Current Topics in Membranes, Volume 51

Aquaporins

Current Topics in Membranes, Volume 51

Series Editors

Dale J. Benos
Department of Physiology and Biophysics
University of Alabama
Birmingham, Alabama

Sydney A. Simon
Department of Neurobiology
Duke University Medical Center
Durham, North Carolina

Current Topics in Membranes, Volume 51

Aquaporins

Edited by

Stefan Hohmann
Department of Cell and Molecular Biology/Microbiology
Göteborg University
Göteborg, Sweden

Søren Nielsen
Department of Cell Biology
University of Aarhus
Aarhus, Denmark

Peter Agre
Departments of Biological Chemistry and Medicine
Johns Hopkins University, School of Medicine
Baltimore, Maryland

ACADEMIC PRESS

A Harcourt Science and Technology Company

San Diego   San Francisco   New York   Boston   London   Sydney   Tokyo

Academic Press
A Harcourt Science and Technology Company
525 B Street, Suite 1900, San Diego, California 92101-4495, USA
http://www.academicpress.com

Academic Press
Harcourt Place, 32 Jamestown Road, London NW1 7BY, UK
http://www.academicpress.com

International Standard Book Number: 0-12-153351-4 (casebound)
International Standard Book Number: 0-12-352095-9 (paperback)

PRINTED IN THE UNITED STATES OF AMERICA
01 02 03 04 05 06 EB 9 8 7 6 5 4 3 2 1

Contents

CHAPTER 9 Future Directions of Aquaporin Research
Stefan Hohmann, Søren Nielsen, Christophe Maurel, and Peter Agre

Contributors

The numbers in parentheses indicate the pages on which authors' contributions begin.

Peter Agre (1, 371), Departments of Biological Chemistry and Medicine, Johns Hopkins University, School of Medicine, Baltimore, Maryland 21205

Eric Beitz (1), Departments of Biological Chemistry and Medicine, John Hopkins University, School of Medicine, Baltimore, Maryland 21205

Mario J. Borgnia (1), Departments of Biological Chemistry and Medicine, John Hopkins University, School of Medicine, Baltimore, Maryland 21205

Roslyn M. Bill (335), Department of Cell and Molecular Biology/Microbiology, Göteborg University, SE-40530 Göteborg, Sweden

Dennis Brown (235), Renal Unit and Program in Membrane Biology, Harvard Medical School, Massachusetts General Hospital East, Charlestown, Massachusetts 02129

Guiseppe Calamita (335), Dipartimento di Fisiologia Generale e Ambiantale, Università degli Studi di Bari, 70126 Bari, Italy

Jennifer Carbrey (1), Departments of Biological Chemistry and Medicine, Johns Hopkins University, School of Medicine, Baltimore, Maryland 21205

Chung-Lin Chou (121), National Heart, Lung and Blood Institute, National Institutes of Health, Bethesda, Maryland 20892

Maarten J. Chrispeels (277), Division of Biology, University of California San Diego, La Jolla, California 92093-0116

Peter M. T. Deen (235), Department of Cell Physiology, University Medical Center of Nijmegen, 6500 HC, Nijmegen, The Netherlands

Andreas Engel (39), M. E. Müller-Institute for Microscopy, Biozentrum, University of Basel, CH-4056 Basel, Switzerland

Jørgen Frøkiaer (155), Department of Cell Biology, University of Aarhus, DK-8000 Aarhus, Denmark

Yoshinori Fuyijoshi (39), Department of Biophysics, Faculty of Science, Kyoto University, Kyoto 606-8502, Japan

Patricia Gerbeau (277), Biochimie et Physiologie Moléculaire des Plantes, ENSAM/INRA/CNRS/UMII, Montpellier, France

Henrik Hager (155), Department of Cell Biology, University of Aarhus, DK-8000 Aarhus, Denmark

Bernard Heymann (39), M. E. Müller-Institute for Microscopy, Biozentrum, University of Basel, CH-4056 Basel, Switzerland

Jason Hoffert (1), Departments of Biological Chemistry and Medicine, Johns Hopkins University, School of Medicine, Baltimore, Maryland 21205

Stefan Hohmann (335, 371), Department of Cell and Molecular Biology/Microbiology, Göteborg University SE-40530 Göteborg, Sweden

Ingela Johansson (277), Department of Plant Biochemistry, University of Lund, Lund, Sweden

Gerald Kayingo (335), Department of Microbiology, University of Stellenbosch, Matieland 7602, South Africa

Landon S. King (1), Departments of Biological Chemistry and Medicine, Johns Hopkins University, School of Medicine, Baltimore, Maryland 21205

Per Kjellbom (277), Department of Plant Biochemistry, University of Lund, Lund, Sweden

Mark A. Knepper (121, 155), National Heart, Lung and Blood Institute, National Institutes of Health, Bethesda, Maryland 20892

David Kozono (1), Departments of Biological Chemistry and Medicine, Johns Hopkins University, School of Medicine, Baltimore, Maryland 21205

Tae-Hwan Kwon (155), Dongguk University, Kyungju, South Korea

Virginia Leitch (1), Departments of Biological Chemistry and Medicine, Johns Hopkins University, School of Medicine, Baltimore, Maryland 21205

Tonghui Ma (185), Departments of Medicine and Physiology, Cardiovascular Research Institute, University of California, San Francisco, California 94143-0521

Geoffrey T. Manley (185), Departments of Medicine and Physiology, Cardiovascular Research Institute, University of California, San Francisco, California 94143-0521

David Marples (155), University of Leeds, Leeds LS2 9JT, United Kingdom

Christophe Maurel (277, 371), Biochimie et Physiologie Moléculaire des Plantes, ENSAM/INRA/CNRS/UMII, Montpellier, France

Alok Mitra (185), Department of Cell Biology, Scripps Research Institute, La Jolla, California 92037

Kaoru Mitsuoka (39), Department of Biophysics, Faculty of Science, Kyoto University, Kyoto 606-8502, Japan

Raphael Morillon (277), Division of Biology, University of California San Diego, La Jolla, California 92093-0116

John D. Neely (1), Departments of Biological Chemistry and Medicine, Johns Hopkins University, School of Medicine, Baltimore, Maryland 21205

Søren Nielsen (121, 155, 371), Department of Cell Biology, University of Aarhus, DK-8000 Aarhus, Denmark

Bernard A. Prior (335), Department of Microbiology, University of Stellenbosch, Matieland 7602, South Africa

William R. Skatch (185), Division of Molecular Medicine, Oregon Health Sciences University, Portland, Oregon 97201

Yuanlin Song (185), Departments of Medicine and Physiology, Cardiovascular Research Institute, University of California, San Francisco, California 94143-0521

Henning Stahlberg (39), M. E. Müller-Institute for Microscopy, Biozentrum, University of Basel, CH-4056 Basel, Switzerland

A. S. Verkman (185), Departments of Medicine and Physiology, Cardiovascular Research Institute, University of California, San Francisco, California 94143-0521

Baoxue Yang (185), Departments of Medicine and Physiology, Cardiovascular Research Institute, University of California, San Francisco, California 94143-0521

Masato Yasui (1), Departments of Biological Chemistry and Medicine, Johns Hopkins University, School of Medicine, Baltimore, Maryland 21205

Preface

For more than 30 years the existence of water-specific transport proteins had been predicted. Ten years ago the characterization of an abundant protein of the erythrocyte membrane called CHIP28 unexpectedly revealed that this protein was such a water channel. Quickly it became apparent that CHIP28, now called aquaporin-1 (AQP-1), was a member of a widespread and ancient family of water and solute-transporting proteins. All eukaryotic organisms and most bacteria possess proteins of the MIP family (MIP for major intrinsic protein, the first protein identified in 1984). In addition to water-specific channels the family comprises glycerol facilitators, which also transport other polyols, urea, and related substances and proteins that transport both water and small solutes.

Since the discovery of AQP-1 as water channel some 600 articles have been published in the scientific literature, mainly on mammalian aquaporins but also more and more on MIP channels from plants and microorganisms. The facts that MIP channels are wide spread and also occur in significant numbers in different organisms (at least 10 in human and more than 30 in higher plants) suggest physiological importance. Over the past 7 years structural, molecular biological, cell biological, physiological, and pathophysiological studies—as well as studies with mutants lacking specific MIP channels—have revealed essential importance of MIP channels in mammalian physiology and pathophysiology as well as in plant and microbial biology. These studies also reveal aquaporins as potential drug targets and as targets for improvement of different crop properties.

In July 2000, more than 200 researchers gathered in Gothenburg, Sweden to discuss the latest advances in research on aquaporins and other MIP channels. The main contributors of this interactive and interdisciplinary conference are the authors of the chapters of this book. Peter Agre, whose group discovered the water channel function of AQP-1, gives an introduction and overview from a historic perspective. Andreas Engel and colleagues as well as Alan Verkman and colleagues discuss structure-function analysis of aquaporins and glycerol facilitators at a point where we are expecting the structure of key members of this protein family at atomic resolution. The physiology and pathophysiology of mammalian aquaporins is discussed in the chapters by Mark Knepper and Søren Nielsen and colleagues and, with a focus on studies on knockout animal models, in the chapter by Alan Verkman and colleagues. Special chapters are dedicated to the many plant MIP channels (by Maarten Chrispeels and colleagues) and the water channels and glycerol facilitators from microorganisms (by Stefan Hohmann and colleagues). The book is then completed by a short account on future perspectives and anticipated developments as well as open questions that probably should be considered in the future.

We hope that this book will stimulate future research on this important protein family. Although book editors usually hope that their product may be a reference for many years to come, we anticipate that the field of aquaporin research will develop rapidly in the future. We especially await novel approaches for the treatment of human diseases based on aquaporin function, or drugs that block aquaporins and strategies to improve crop properties on the basis of MIP channel expression or knockout.

Stefan Hohmann
Søren Nielsen
Peter Agre

Previous Volumes in Series

Current Topics in Membranes and Transport

* Part of the series from the Yale Department of Cellular and Molecular Physiology

Current Topics in Membranes

CHAPTER 1

Discovery of the Aquaporins and Their Impact on Basic and Clinical Physiology

Peter Agre, Mario J. Borgnia, Masato Yasui, John D. Neely, Jennifer Carbrey, David Kozono, Eric Beitz, Jason Hoffert, Virginia Leitch, and Landon S. King

Departments of Biological Chemistry and Medicine, Johns Hopkins University, School of Medicine, Baltimore, Maryland 21205

I. PRE-AQUAPORIN ERA

Water channels have become increasingly well recognized during the past 10 years; however, the basic issues of membrane water permeability were recognized much earlier.

A. *Water Permeation of Cell Membranes*

The combined efforts of multiple laboratories indicated that water permeation is a feature of some but not all cells and tissues. Since the time of August Krogh and his studies of capillary permeability, physiologists at the University of Copenhagen have addressed the problem of how water and solutes are transported across epithelia. Hans Ussing and colleagues studied amphibian skin and recognized that certain epithelia are exceedingly permeable, hence "leaky" (reviewed by Ussing, 1965). Other investigators followed these studies with electron microscopic analyses, which indicated that water channels ("aggrephores") must reside within tissues such as amphibian bladder (reviewed by Kachadorian *et al.*, 2000), and when stimulated appropriately, the vesicles were believed to traffic to the cell surface, inducing water permeability. This process was reversed by reinternalization of the putative water channels, and the "shuttle hypothesis" became widely recognized (Wade *et al.*, 1981).

A. K. Solomon and colleagues concentrated their attention on red blood cells. They measured large and selective water fluxes through the membranes with low Arrhenius activation energies, indicating that porelike molecules must reside in the plasma membranes (reviewed by Solomon, 1968). Taking this further, Robert Macey and colleagues observed that membrane water permeability is selectively inhibited by $HgCl_2$ and certain organomercurials and that water permeability could be restored by treating the red cells with reducing agents (reviewed by Macey, 1984). Thus, the water pores must be formed from proteins with free sulfhydryls that react with mercurials.

Fortunately, these observations were integrated by Alan Finkelstein in a monograph titled *Water Movements through Lipid Bilayers, Pores, and Plasma Membranes—Theory and Reality* (1987). In this comprehensive treatise, Finkelstein posed the question "Who are the water pores?" To answer this, he considered the possibility that one of the known red cell proteins must be the long-sought water channel, because his biochemist friends (stated to be "few in number") assured him that all the major membrane proteins must certainly have already been discovered by then. In retrospect, Finkelstein pointed out that the identity of the membrane water pores was certainly not known. Attempts to identify molecular water channels were made by several laboratories using various approaches: labeling of red cell anion transporter with isotopic mercury (Solomon *et al.*, 1983); measuring increased osmotic water permeability of oocytes expressing glucose transporter protein (Fischbarg *et al.,* 1990); functional evaluation of candidate proteins from amphibian bladder (Harris *et al.,* 1992); and attempts to clone by expression (Zhang *et al.,* 1990). Unfortunately, none of these direct approaches was successful.

B. Red Cell 28-kDa Membrane Protein

It has been said that "hypothesis-driven research is highly overrated" (A. G. Gilman, personal communication, 2000), suggesting that unexpected results may sometimes yield insights that are much greater than anticipated and may change a field altogether. The surprising discovery of the 28-kDa red cell membrane protein, now known as AQP1, is a clear demonstration of the importance that a chance observation can make in a field stymied by logical approaches.

The Rh blood group antigens are of large clinical importance. Their molecular identities, however, have long resisted identification. Utilizing radiochemically labeled red cell membranes from Rh(D) positive red cells, a 32-kDa membrane protein was isolated by hydroxylapatite chromatography (Agre *et al.,* 1987; Saboori *et al.,* 1988). This protein is virtually invisible to staining with Coomassie blue; however, in pure form the protein is well visualized with silver reagent. A 28-kDa polypeptide was also unstained by Coomassie but was visualized with silver reagent (explaining why Professor Finkelstein's biochemist friends failed to see it). The 28-kDa polypeptide was initially believed to be a proteolytic fragment of the 32-kDa Rh polypeptide. Surprisingly, antibodies raised in rabbits only reacted with the 28-kDa band, which was subsequently characterized and found to be unrelated to Rh (Denker *et al.,* 1988). Although initially believed to be linked with the membrane skeleton, the 28-kDa polypeptide could be solubilized in higher concentrations of nondenaturing detergent. The relative insolubility of the 28-kDa polypeptide in the detergent *N*-lauroylsarcosine, a substance once well known to the baby boomer generation as Gardol, the secret ingredient of Colgate's toothpaste, proved to be extremely helpful. Virtually all other red cell membrane proteins can be removed in one step, leaving nearly pure 28-kDa polypeptide. This procedure permitted isolation of pure 28-kDa polypeptide in milligram concentrations (Fig. 1).

Several features of the 28-kDa polypeptide suggested that it may have a channel-like structure and resulted in the temporary designation "CHIP28," for channel-like integral protein of 28 kDa (Smith and Agre, 1991). When analyzed in nonionic detergent, the protein exhibited filtration and velocity sedimentation properties indicating that it is a multisubunit oligomer, predicted to be a homotetramer. Most of the protein resides between the leaflets of the lipid bilayer, and a short N terminus and the C terminus were localized to the cytoplasmic space. Unlike other membrane proteins, a complex polylactosaminoglycan was attached to only one of the

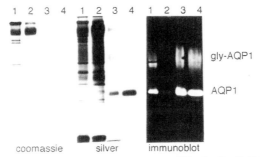

FIGURE 1   Purification of AQP1 protein from human red blood cells. (*Left*) SDS–PAGE stained with Coomassie; (*middle*) same stained with silver; (*right*) immunoblot reacted with anti-AQP1. Lane 1, whole red cell membranes; lane 2, *N*-lauroylsarcosine-soluble proteins; lane 3, *N*-lauroylsarcosine-insoluble proteins; lane 4, FPLC purified AQP1. Note 28-kDa unglycosylated and 40- to 60-kDa glycosylated protein. Reproduced with permission from *Methods in Enzymology* (Agre *et al.*, 1998b).

four subunits. The amino acid sequence of the first 35 N-terminal residues was determined by Edman degradation protein sequencing. Interestingly, the sequence was observed to be related to major intrinsic protein of lens (MIP), an abundant but functionally undefined protein (Gorin *et al.*, 1984). The 28-kDa protein was noted to be extremely abundant in red blood cells (approximately 200,000 monomers per membrane), in apical brush border of renal proximal tubules (approximately 4% of the total membrane protein), and in descending thin limbs of Henle's loop (Denker *et al.*, 1998). Recognizing that these proteins are all highly permeable to water, the late John C. Parker (1936–1993), a noted membrane physiologist at the University of North Carolina at Chapel Hill, first suggested that the 28-kDa polypeptide may be the long-sought water channel (Parker and Agre, personal discussions, 1990). As will be documented in this volume, Parker's suggestion was to have major consequences for the direction of water transport research.

II. THE FIRST RECOGNIZED WATER CHANNEL PROTEIN

As with other physiologically important molecules, identification of the first member of the water channel family led to increased interest in this area of biology.

A. cDNA Cloning

The N-terminal sequence of CHIP28 was utilized to design oligonucleotides that were used clone the cDNA from an erythroid library (Preston and Agre, 1991). This was done in stages, and confirmation that the product achieved was obtained by demonstrating that the expressed protein reacted with the specific antibody

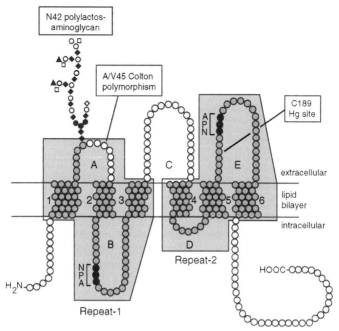

FIGURE 2 Schematic diagram of AQP1 monomer showing sequence repeats and selected structural features. Reproduced with permission from *Journal of Structural Biology* (Heymann *et al.,* 1998).

reactive with the C terminus. Sequence analysis demonstrated that the protein contains a tandem repeat of two related peptides, each encoding three presumed bilayer-spanning domains (Fig. 2). A potential glycosylation site was noted in the extracellular loop A, and two highly conserved loops were noted to contain the preserved motif Asn-Pro-Ala (NPA).

Other clues to possible function of this protein were observed at this time. The GenBank contained related sequences encoding a microbial homolog and sequences encoding four plant homologs. Of particular interest, expression of a homolog from pea plants (TUR) was noted to become induced by water deprivation (Guererro *et al.,* 1990). Moreover, the size of the 28-kDa polypeptide was close to the size of the membrane water channel independently determined by radiation inactivation studies (van Hoek *et al.,* 1991).

B. Functional Studies

Unlike measurement of ion currents or transport of radioisotopically labeled solutes, the measurement of membrane water transport poses special difficulties (Solomon, 1989; Verkman, 1992). Because all membranes exhibit some degree of diffusional water permeability, significant background levels of water permeability are always present. Likewise, the movement of water across cell membranes is very fast and reaches equilibrium in milliseconds. Thus, the use of isotopically labeled water (^3H_2O) is not generally helpful because hydrogen ions are freely exchangeable with other water molecules and isotopic studies are not amenable to fast measurements. Expression of transport proteins in *Xenopus laevis* oocytes had been successfully developed by E. M. Wright and colleagues to clone other transport processes (Hediger *et al.*, 1987). Recognizing that these cells are relatively easy to manipulate and have particularly low water permeability, Eric Windhager at Cornell Medical School first proposed oocyte expression as a mechanism to study water transporters.

1. Expression in Oocytes

The function of CHIP28 was first assessed by expression in *X. laevis* oocytes (Preston *et al.*, 1992). By cloning the cDNA into an expression vector, complementary RNA was prepared and injected into oocytes. After 3 days in culture, the oocytes were transferred from isotonic modified Barth's solution (200 mos*M*) to diluted solution (70 mos*M*). This resulted in a dramatic swelling and explosion in the test oocytes, whereas control water-injected oocytes remained virtually unchanged (Fig. 3). When quantitated, the coefficient of osmotic water permeability (P_f) of the oocytes expressing CHIP28 rose 20-fold (to ~ 200 cm/s $\times 10^{-4}$). Similar to water channels in native membranes, the Arrhenius activation energy was low, $E_a < 3$ kcal/mol (equivalent to diffusion of water in bulk solution); this was distinct from that of water-injected control oocytes, $E_a \gg 10$ kcal/mol. Also notable was the inhibition by 1 m*M* $HgCl_2$, which was reversed by the reducing

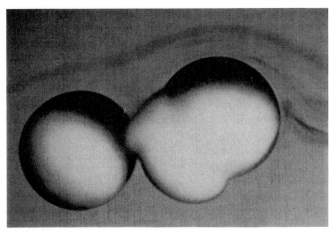

FIGURE 3   Osmotic swelling and rupture of *X. laevis* oocyte expressing AQP1 protein (*right*) vs water-injected control oocyte (*left*). Reproduced with permission from *Science* (Preston *et al.*, 1992).

agent 2-mercaptoethanol. The possibility that this permeability represented a non-specific leak was ruled out, because oocytes retained their normal resting potentials and failed to exhibit increased membrane currents. The lack of ion conductance by AQP1 was disputed by a single group of investigators who reported Forskolin-induced cation permeation by AQP1 expressed in oocytes (Yool *et al.*, 1996); however, this was not reproduced by other laboratories (Agre *et al.*, 1997). More recently, the originators of the AQP1–ion channel hypothesis have reported a cGMP gated conductance when AQP1 is expressed under special circumstances (Anthony *et al.*, 2000), an interesting observation that awaits confirmation. Other permeability issues are also still unresolved, including the interesting possibility that AQP1 is permeated by CO_2 or other gases (Nakhoul *et al.*, 1998; Yang *et al.*, 2000).

2. Reconstitution into Membranes

The initial oocyte expression studies were viewed as strong evidence that the CHIP28 protein is a membrane water channel, but the possibility remained that it could be an activator of endogenous water channels. Thus, highly purified protein was reconstituted at different concentrations into membrane proteoliposomes (Fig. 4), which were used for biophysical analysis of membrane water permeability (Zeidel *et al.*, 1992, 1994). Purified CHIP28 protein was reconstituted with pure phospholipid (protein:lipid ranging from 1:10 to 1:100), yielding small unilamellar vesicles of ~100 nm in diameter. These CHIP28 proteoliposomes were loaded with carboxyfluorescein and analyzed by stopped-flow transfer from iso-osmolar buffer to hyperosmolar solution, resulting in shrinkage of the vesicles with a

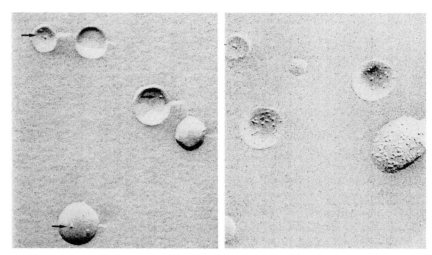

FIGURE 4 AQP1 reconstituted into proteoliposomes. Freeze fracture electron micrograph of membranes reconstituted with lower (left) or higher (right) concentrations of AQP1 protein. Reproduced with permission from *Biochemistry* (Zeidel *et al.*, 1994).

timescale of milliseconds. By comparisons with control liposomes (reconstituted membrane vesicles lacking CHIP28), the background water permeability was determined. The CHIP28 proteoliposomes exhibited up to a 50-fold increase in water permeability with low Arrhenius activation energy and mercurial inhibition. The unit permeability was calculated to be $\sim 3 \times 10^9$ water molecules per subunit per second. Moreover, permeability to H^+ and urea was not detected. Proteolytic removal of the 4-kDa C-terminal peptide did not diminish water permeability. Together these studies confirmed that CHIP28 is a constitutively active membrane water channel and quantitatively sufficient to explain the water permeability of red blood cells. Using partially purified CHIP28 protein, similar studies were undertaken by other investigators with similar interpretations (van Hoek and Verkman, 1992).

C. Aquaporins, a New Name for an Ancient Protein Family

As part of the Human Genome Project, the Genome Nomenclature Committee was charged with the issue of classifying genes that had been given different names by different laboratories. Lack of uniform nomenclature is a problem that has been present since ancient times. In *Genesis*, the effort to build a tower with its top in the heavens suddenly stopped when God caused multiple languages to be spoken by the workers, hence the Tower of Babel. Even in modern times, workers are often reluctant to accept and use nomenclature proposed by competing investigators, so the Human Genome Organization appointed a nomenclature committee to assign gene symbols. Several observations from the Human Genome Nomenclature Committee seem to reflect basic human nature:

> It is always the people with the worst data that make the most fuss about the gene symbol.
> —Phyllis McAlpine, University of Winnipeg

> People will accept any gene symbol, if it is No. 1.
> —Sue Povey, University of London

> The use of aliases will continue, but only by the individuals who invented them.
> —Alan Scott, Johns Hopkins University

The first molecular characterization of a molecular water channel quickly led to the identification of other mammalian homologs and plant homologs. To aid in recognizing these related proteins, the name "aquaporin" was proposed (Agre *et al.*, 1993). This has now been adopted as the name for all sequence-related water transporters, and the gene symbol *AQP* has been adopted by the Human Genome Nomenclature Committee (Agre, 1997). CHIP28 is now officially referred to as AQP1, and this nomenclature is used throughout the remainder of this volume.

D. Structure of AQP1 Protein

Determination of protein structures is a major step in the development of pharmacological agents. Soon after its discovery, the structure of AQP1 was pursued
by multiple scientific groups.

1. Structure Deduced from Sequence

Hydropathy analysis of the deduced amino acid sequence of lens MIP protein
predicted it to be an integral membrane protein with six bilayer-spanning domains
(Gorin *et al.,* 1984). Closer inspection revealed two tandem repeats (Figs. 2 and 5),
each containing three bilayer-spanning domains with two perfectly conserved motifs in the connecting loops, Asn-Pro-Ala (NPA) (Reizer *et al.,* 1993; Wistow *et al.,*
1991). The demonstration of water channel function made it possible to engineer
recombinants with the goal of establishing the sites of mercurial inhibition and the
membrane topology of biologically active recombinants.

Mercurials are known to react with free sulfhydryls, and the primary sequence
of AQP1 contains four cysteines. In individual recombinants, each cysteine was
replaced by a serine, an amino acid with the same structure except where the sulfur
atom is replaced by an oxygen (Preston *et al.,* 1993). These studies implicated the
residue Cys 189 in loop E as the site of mercurial inhibition and suggested that this
domain may form a critical narrowing in the aqueous pore. In subsequent studies,
the mercury-insensitive mutant Cys 189 Ser was further mutated by introducing
a cysteine at the corresponding residue in loop B, Ala 73 Cys. This restored
mercurial sensitivity, indicating that loops B and E are both functionally important.
Recombinants were prepared in which monomers, containing or lacking the Hg^{2+}
inhibitory site, were linked in tandem as 55-kDa dimers. Analyses were undertaken
of the dimers as well as immunoprecipitation of full-length and truncated forms
of the AQP1 monomers. Together, all of these studies indicated that each of the
four monomers in the AQP1 tetramer carries an individual aqueous pore (Jung
et al., 1994a).

*Bam*HI restriction sites were cloned into the AQP1 cDNA at different sites.
DNA encoding a 31-residue peptide epitope from the coronavirus E1 glycoprotein
was then inserted at these different sites, and the water transport was assessed by
expression in oocytes (Preston *et al.,* 1993). This epitope was helpful, because
specific antibodies were available and because the epitope contains proteolytic
cleavage sites, whereas the intact AQP1 protein does not.

Insertions into loops B or E usually caused loss of function when the recombinants were expressed in oocytes (Jung *et al.,* 1994a; Preston *et al.,* 1993). Nevertheless, insertions into the N or C termini or connecting loops A, C, or D encoded functional molecules which were then evaluated to see whether the E1
epitope was at the extracellular surface or the cytoplasmic surface of the oocyte
plasma membrane. By antibody binding or by selective proteolysis, the epitope was

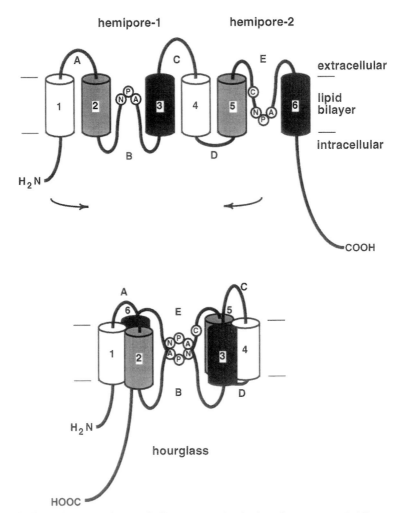

FIGURE 5   Schematic diagram of AQP1 monomer showing hourglass structure. (*Top*) Two repeats of three bilayer-spanning domains are oriented in obverse symmetry. Loops B and E are believed to fold back into the membrane. (*Bottom*) When folded together, the repeats form a single aqueous pathway through the membrane. Reproduced with permission from *The Journal of Biological Chemistry* (Jung *et al.*, 1994a).

confirmed to reside at the previously predicted positions. Of greatest importance was the location of loop C at the extracellular side of the membrane. This placed a constraint on the AQP1 topology, indicating that the two repeats must be obversely symmetric with a structure resembling the ancient hourglass (Fig. 5). Another group of investigators approached the topology problem by preparing truncated molecules (Skach *et al.,* 1994). They initially concluded that the water channel is composed of four transmembrane domains. It is now believed, however, that this may represent an intermediate state (see Chapter 5). Thus, the hourglass model was generally believed to be the most likely structure (Jung *et al.,* 1994a).

2. Protein Structural Determinations

At the same time reconstitution of purified AQP1 protein was being functionally characterized (see above), the protein was being analyzed for structure. The hydrodynamic properties of the purified AQP1 protein revealed it to be a homotetramer composed of four 28-kDa subunits embedded in the membrane with only a 4-kDa C terminus extending into the cytoplasm (Smith and Agre, 1991). Cultured cells transfected with the AQP1 cDNA as well as proteoliposomes were shown by negative staining electron microscopy to have a multisubunit structure interpreted as being a tetramer (Verbavatz *et al.,* 1993).

Given the serious difficulties associated with crystallizing integral membrane proteins for X-ray diffraction crystallography, the structure of the AQP1 protein was undertaken by cryo-electron microscopy of membrane crystals. Reconstituting purified AQP1 protein at high concentrations relative to the concentration of phospholipid (approximately 1:1) yielded large, highly ordered membrane sheets and resealed membrane vesicles of ∼300 μm in diameter. Measuring the water permeability of the membrane vesicles confirmed that the reconstituted protein was 100% biologically active (Walz *et al.,* 1994). A series of studies revealed the 2D structure at increasing resolution, and by studying the membrane crystals at tilts of up to 60°, the 3D structure was established at 6Å (Cheng *et al.,* 1997; Li *et al.,* 1997; Walz *et al.,* 1997). The temperature of the electron diffractions is critical to the measurements, and the use of a special electron microscope with a liquid-helium-cooled stage, has now established the 3D structure at 3.8 Å. Short pore helices are now identified after the NPA motifs in loops B and E, and protrusions representing side chains are visualized (Mitsuoka *et al.,* 1999). Merging these determinations with model building has yielded the structure at the atomic level (Murata *et al.,* 2000). A narrow aqueous pore through the center of AQP1 is flanked by critical residues (Fig. 6).

The structural studies have concentrated on the membrane domains, because this part of the molecule confers water transport (Zeidel *et al.,* 1994). The importance of the N and C termini is not understood, and these domains are least well conserved when species orthologs and other aquaporin paralogs are compared. The N and C termini are believed to be important in membrane targeting, but recent work suggests that in some tissues, AQP1 may associate with other proteins by being part of a large complex associated through a PDZ binding motif (Cowan *et al.,* 2000). Certainly this area of research awaits further investigation.

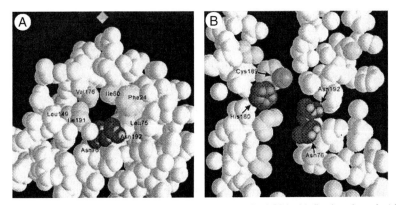

FIGURE 6 Space-filled diagrams of the aqueous pore in AQP1. A) Section through AQP1 monomer parallel to the plane of the membrane showing functionally important hydrophilic residues (darkly shaded–Asn76 and Asn192). B) Section showing side view of narrow aqueous channel flanked by Hg^{++} inhibitory site (Cys189). Reproduced with permission from *Nature* (Murata *et al.,* 2000).

E. Tissue Distribution

Prior to cloning of the cDNA encoding AQP1, it had been recognized that multiple tissues exhibit high levels of water permeability, and these tissues were believed to contain molecular water channels. With the development of high-affinity antibodies specific for the N and C termini of AQP1, it became possible to precisely locate the cellular and subcellular sites where the protein is expressed. These studies provided great insight into the physiological and pathophysiological roles for the AQP1 protein.

1. AQP1 in Kidney

Beginning with Homer Smith in the 1930s, transport physiologists have given the kidney more attention than any other tissue. Despite its great importance, Smith recognized the limitations of kidney research:

> The history of renal physiology has erred, more often than not, by attempts at oversimplification. The problems of water and salt excretion appear to be extremely complex, and especially liable to this danger.
> —Homer W. Smith, 1937, *The Physiology of the Kidney*

The kidney is a major site where nitrogenous wastes are excreted, pH is adjusted, and body water balance is achieved. Each kidney contains roughly one million individual nephron units: glomerulus, proximal convoluted and straight tubules, descending thin limbs of Henle's loop, thin and thick ascending limbs, and connecting tubules that join multiple nephrons to the collecting ducts. The average adult human generates almost 200 liters of glomerular filtrate daily, but almost 90% of the water is reabsorbed in the proximal tubules and descending thin limbs. These tissues are known to be highly permeable to water, and water channels were predicted to reside at these locations. Using affinity-purified poly-clonal antibodies, AQP1 was identified at the apical brush border and at the basal and lateral membranes of both proximal tubules and descending thin limbs of rat kidney (Denker *et al.*, 1988; Nielsen *et al.*, 1993a) (Fig. 7A,B).

Transcellular water movements occur through AQP1. Water enters through AQP1 in the apical brush border, and water is released into the interstitium from AQP1 in the basolateral membrane (Fig. 7C). Moreover, the appearance of this pro-tein coincides in rat kidney with the development of concentrating mechanisms that occur at the time of weaning (Smith *et al.*, 1993). This distribution was confirmed in rat kidney by immunogold electron microscopy with a second affinity-purified antibody specific for the N terminus of AQP1 (Nielsen *et al.*, 1993a) and with whole antiserum (Sabolic *et al.*, 1992). A similar pattern was observed in human kidney (Maunsbach *et al.*, 1997). A very important negative was established when AQP1 was not observed over renal collecting duct principal cells (Fig. 7A,B), the site where water permeability was known to be regulated by antidiuretic hormone (vasopressin). This first indicated the likelihood of a second member of the aqua-porin family (Nielsen *et al.*, 1993a). A second area of AQP1 expression was observed over the capillary endothelia, the descending vasa recta (Nielsen *et al.*, 1995b; Pallone *et al.*, 1997).

2. AQP1 in Other Tissues

In situ hybridizations indicated that AQP1 is expressed in complex patterns (Bondy *et al.*, 1993). AQP1 mRNA transiently appears in some tissues near the time of birth (e.g., corneal endothelium, periosteum, and heart). AQP1 mRNA first appears in other tissues around the time of birth and throughout adult life (e.g., kidney, lung, and blood-forming tissues). AQP1 mRNA was observed in still other tissues throughout fetal and postnatal life (e.g., choroid plexus). The affinity-purified antibodies were used to establish that AQP1 protein appeared in each of these tissues (Nielsen *et al.*, 1993b).

Of particular importance, AQP1 was found to be abundant in capillary endothelia in many distributions throughout the body where the protein was shown by light microscopy and immunogold electron microscopy to reside in both lumenal and ablumenal membranes. In lung, this was especially prominent in the capillary endothelia in peribronchiolar regions, indicating a special need for this protein in the reabsorption of fluid just after birth or in settings of interstitial edema (King *et al.*, 1996; Umenishi *et al.*, 1996). Premature human infants suffer significant morbidity due to immature lung, and glucocorticoid therapy is known to improve lung function. A classical glucocorticoid response element is located in the proximal promoter region of the human *AQP1* gene (Moon *et al.*, 1997), and treatment of pregnant rats with glucocorticoids results in a 5- to 10-fold boost in AQP1 expression in the fetal lungs. Other sites where AQP1 is expressed include liver

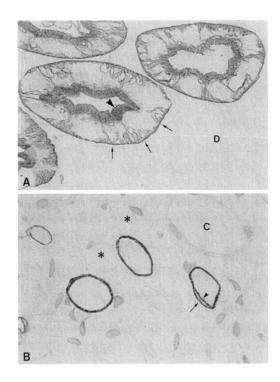

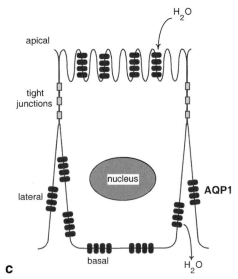

cholangiocytes (Roberts *et al.,* 1994) and fibroblasts beneath airway epithelium (King *et al.,* 1996, 1997; Nielsen *et al.,* 1997a). The protein is also expressed in choroid plexus, where it may participate in generation of cerebrospinal fluid; in nonpigment epithelia in eye, where it may participate in formation of aqueous humor; and in corneal endothelium and lens epithelium (Nielsen *et al.,* 1993b).

F. AQP1-Null Humans

The presence of AQP1 in diverse tissues suggested a fundamental role for this protein; however, the clinical consequences of AQP1 deficiency were initially unclear.

Polymorphisms in red cell membrane proteins are known to be the molecular basis of blood group antigens. Although hundreds of different human red cell antigens have been described serologically, these belong to approximately 24 different blood groups, with each group representing a single genetic locus (Agre and Cartron, 1992). Previous discovery of some blood groups occurred when individuals totally lacking a certain antigen were sensitized with the blood of an individual who expresses the antigen. Although sensitization may occur after blood transfusions, relatively few individuals ever receive blood transfusions. Sensitization also occurs when maternal and fetal blood mixing occurs during birth. Because of potentially severe clinical manifestations, this is best recognized in the setting of maternal–fetal Rh incompatibilities; however, other blood group antigens can also lead to the appearance of circulating antibodies.

By classic linkage analyses, the human Colton blood group antigens were linked to the short arm of human chromosome 7 (Zelinkski *et al.,* 1990). Demonstration of the *AQP1* gene locus at the short arm of human chromosome 7 provided a strong clue that AQP1 protein is the structural basis of Co antigens (personal discussions with Marion Reid and Colvin Redman, New York Blood Center, 1993). Human Co antigens are generally not clinically important; 99.8% of the population is Co(a+), and ~8% is Co(b+). The International Blood Group Reference Laboratory in Bristol, United Kingdom, maintains a registry of all rare blood types, and DNA from peripheral blood leukocytes was obtained from multiple individuals with

FIGURE 7 AQP1 in kidney. (A) Immunohistochemical staining of AQP1 in rat renal cortex. Specific staining reveals abundant AQP1 in apical brush border (arrowhead), basal, and lateral membrane (small arrows) in proximal convoluted tubule. Distal convoluted tubules (D) fails to label. (B) Immunohistochemistry of AQP1 in apical and basolateral membrane in descending thin limb of Henle's loop. Ascending thin limbs (*) and collecting ducts (C) fail to label. Reproduced with permission from *The Journal of Cell Biology* (Nielsen *et al.,* 1993a). (*Bottom*) Schematic model showing transcellular water permeation of proximal tubule through AQP1. Reproduced with permission from *Kidney International* (Nielsen and Agre, 1995).

each phenotype. Polymerase chain amplifications of genomic DNA revealed that a polymorphism at residue 45 of AQP1 is the basis of the blood group. Ala 45 occurs in Co(a+), and Val 45 occurs in Co(b+) (Fig. 2); this specificity was confirmed by multiple immunoprecipitations and immunoblotting (Smith *et al.*, 1994).

Worldwide referencing has yielded only six individuals who lack the Co blood group antigens, and we were able to obtain blood and urine from three unrelated Co-null individuals (Preston *et al.*, 1994b). Immunoblotting revealed a total lack of AQP1 protein in the red cells from Probands 1 and 2, and only extremely low levels of the AQP1 protein in red cells from Proband 3. This was accompanied by a marked diminution in the red cell osmotic water permeability (Mathai *et al.*, 1996; Preston *et al.*, 1994b). Immunoblotting of urine sediment confirmed the lack of AQP1 protein in renal brush border membranes. Genomic DNA analyses demonstrated disruptions in *AQP1* (Preston *et al.*, 1994b). Proband 1 was homozygous for complete deletion of the first exon of *AQP1,* which normally encodes half of the AQP1 protein. Proband 2 was homozygous for a frameshift in the first exon. Proband 3 is homozygous for a missense mutation at the end of the first transmembrane helix where Pro 38 is replaced by Leu. Presumably this individual expresses an unstable form of AQP1, because when the P38L protein is expressed in oocytes, it is degraded.

All three Co-null individuals were identified because they developed anti-Co antibodies after pregnancy; the clinical courses, however, were not identical. Proband 2 successfully bore four children without incident. In contrast, the third pregnancy of Proband 1 was complicated by a life-threatening fetal hemolytic syndrome, which required intrauterine transfusions of the fetus with Co-null red cells from the mother. Because of persisting anti-Co antibodies, all three probands are at risk for hemolytic reactions if transfusions should ever be given. Thus, units of their own blood have been cryopreserved at nearby hospitals should emergency transfusions ever be needed.

Surprisingly, none of these Co-null individuals (referred to here as "AQP1 null") was aware of any other significant clinical difficulties. To ascertain whether sub-clinical deficiencies may exist, careful clinical evaluations were undertaken in the Clinical Research Unit at the Johns Hopkins Hospital. Because they totally lack AQP1 protein, Probands 1 and 2 were studied. Baseline evaluations of both were entirely normal, but after 24-h fluid deprivation, neither could concentrate her urine above 450 mos$M$, whereas normal individuals all concentrated their urine to $\sim$1000 mos$M$ after only 8 h. Likewise, hyperosmolar saline infusions failed to raise the urinary osmolality, even though it caused a significant rise in serum osmolality. Although the individuals do not suffer as long as water is available, it is easy to see why AQP1 deficiency would be problematic for individuals with limited access to water.

Other evaluations of the AQP1-null individuals were undertaken, including high-resolution computer axial tomography of lung before and after intravenous fluid challenge. Normal individuals experience a 2- to 3-fold increase in the thickness of airways after infusions, whereas neither of the two AQP1-null individuals exhibited a change in airway thickness. Thus, rate of fluid exchange between blood vessels and peribronchiolar interstitium is reduced when AQP1 is absent. Although no evidence suggests that AQP1-null families have increased problems with clearing perinatal lung edema, this remains a distinct possibility. By traditional gene disruption techniques, mice lacking AQP1 protein have been produced. Similar to AQP1-null humans, these mice exhibit multiple subtle defects (see Chapter 5).

III. OTHER MAMMALIAN AQUAPORINS

Discovery of AQP1 rapidly led to homology cloning efforts in multiple laboratories. Curiously, each new homolog was simultaneously reported by different research groups. This mini-industry has identified 10 mammalian homologs, and several have been partially characterized (Fig. 8).

A. Orthodox Aquaporins

The first characterized member of the family, AQP1, is freely permeated by water but not by other solutes. Several other members of the family were also found to be highly water selective.

1. AQP2, the Vasopressin-Regulated Water Channel

First attention was directed toward the kidney, where a cDNA encoding the anticipated collecting duct homolog was amplified using degenerate primers designed from the sequence of AQP1 (Fushimi *et al.,* 1993). A series of reports defined this protein in various physiologic and clinical states (see Chapters 3, 4, and 6). Although it was predicted to be the vasopressin-regulated water channel, formal demonstration was achieved by an elegant report in which collecting ducts were isolated from rat kidney and examined for water permeability and the distribution of AQP2 was established by immunogold electron microscopy (Fig. 9A) (Nielsen and Agre, 1995). In summary, AQP2 is rapidly targeted to the plasma membrane and reinternalized, thereby confirming the "shuttle hypothesis" and demonstrating a rapid form of regulation (Fig. 9). In addition, chronic thirsting was shown to lead to an increase in AQP2 biogenesis.

Large interest in AQP2 resulted in the identification of humans with mutations in the gene encoding AQP2 and a severe clinical phenotype (Deen *et al.,* 1994). Vasopressin-deficient individuals are unable to concentrate their urine and suffer from diabetes insipidus (DI). Subsequently, a group of patients with

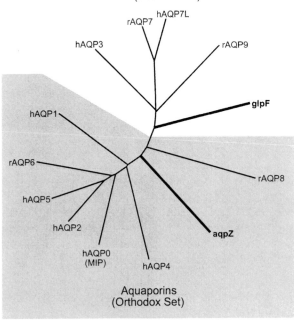

Aquaglyceroporins (Cocktail Set) — hAQP7L, rAQP7, hAQP3, rAQP9, glpF

Aquaporins (Orthodox Set) — hAQP1, rAQP6, hAQP5, hAQP2, hAQP0 (MIP), hAQP4, aqpZ, rAQP8

```
AqpZ  -------------------MFRKLAAECFGTFWLVFGGCGSAVLAAGFP-----ELGIG   35
AQP1  ---------MASEFKK--KLFWRAVVAEFLATTLFVFISIG-SALGFKYPVGNNQTAVQD   48
AQP3  MGRQKELVSRCGEMLHIRYRLLRQALAECLGTLILVMFGCG-SVAQVVLSR----GTHGG   55
GlpF  ------------MSQT-STLKGQCIAEFLGTGLLIFFGVG-CVAALKVA-----GASFG   40

AqpZ  FAGVALAFGLTVLTMAFAVGHISGGHFNPAVTIGLWAGGRFPAKEVVGYVIAQVVGGIVA   95
AQP1  NVKVSLAFGLSIATLAQSVGHISGAHLNPAVTLGLLLSCQISIFRALMYIIAQCVGAIVA  108
AQP3  FLTINLAFGFAVTLGILIAGQVSGAHLNPAVTFAMSFLARDPWIKLPIYTLAQTLGAFLG  115
GlpF  QWEISVIWGLGVAMAIYLTAGVSGAHLNPAVTIALWLFACFDKRKVIPFIVSQVAGAFCA  100

AqpZ  AALVYLIASGKT-GFD---AAASG----FASNGYGEHSPGGYS-MLSALVVELVLSAGFL  145
AQP1  TAILSGITSS----------LTGNS-LGRNDLADGVNSGQ-----GLGIEIIGTLQLV  150
AQP3  AGIVFGLYYDAIWHFADNQLFVSGPN---GTAGIFATYPSGHLDMINGFFDQFIGTASLI  171
GlpF  AALVYGLYYNLFFDFEQTHHIVRGSVESVDLAGTFSTYPNPHINFVQAFAVEMVITAILM  159

AqpZ  LVIHGATDKFA----PAG-FAPIAIGLALTLIHLISIPVTNTSVNPARSTAV---AIFQG  197
AQP1  LCVLATTDRRRRD--LGG-SAPLAIGLSVALGHLLAIDYTGCGINPARSFGS---AVITH  204
AQP3  VCVLAIVDPYNNPG-PRG-LEAFTVGLVVLVIGTSMGFNSGYAVNPARDFGPRLFTALAG  229
GlpF  GLILALTDDGN--GVPRGPLAPLLIGLLIAVIGASMGPLTGFAMNPARDFGPKVFAWLAG  217

AqpZ  G----WALE----QLWFFWVVPIVGGIIGGLIYRTLL------------------EKRD  230
AQP1  N----FS-N-----HWIFWVGPFIGGALAVLIYDFILAPRSSDLTDRVKVWTSGQVEEYD  254
AQP3  WGSAVFTTGQ--HWWWVPIVSPLLGSIAGVFVYQLMIGCHLE-------QPPPSNEEEN  279
GlpF  WGNVAFTGGRDIPYFLVPLFGPIVGAIVGAFAYRKLIGRHLP--------CDICVVEEKE  269

AqpZ  ---------------                                            231
AQP1  LDADDINSRVEMKPK                                            269
AQP3  VKLAHVKHKEQI---                                            292
GlpF  TTTP-SEQKASL---                                            281
```

an X-linked disorder in which patients' kidneys fail to respond to vasopressin (nephrogenic diabetes insipidus, NDI) was identified, and several of these kindreds were found to carry mutations in the gene encoding the vasopressin V2 receptor on the X chromosome (reviewed by Bichet, 1998). Once the AQP2 water channel was identified, genomic DNA from NDI kindreds lacking V2 receptor mutations were found to contain mutations in *AQP2* causing recessive (Van Lieburg *et al.*, 1994) or dominantly inherited disorders (Mulders *et al.*, 1998).

Although studies thus far have been restricted to rodent models, several common clinical disorders of water balance are now believed to reflect secondary abnormalities in AQP2 expression. These provocative studies suggest that the human counterparts of these disorders will be linked to AQP2 expression. Important advances include the recognition that expression of AQP2 protein is reduced after lithium administration, a commonly used agent in the treatment of manic depressive disease that is known to produce polyuria (Marples *et al.*, 1995). Polyuria is a frequent side effect after removal of urinary obstruction (Frokiaer *et al.*, 1996), in diuretic-induced hypokalemia (Marples *et al.*, 1996), and in nocturnal enuresis (Frokiaer and Nielsen, 1997); AQP2 down-regulation has been described in each of these settings. Overexpression of AQP2 may contribute to the excessive water retention that frequently complicates congestive heart failure (Nielsen *et al.*, 1997c; Xu *et al.*, 1997). Most recently, overexpression of AQP2 has been implicated in the fluid retention during pregnancy (Ohara *et al.*, 1998), which may precipitate eclampsia.

2. AQP0 (MIP) from Lens Fiber Cells

Although the function of this protein is still unsettled, it was the first member of the family to be isolated and cloned (Gorin *et al.*, 1984). When expressed in oocytes, AQP0 confers only a modest increase in membrane permeability (Chandy *et al.*, 1997; Kushmerick *et al.*, 1995; Mulders *et al.*, 1995); however, it was recently noted that water permeability is elevated under slightly acidic conditions, pH 6 (Nemeth-Cahalan and Hall, 2000). Although the protein was reported to conduct ions when reconstituted into planar lipid bilayers (Ehring *et al.*, 1990), increased membrane currents were not detected when AQP0 is expressed in oocytes (Mulders *et al.*, 1995).

AQP0 protein is expressed only in fiber cells of lens, a site not known to have large levels of fluid transport, and a secondary structural role as a cell-to-cell

---

FIGURE 8 Phylogenetic and sequence analysis of mammalian and *Escherichia coli* aquaporin homologs. (*Top*) Mammalian aquaporins (*shaded*) and aquaglyceroporins (*unshaded*). Reproduced with permission from *The Journal of Biological Chemistry* (Agre *et al.*, 1998a). (*Bottom*) Sequence alignment of mammalian AQP1 and AQP3 with *E. coli* AqpZ and GlpF. Reproduced with permission from *The Annual Review of Biochemistry* (Borgnia *et al.*, 1999).

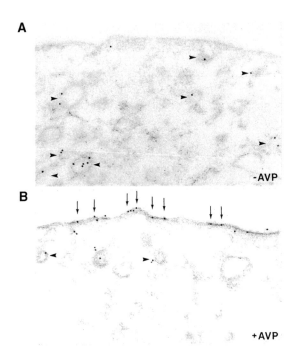

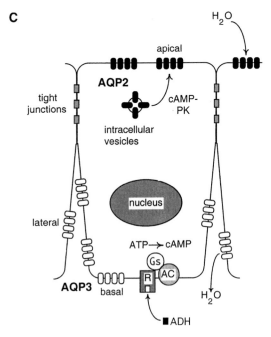

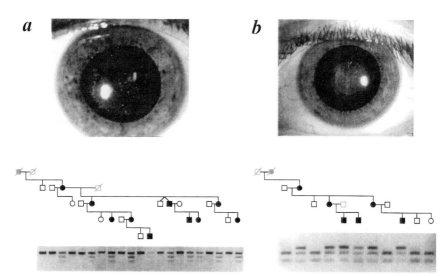

FIGURE 10 Cataracts in patients from two unrelated families with mutations in the gene encoding AQP0. (*a*) Lens examination of patient with T138R mutation showing multiple opacities throughout. (*b*) Lens examination of patient with E134G mutation showing single lamellar opacity. (*Bottom*) Restriction fragment analyses demonstrates cosegregation of patterns with cataract in both families (black symbols). Reproduced with permission from *Nature Genetics* (Berry *et al.*, 2000).

adhesion protein has been proposed (Fotiadis *et al.*, 2000). The protein is essential for lens homeostasis, because mice bearing mutations in the gene suffer from congenital cataracts (Shiels and Bassnett, 1996). This is now known to be responsible for some forms of congenital cataract in humans (Berry *et al.*, 2000; Francis *et al.*, 2000). Interestingly, in mice and humans, the cataracts are a dominant trait, consistent with a structural defect, and the different mutations cause clinically distinct forms of cataract (Fig. 10).

3. AQP4, the Brain Water Channel

The homolog was cloned from brain (Jung *et al.*, 1994b) and lung (Hasegawa *et al.*, 1994). Unlike most other aquaporins, AQP4 is not inhibited by mercurials. Also unlike other aquaporins, AQP4 has two translation initiation sites encoding

FIGURE 9 AQP2 in principal cells of renal collecting duct. (A) Immunogold electron microscopy of rat collecting duct principal cell in unstimulated state. (B) Same after stimulation with vasopressin. Reproduced with permission from *Proceedings of the National Academy of Science USA* (Nielsen *et al.*, 1995d). (C) Schematic model for transcellular water movement through collecting duct principal cell as regulated by exocytosis of AQP2. Reproduced with permission from *Kidney International* (Nielsen and Agre, 1995).

two isoforms of 301 and 323 amino acids (Lu *et al.*, 1996). Although the biological need for this is uncertain, AQP4 homotetramers formed from the 301 residue monomers exist, as well as AQP4 heterotetramers containing both 301 and 323 residue monomers (Neely *et al.*, 1999).

The distribution of AQP4 has been defined in multiple tissues, including renal collecting duct principal cells and airway epithelia (Frigeri *et al.*, 1995; Nielsen *et al.*, 1997a; Terris *et al.*, 1995). AQP4 is most abundantly expressed in brain where it resides in the astroglial cells surrounding blood vessels (Fig. 11A) (Nagelhus *et al.*, 1998; Nielsen *et al.*, 1997b). Square arrays have been observed in the plasma membranes of astroglial and ependymal cells, but their molecular identities were uncertain. Similar arrays were observed in cultured cells transfected with *AQP4* (Yang *et al.*, 1996), and direct labeling of fracture replicas with anti-AQP4 established that square arrays contain AQP4 (Rash *et al.*, 1998) (Fig. 11B). This distribution predicted that AQP4 is involved in dissipation of brain edema or possibly disturbances of brain water homeostasis such as pseudotumor cerebrii (Lee *et al.*, 1997). Human mutants have not yet been identified, but mice bearing *AQP4* gene disruption were found to have improved survival in acute brain edema models (Manley *et al.*, 2000) (see Chapter 5). AQP4 is also expressed in fast twitch skeletal muscle fibers and is deficient in the mouse model of Duchenne's muscular dystrophy (Frigeri *et al.*, 1998). Regulation of AQP4 water permeability by protein kinase C was proposed after a study demonstrated that high concentrations of phorbol diesters partially reduce the water permeability of AQP4 oocytes (Han *et al.*, 1998); however, these agents are known to induce nonspecific perturbations of surface membrane.

4. AQP5 from Secretory Glands

This homolog was cloned from a submandibular gland cDNA library (Raina *et al.*, 1995). Immunohistochemical and immunogold electron microscopic studies also demonstrated the protein in the apical membranes of secretory glands, type I pneumocytes, and corneal epithelium (Hamann *et al.*, 1998; Nielsen *et al.*, 1997a). Although regulation of AQP5 has been proposed, no convincing evidence suggests that protein trafficking or gating occurs. The distribution implicates AQP5 in human dry disorders. Recent studies of lacrimal glands and salivary glands from patients with Sjögren's syndrome show abnormal trafficking of AQP5 (Steinfeld *et al.*, 2001; Tsubota *et al.*, 2001). In support of this, mice with disruptions in the gene encoding AQP5 were found to have reduced salivary gland secretion (Ma *et al.*, 1999).

B. *Mammalian Aquaglyceroporins*

AQP3, AQP7, AQP9, and possibly AQP8 constitute a subgroup of aquaporins with a broader permeation range that includes glycerol, hence the name

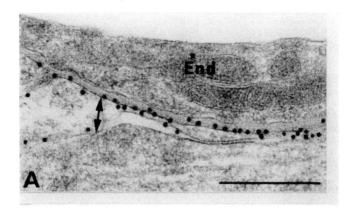

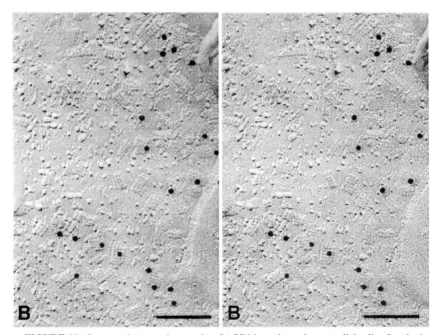

FIGURE 11 Immunoelectron micrographs of AQP4 in perivascular astroglial cells of rat brain. (A) Ultrathin section of rat cerebellum reacted with anti-AQP4. Reproduced with permission from *The Journal of Neuroscience* (Nielsen *et al.*, 1997b). (B) Stereoscopic view of end-feet in freeze fracture replica. Reproduced with permission from *The Proceedings of the National Academy of Science* (Rash *et al.*, 1998).

"aquaglyceroporins." Three groups cloned AQP3 at the same time (Echevarria *et al.*, 1994; Ishibashi *et al.*, 1994; Ma *et al.*, 1994). Although one group of scientists initially disputed the permeation by water, all agreed that the protein transports glycerol. Likewise, it seems most likely that the protein transports water and glycerol through the same pore (Kuwahara *et al.*, 1997) although evidence for two distinct pathways has been reported (Echevarria *et al.*, 1996). The distribution in multiple sites, including the basolateral domains of renal collecting duct principal cells, airway epithelia, and secretory glands, suggests several functions (Ecelbarger *et al.*, 1995; Frigeri *et al.*, 1995; Nielsen *et al.*, 1997a). Mice engineered with disruptions of the *AQP3* gene exhibit a nephrogenic diabetes insipidus phenotype (Chapter 5); however, human mutants have not yet been reported.

Curiously, AQP7, AQP8, and AQP9 were each simultaneously cloned by multiple groups. AQP7 is permeated by water and glycerol. The permeation of AQP8 and AQP9 is disputed, but strong evidence supports the permeation of AQP9 by a range of solutes (Tsukaguchi *et al.*, 1998). The subcellular distribution of these homologs is still uncertain and species differences have been reported. Although rigorous tissues and subcellular localization studies are not yet available, AQP7 is expressed in testis (Ishibashi *et al.*, 1997); AQP8 is expressed in pancreas and liver (Koyama *et al.*, 1997); and AQP9 is expressed in liver and leukocytes (Tsukaguchi *et al.*, 1998).

C. AQP6, a Very Strange Member of the Family

Another homolog (AQP6) was cloned from kidney but evoked marginal scientific interest because it was reported to bear only minimal water permeability, which was allegedly inhibited by Hg^{2+} (Ma *et al.*, 1995). The AQP6 protein was subsequently demonstrated to reside within intracellular vesicles in three kidney epithelia: glomerular podocytes and proximal straight tubules. The AQP6 protein is most abundantly expressed in the α-intercalated cells of collecting duct (Yasui *et al.*, 1999a). Note that these cells lie alongside the principal cells of collecting duct, where the AQP2 resides in intracellular vesicles traffic to the plasma membrane and are then reinternalized. In contrast, AQP6 appears exclusively in the membrane of intracellular vesicles alongside H^+-ATPase (Yasui *et al.*, 1999b). The α-intercalated cells are known to play an important function in acid–base metabolism by releasing acid into the urinary lumen.

The function of AQP6 was recently reevaluated by expression in oocytes. Surprisingly, the low level of water permeability was significantly increased by treatment with Hg^{2+} (Yasui *et al.*, 1999b), and two cysteine residues were demonstrated to be required for this effect (Fig. 12A). Moreover, the Hg^{2+}-treated AQP6 oocytes

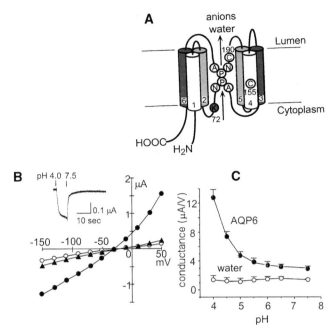

FIGURE 12 AQP6 model and electrophysiologic analyses in *X. laevis* oocytes. (A) Membrane topology of AQP6 showing positively charged residue (K72) at cytoplasmic mouth of pore, juxtaposed NPA motifs, and cysteines critical to Hg^{2+} activation (C155 and C190). (B) IV plot of AQP6 oocyte at pH 7.5 (open circles), after 1 min at pH 4.0 (solid circles), or 1 min after return to pH 7.5 (triangles). (C) Ion conductance of AQP6 oocytes and control water-injected oocytes measured at different pH. Reproduced with permission from *Nature* (Yasui *et al.*, 1999b).

were found to have a large membrane current that exhibited intermediate selectivity for anions. Because AQP6 is expressed in acid-secreting cells, it was questioned whether these effects may be triggered by low pH. A shift to pH 4.0 rapidly induced the ion permeation and swelling, which were rapidly reversed by return to pH 7.0 (Fig. 12B,C). AQP6 has several residues that are not present in other aquaporins. In particular, a charged residue Lys 72 lies at the cytoplasmic mouth of the pore (Fig. 12A), and mutagenesis to the negatively charged Glu led to loss of ion selectivity.

The presence of AQP6 in intracellular vesicles in α-intercalated cells suggests that the H^+-ATPase creates an intravesicular acid environment, opening AQP6 to maintain electroneutrality. The presence of AQP6 in proximal straight tubules suggests that the protein may participate in the acidification of endosomal vesicles. These features are distinct from those of all other members of the aquaporin family and suggest that the biophysical properties of aquaporins may be much more complex than previously believed.

D. Multiple Aquaporins in a Tissue

The initial studies of AQP1 suggested that multiple homologs must exist to explain water permeation of tissues lacking the protein (Nielsen *et al.,* 1993a,b). Although not originally foreseen, multiple homologs have been identified in certain tissues. Indeed the assembly of several aquaporins creates a complex intracellular plumbing system. The biological need for multiple aquaporins seemingly reflects multiple needs: (1) the need for regulated and unregulated aquaporins; (2) the need for channels exclusively permeated by water or the need for channels permeated by water plus small solutes or even ions; and (3) the need for specific trafficking signals that target aquaporins to specific plasma membrane domains or intracellular sites in polarized epithelia.

As discussed earlier, the kidney contains at least six aquaporins in nonoverlapping distributions: AQP1, AQP2, AQP3, AQP4, AQP6, and AQP7. The collecting duct exhibits the greatest complexity, with principal cells containing AQP2 in intracellular vesicles and at the plasma membrane and AQP3 or AQP4 at the basolateral domains (Fig. 9C). The adjacent α-intercalated cells are only known to contain AQP6, whose primary function may be to provide anion transport.

The epithelia of lung tissues are also complex (Nielsen *et al.,* 1997a). Distal lung features AQP5 in the apical membrane of type I epithelia, a flattened cell type that provides most of the gas exchange surface area. Interestingly, no aquaporin has yet been identified on the basolateral membrane of these cells (Fig. 13A). In the airway epithelia, even more complexity is present with AQP3 in basal cells and AQP4 in the lateral membranes of surface epithelia. AQP1 is present in subepithelial fibroblasts and capillary endothelia (Fig. 13B). Secretory glands contain two different aquaporins in the basolateral membranes (AQP3 or AQP4) and AQP5 in the apical membrane (Fig. 13C).

The eye is also complex, with at least five different aquaporins present (Hamann *et al.,* 1998) (Fig. 13D). As noted previously, AQP1 exists in nonpigment ciliary epithelia, outflow tracks, lens epithelium, and corneal endothelium. It is noteworthy that AQP1-null humans are not known to exhibit ocular dysfunction. However, it is very possible that the protein is necessary during injury or other stress. AQP0 resides in lens, where its absence causes cataract. AQP3 resides in the conjunctivae, AQP5 in corneal epithelium and lacrimal glands, and AQP4 in retinal Müller cells. Presumed functions may be inferred from the sites of expression; however, demonstration of precise physiological or pathophysiological roles has not yet occurred.

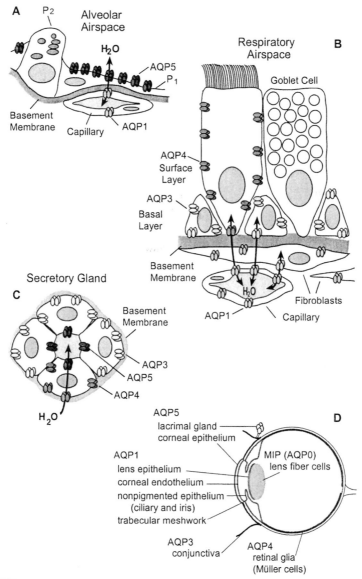

FIGURE 13 Schematic diagrams of multiple aquaporins in complex tissues of lung and eye. (A) Alveolus, (B) trachea, (C) secretory acinus, and (D) eye. Reproduced with permission from *The American Journal of Physiology* (Hamann *et al.,* 1998; Nielsen *et al.,* 1997a).

IV. NONMAMMALIAN HOMOLOGS

The recent elucidation of multiple genomes is revealing aquaporins and aqua-glyceroporins in virtually all life forms. It is anticipated that aquaporins will play multiple physiological functions in these diverse species.

A. Insect Homologs

The big brain (Bib) *Drosophila* mutant was recognized before the function of AQP1 was established (Rao *et al.*, 1990). The mutant phenotype is a distinctive perturbation during the developmental switch that occurs when neural progenitors and epithelial progenitors diverge from a common cell type. Although the Bib protein must be performing an important function, no biophysical demonstration of water permeation or another transport event has yet been shown.

A second insect homolog was found in the sap-sucking insect *Cicadella viridis* (AQP$_{CIC}$). These proteins are extremely abundant in the filter plate of the midgut, where large amounts of fluid are taken in and transferred to Malpighian tubules (Beuron *et al.*, 1995). Interesting structural information is being obtained by studying this native protein.

B. Microbial Homologs

The *Escherichia coli* glycerol facilitator protein (GlpF) was discovered long before aquaporins were recognized (Heller *et al.*, 1980), but the second homolog (AqpZ) was discovered more recently (Calamita *et al.*, 1995). *E. coli* genome sequencing has confirmed that these are the only two aquaporin homologs in that bacteria. As noted previously, phylogenetic analysis of mammalian aquaporins splits the superfamily into two subfamilies—orthodox aquaporins and aquaglycero-porins (Fig. 8A). When compared to these sequences, AqpZ aligns with the mammalian orthodox aquaporins, and oocytes expressing this protein are permeated by water but not glycerol (Calamita *et al.*, 1995). In contrast, GlpF aligns with mammalian aquaglyceroporins, and oocytes expressing GlpF are permeated by glycerol, whereas water permeation was not detected (Maurel *et al.*, 1994).

Structural explanations for these functional differences are being sought. The primary amino acid sequences are highly homologous, yet a few sequence differences are present (Fig. 8B). Of note, AqpZ and orthodox aquaporins have a Ser following the second NPA motif in loop E (NPARS), whereas GlpF and aquaglyceroporins have an Asp at this site (NPARD) as well as a motif (GLYY) at the extracellular end of the third transmembrane domain. Electron microscopic studies of GlpF expressed in oocytes were interpreted as showing that the protein is a monomer (Bron *et al.*, 1999), whereas electron crystallographic studies of purified, reconstituted GlpF at 7 Å showed a tetrameric assembly with a central hourglass geometry, suggesting a larger pore (Braun *et al.*, 2000). Moreover, two residues at the extracellular end of the sixth transmembrane domain have been reported to determine whether AQP_{CIC} transports water like an orthodox aquaporin (YW) or glycerol like an aquaglyceroporin (PL) (Lagrée *et al.*, 1998, 1999); however, preliminary studies from other laboratories have not confirmed the general significance of these observations.

Why a simple gram-negative organism would need two functionally related members of this protein family is puzzling. Presumably, the specificity of AqpZ for water but not glycerol and GlpF is important to the bacterium for glycerol transport. Nevertheless, null phenotypes are not strikingly obvious. AqpZ-null bacteria exhibit relatively less growth than the parent strain when cultured under conditions of maximal growth (Calamita *et al.*, 1998). The genomes of most bacteria contain only one member of the aquaporin family, and the functional importance is yet undefined.

Saccharomyces cerevisiae is a favorite organism for genetic manipulations, and the recent genome sequencing has revealed genes for four aquaporin homologs. Two encode proteins with sequences more closely related to aquaglyceroporins, of which the FPS1 homolog has been studied extensively (see Chapter 8). Two other sequences similar to orthodox aquaporins have been identified (Bonhivers *et al.*, 1998; Laizé *et al.*, 1999). In lab strains, *AQY1* encodes a protein that is not functional when expressed in oocytes. Presumably, this reflects negative pressure, because *AQY1* from wild-type strains encodes a functional water channel (Bonhivers *et al.*, 1998). The second gene, *AQY2*, is fragmented in the sequenced genome; however, other strains contain an open reading frame that is nonfunctional in oocytes (Laizé *et al.*, 2000). The physiological significance of these two yeast aquaporin proteins is still being sought, and it remains unknown whether microbial aquaporins are involved in invasiveness or other clinically important processes.

C. Plant Homologs

The plant kingdom contains a large and complex group of organisms that need careful water balance. Hydraulic engineering is involved in many physiological functions of plants. Not surprisingly, numerous aquaporins have been identified in plants, and each species may carry genes for several dozen aquaporin homologs. The functions of some of these proteins are being uncovered, and a complex and fascinating story is emerging (reviewed by Maurel, 1997) (see Chapter 7).

Briefly, three types of plant homologs have been identified: tonoplast intrinsic proteins (TIPs), plasma membrane intrinsic proteins (PIPs), and the symbiosome protein in the roots of legumes colonized with nitrogen-fixing bacteria (Nod26).

When expressed in oocytes, some of these proteins (γ-TIP) have been shown to transport water (Maurel *et al.,* 1993), whereas the plant homolog Nod26 has broader solute permeation (Dean *et al.,* 1999). Why many plant homologs do not exhibit any function when expressed in oocytes is not understood (M. J. Chrispeels, personal communication, 2000).

Some plant aquaporins have been linked to physiologically important processes. As noted previously, the existence of turgor-responsive genes in plants included a protein with a sequence related to AQP1 (Guerrero *et al.,* 1990), suggesting that some members of the plant aquaporins sense water deprivation. By genetically reducing expression of the PIP1b homolog in *Arabidopsis thaliana* (Kaldenhoff *et al.,* 1998), the plant was still able to maintain stem turgor by increasing the arborization of rootlets (Fig. 14). This provides an interesting lesson on the need for care during phenotype determinations, because the plants appeared similar above the surface. Curiously, another aquaporin homolog TobRB7 is up-regulated

FIGURE 14   Functional consequence of aquaporin PIP1b expression in *A. thaliana. (Left)* Plant overexpressing antisense RNA exhibits low PIP1b protein content and compensates with increased arborization of rootlets. *(Right)* Genetically unmodified plant. Reproduced with permission from *The Plant Journal* (Kaldenhoff *et al.,* 1998).

in the roots of tobacco plants during nematodal infestation, permitting the parasite to suck fluid from the host organism (Opperman *et al.,* 1994). To preserve the well-recognized benefits of genetic diversity (hybrid vigor), most plants defend against self-pollination. An interesting aquaporin mutant in the crucifer family eliminates a checkpoint, suggesting that water transport between stigma and pollen is involved in this process (Ikeda *et al.,* 1997). Other physiological and pathophysiological processes are being sought, and it is likely that aquaporins will attract the continued interest of plant physiologists.

V. PERSPECTIVE

Discovery of the aquaporins has dramatically changed the field of membrane water transport, and the impact of this discovery on basic and clinical physiology is becoming obvious. Since publication of the first molecular water channel (Preston *et al.,* 1992), more than 700 reports have appeared in the scientific and clinical literature, and it is clear that aquaporin research will advance our understanding of diverse physiological and pathophysiological processes in humans, other vertebrates, invertebrates, microbials, and plants. It is humbling to see that a single, unexpected observation can still provide new insights that alter the course of biological research. Nevertheless, the greater importance of our work is often not obvious, and it is worth remembering a gentle caution issued a century ago:

In summary, there are no small problems. Problems that appear small are large problems that are not understood. Nature is a harmonious mechanism where all parts, including those appearing to play a secondary role, cooperate in the functional whole. No one can predict their importance in the future. All natural arrangements, however capricious they may seem, have a function.
—Santiago Ramón y Cajal, 1897, *Advice for a Young Investigator*

Acknowledgments

We thank our colleagues in these studies. In addition, we are grateful for support from the National Institutes of Health, the Cystic Fibrosis Foundation, and the Human Frontier Science Program.

References

Agre, P. (1997). Aquaporin nomenclature workshop: Mammalian aquaporins. *Biol. Cell* **89,** 321–329.

Agre, P. (1998). Aquaporin null phenotypes: The importance of classical physiology. *Proc. Natl. Acad. Sci. USA* **95,** 9061–9063.

Agre, P., and Cartron, J. P., eds. (1992). "Protein Blood Group Antigens of the Human Red Cell". Johns Hopkins University Press, Baltimore.

Agre, P., Saboori, A. M., Asimos, A., and Smith, B. L. (1987). Purification and partial characterization of the M_r 30,000 integral membrane protein associated with the erythrocyte Rh(D) antigen. *J. Biol. Chem.* **262,** 17497–17503.

Agre, P., Sasaki, S., and Chrispeels, M. J. (1993). Aquaporins, a family of membrane water channels. *Am. J. Physiol.* **265,** F461.

Agre, P., Lee, M. D., Devidas, S., Guggino, W. B., Sasaki, S., Uchida, S., Kuwahara, M., Fushimi, K., Marumo, F., Verkman, A. S., Yang, B., Deen, P. M. T., Mulders, S. M., Kansen, S. M., and van Os, C. H. (1997). Aquaporins and ion conductance. *Science* **275,** 1490–1492.

Agre, P., Bonhivers, M., and Borgnia, M. J. (1998a). The aquaporins, blueprints for cellular plumbing systems. *J. Biol. Chem.* **273,** 14659–14662.

Agre, P., Mathai, J. C., Smith, B. L., and Preston, G. M. (1998b). Functional analyses of the aquaporin water channels. *Methods Enzymol.* **294,** 550–572.

Anthony, T. L., Brooks, H. L., Boassa, D., Leonov, S., Yanochko, G. M., Regan, T. W., and Yool, A. J. (2000). Cloned human aquaporin-1 is a cyclic GMP gated ion channel. *Mol. Pharmacol.* **57,** 576–588.

Berry, V., Francis, P. F., Kaushal, S., Moore, A., and Bhattacharya, S. (2000). Missense mutations in MIP underlie autosomal dominant 'polymorphic' and lamellar cataracts linked to 12q. *Nat. Genet.* **25,** 15–17.

Beuron, F., Le Caherec, F., Guillam, M. T., Cavalier, A., Garret, A., Tassan, J. L., Delamarche, C., Schultz, P., Mallouh, V., Rolland, J. P., and Thomas, D. (1995). Structural analysis of a MIP family protein from the digestive tract of Cicadella viridis. *J. Biol. Chem.* **270,** 17414–17422.

Bichet, D. G. (1998). Nephrogenic diabetes insipidus. *Am. J. Med.* **105,** 431–442.

Bondy, C., Chin, E., Smith, B. L., Preston, G. M., and Agre, P. (1993). Developmental gene expression and tissue distribution of the CHIP28 water channel protein. *Proc. Natl. Acad. Sci. USA* **90,** 4500–4504.

Bonhivers, M., Carbrey, J., Gould, S. J., and Agre, P. (1998). Aquaporins in *Saccharomyces:* Genetic and functional distinctions between genetic and wild-type strains. *J. Biol. Chem.* **273,** 27565–27572.

Borgnia, M., Nielsen, S., Engel., A., and Agre, P. (1999). Cellular and molecular biology of the aquaporin water channels. *Annu. Rev. Biochem.* **68,** 425–458.

Braun, T., Philippsen, A., Wirtz, S., Borgnia, M. J., Agre, P., Kühlbrandt, W., Engel, A., and Stahlberg, H. (2000). Projection structure of the glycerol facilitator at 3.5 Å resolution. *EMBO Rep.* **1,** 183–189.

Bron, P., Lagree, V., Froger, A., Rolland, J. P., Hubert, J. F., Delamarche, C., Deschamps, S., Pellerin, I., Thomas, D., and Haase, W. (1999). Oligomerization state of MIP proteins expressed in Xenopus oocyte revealed by freeze-fracture electron-microscopy analysis. *J. Struct. Biol.* **30,** 287–296.

Calamita, G., Bishai, W. R., Preston, G. M., Guggino, W. B., and Agre, P. (1995). Molecular cloning and characterization of aquaporin Z: A water channel from *E. coli. J. Biol. Chem.* **270,** 29063–29066.

Calamita, G., Kempf, B., Bonhivers, M., Bishai, W. R., Bremer, E., and Agre, P. (1998). Regulation of the *E. coli.* water channel gene, AqpZ. *Proc. Natl. Acad. Sci. USA* **95,** 3627–3631.

Chandy, G., Zampighi, G. A., Kreman, M., and Hall, J. E. (1997). Comparison of the water transporting protperited of MIP and AQP1. *J. Membr. Biol.* **159,** 29–39.

Cheng, A., van Hoek, A. N., Yeager, M., Verkman, A. S., and Mitra, A. K. (1997). Three-dimensional organization of a human water channel. *Nature* **387,** 627–630.

Cowan, C. A., Yokoyama, N., Bioanci, L. M., Henkemeyer, M., and Fritzsch, B. (2000). EphB2 guides axons at the midline and is necessary for normal vestibular function. *Neuron* **26,** 417–430.

Dean, R. M., Rivers, R. L., Zeidel, M. L., and Roberts, D. M. (1999). Purification and functional reconstitution of soybean nodulin 26. An aquaporin with glycerol transport properties. *Biochemistry* **38,** 347–353.

Deen, P. M., Verdijk, M. A., and Knoers, N. V. *et al.* (1994). Requirement of human renal water channel aquaporin-2 for vasopressin-dependent concentration of urine. *Science* **264,** 92–95.

Denker, B. M., Smith, B. L., Kuhajda, F. P., and Agre, P. (1988). Identification, purification, and characterization of a novel *Mr* 28,000 integral membrane protein from erythrocytes and renal tubules. *J. Biol. Chem.* **263,** 15634–15642.

Ecelbarger, C. A., Terris, J., Frindt, G., Echevarria, M., Marples, D., Nielsen, S., and Knepper, M. A. (1995). Aquaporin-3 water channel localization and regulation in rat kidney. *Am. J. Physiol.* **269,** F663–F672.

Echevarria, M., Windhager, E. E., Tate, S. S., and Frindt, G. (1994). Cloning and expression of AQP3, a water channel from medullary collecting duct of rat kidney. *Proc. Natl. Acad. Sci. USA* **91,** 10997–11001.

Echevarria, M., Windhager, E. E., and Frindt, G. (1996). Selectivity of the renal collecting duct water channel aquaporin-3. *J. Biol. Chem.* **271,** 25079–25082.

Ehring, G. R., Zampighi, G., Horwitz, J., Bok, D., and Hall, J. (1990). Properties of channels reconstituted from the major intrinsic protein of lens fiber membranes. *J. Gen. Physiol.* **96,** 631–664.

Finkelstein, A. (1987). Water Movement through Lipid Bilayers, Pores, and Plasma Membranes. Theory and Reality. Wiley, New York.

Fischbarg, J., Kuang, K., Vera, J. C., Arant, S., Silverstein, S. C., Loike, J., and Rosen, O. M. (1990). Glucose transporters serve as water channels. *Proc. Natl. Acad. Sci. USA* **87,** 3244–3247.

Fotiadis, D., Hasler, L., Müller, D. J., Stahlberg, H., Kistler, J., and Engel, A. (2000). Surface topography of lens MIP (AQP0) supports two functions. *J. Mol. Biol.* **300,** 779–789.

Francis, P., Chung, J.-J., Yasui, M., Berry, V., Moore, A., Wyatt, M. K., Wistow, G., Bhattacharya, S. S., and Agre, P. (2000). Functional impairment of lens aquaporin in two families with dominantly inherited cataract. *Human Mol. Genet.* **9,** 2329–2334.

Frigeri, A., Gropper, M., Turck, C. W., and Verkman, A. S. (1995). Immunolocalization of the mercurial-insensitive water channel and glycerol intrinsic protein in epithelial cell plasma membranes. *Proc. Natl. Acad. Sci. USA* **92,** 4328–4331.

Frigeri, A., Nicchia, G. P., Verbavatz, J. M., Valenti, G., and Svelto, M. (1998). Expression of aquaporin-4 in gast-twitch fibers of mammalian skeletal muscle. *J. Clin. Invest.* **102,** 695–703.

Frokiaer, J., and Nielsen, S. (1997). Do aquaporins have a role in nocturnal enuresis? *Scan. J. Urol. Nephrol. Suppl.* **183,** 31–32.

Frokiaer, J., Marples, D., and Knepper, M. A. *et al.* (1996). Bilateral ureteral obstruction downregulates expression of vasopressin-sensitive AQP-2 water channel in rat kidney. *Am. J. Physiol.* **270,** F657–F668.

Fushimi, K., Uchida, S., Hara, Y., Hirata, Y., Marumo, F., and Sasai, S. (1993). Cloning and expression of apical membrane water channel of rat kidney collecting tubule. *Nature* **361,** 549–552.

Gorin, M. B., Yancey, S. B., Cline, J., Revel, J. P., and Horwitz, J. (1984). The major intrinsic protein (MIP) of the bovine lens fiber membrane. *Cell* **39,** 49–59.

Guerrero, F. D., Jones, J. T., and Mullet, J. E. (1990). Turgor-responsive gene transcription and RNA levels increase rapidly when pea shoots are wilted. *Plant Mol. Biol.* **15,** 11–26.

Hamann, S., Zeuthen, T., la Cour, M., Nagelhus, E., Ottersen, O. P., Agre, P., and Nielsen, S. (1998). Aquaporins in complex tissues: III. Distribution of aquaporins 1–5 in human and rat eye. *Am. J. Physiol.* **274,** C1332–C1345.

Han, Z., Wax, M. B., and Patil, R. V. (1998). Regulation of aquaporin-4 water channels by phorbol ester-dependent protein phosphorylation. *J. Biol. Chem.* **273,** 6001–6004.

Harris, H. W., Hosselet, C., Guay-Woodford, L., and Zeidel, M. L. (1992). Purification and partial characterization of candidate antidiuretic water channel proteins of Mr 55,000 and 53,000 from toad bladder. *J. Biol. Chem.* **267,** 22115–22121.

Hasegawa, H., Ma, T., Skach, W., Matthay, M. A., and Verkman, A. S. (1994). Molecular cloning of a mercurial sensitive water channel expressed in selected water-transporting tissues. *J. Biol. Chem.* **269,** 5497–5500.

Hediger, M. A., Coady, J. M., Ikeda, T. S., and Wright, E. M. (1987). Expression cloning and cDNA sequencing of the Na+/glucose co-transporter. *Nature* **330,** 379–381.

Heller, K. B., Lin, E. C., and Wilson, T. H. (1980). Substrate specificity and transport properties of the glycerol facilitator of *Escherichia coli. J. Bacteriol.* **144,** 274–278.

Heymann, J. B., Agre, P., and Engel, A. (1998). Progress on the structure and function of aquaporin-1. *J. Struct. Biol.* **121,** 191–206.

Ikeda, S., Nasrallah, J. B., Dixit, R., Preiss, S., and Nasrallah, M. E. (1997). An aquaporin-like gene required for the Brassica self-incompatibility response. *Science* **276,** 1564–1566.

Ishibashi, K., Sasaki, S., Fushimi, K., Uchida, S., Kuwahara, M., and Marumo, F. (1994). Molecular cloning and expression of a member of the aquaporin family with permeability to glycerol and urea in addition to water expressed at the basolateral membrane of kidney collecting duct cells. *Proc. Natl. Acad. Sci. USA* **91,** 6269–6273.

Ishibashi, K., Kuwahara, M., Gu, Y., Kageyama, Y., Tohsaka, A., Suzuki, F., Marumo, F., and Sasaki, S. (1997). Cloning and functional expression of a new water channel abundantly expressed in testis and permeable to water, glycerol, and urea. *J. Biol. Chem.* **272,** 20782–20786.

Jung, J. S., Preston, G. M., Smith, B. L., Guggino, W. B., and Agre, P. (1994a). Molecular structure of the water channel through aquaporin CHIP: The hourglass model. *J. Biol. Chem.* **269,** 14648–14654.

Jung, J. S., Bhat, R. V., Preston, G. M., Baraban, J. M., and Agre, P. (1994b). Molecular characterization of an aquaporin cDNA from brain: Candidate osmoreceptor and regulator of water balance. *Proc. Natl. Acad. Sci. USA* **91,** 13052–13056.

Kachadorian, W. A., Wade, J. B., and DiScala, V. A. (2000). Vasopressin: Induced structural changes in toad bladder luminal membrane. *J. Am. Soc. Neph.* **11,** 376–380.

Kaldenhoff, R., Grote, K., Zhu, J. J., and Zimmermann, U. (1998). Significance of plasmalemma aquaporins for water-transport in arabidopsis thaliana. *Plant J.* **14,** 121–128.

King, L. S., Nielsen, S., and Agre, P. (1996). Aquaporin-1 water channel protein in lung: Ontogeny, steroid-induced expression, and distribution in rat. *J. Clin. Invest.* **97,** 2183–2191.

King, L. S., Nielsen, S., and Agre, P. (1997). Aquaporins in complex tissues: I. Developmental patterns in respiratory tract and glandular tissue of rat. *Am. J. Physiol.* **273,** C1541–C1548.

Koyama, Y., Yamamoto, T., Kondo, D., Funaki, H., Yaoita, E., Kawasaki, K., Sato, H., Hatakeyama, K., and Kihara, I. (1997). Molecular cloning of a new aquaporin from rat pancreas and liver. *J. Biol. Chem.* **272,** 30329–30333.

Kushmerick, C., Rice, S. J., Baldo, G. J., Haspel, H. C., and Mathias, R. T. (1995). Ion, water, and neutral solute transport in Xenopus oocytes expressing frog lens MIP. *Exp. Eye Res.* **61,** 351–362.

Kuwahara, M., Gu, Y., Ishibashi, K., Marumo, F., and Sasaki, S. (1997). Mercury-sensitive residues and pore site in AQP3 water channel. *Biochemistry* **36,** 13973–13978.

Lagrée, V., Froger, A., Deschamps, S., Pellerin, I., Delamarche, C., Bonnec, G., Gouranton, J., Thomas, D., and Hubert, J. F. (1998). Oligomerization state of water channels and glycerol facilitators: Involvement of loop E. *J. Biol. Chem.* **273,** 33949–33953.

Lagrée, V., Froger, A., Deschamps, S., Hubert, J. F., Delamarche, C., Bonnec, G., Thomas, D., Gouranton, J., and Pellerin, I. (1999). Switch from an aquaporin to a glycerol channel by two amino acids substitution. *J. Biol. Chem.* **274,** 6817–6819.

Laizé, V., Gobin, R., Rousselet, G., Badier, C., Hohmann, S., Ripoche, P., and Tacnet, F. (1999). Molecular and functional study of AQY1 from Saccharomyces cerevisiae. Role of the C-terminal domain. *Biochem. Biophys. Res. Commun.* **257,** 139–144.

Laizé, V., Tacnet, F., Ripoche, P., and Hohmann, S. (2000). Polymorphism of yeast aquaporins. *Yeast* **16,** 897–903.

Lee, M. D., King, L. S., and Agre, P. (1997). The aquaporin family of water channel proteins in clinical medicine. *Medicine* **76,** 141–156.

Li, H., Lee, S., and Jap, B. K. (1997). Molecular design of aquaporin-1 water channel as revealed by electron crystallography. *Nature Struct. Biol.* **4,** 263–265.

Lu, M., Lee, M. D., Smith, B. L., Jung, J. S., Agre, P., Verdijk, M. A. J., Merkx, G., Rijss, J. P. L., and Deen, P. (1996). The human aquaporin-4 gene: Definition of the locus encoding two water channel polypeptides in brain. *Proc. Natl. Acad. Sci. USA* **93,** 10908–10912.

Ma, T., Frigeri, A., Hasegawa, H., and Verkman, A. S. (1994). Cloning of a water channel homolog expressed in brain meningeal cells and kidney collecting duct that functions as a stilbene-sensitive glycerol transporter. *J. Biol. Chem.* **269,** 21845–21849.

Ma, T., Yang, B., Kuo, W. L., and Verkman, A. S. (1995). cDNA cloning and gene structure of a novel water channel expressed exclusively in human kidney: Evidence for a gene cluster of aquaporins at chromosome locus 12q13. *Genomics* **35,** 543–550.

Ma, T., Song, Y., Gillespie, A., Carlson, E. J., Epstein, C. J., and Verkman, A. S. (1999). Defective secretion of saliva in transgenic mice lacking aquaporin-5 water channels. *J. Biol. Chem.* **274,** 20071–20074.

Macey, R. I. (1984). Transport of water and urea in red blood cells. *Am. J. Physiol.* **246,** C195–203.

Manley, G. T., Fujimura, M., Ma, T., Feliz, F., Bollen, A., Chan, P., and Verkman, A. S. (2000). Aquaporin-4 deletion in mice reduces brain edema following acute water intoxication and ischemic stroke. *Nature Med.* **6,** 159–163.

Marples, D., Christensen, S., Christensen, E. I., Ottosen, P. D., and Nielsen, S. (1995). Lithium-induced downregulation of aquaporin-2 water channel expression in rat kidney medulla. *J. Clin. Invest.* **95,** 1838–1845.

Marples, D., Frokiaer, J., Dorup, J., Knepper, M. A., and Nielsen, S. (1996). Hypokalemia-induced downregulation of aquaporin-2 water channel expression in rat kidney medulla and cortex. *J. Clin. Invest.* **97,** 1960–1968.

Mathai, J. C., Mori, S., Smith, B. L., Preston, G. M., Mohandas, N., Collins, M., van Zijl, P. C. M., Zeidel, M. L., and Agre, P. (1996). Functional analysis of aquaporin-1 deficient red cells: the Colton-null phenotype. *J. Biol. Chem.* **271,** 1309–1313.

Maunsbach, A. B., Marples, D., Chin, E., Ning, G., Bondy, C., Agre, P., and Nielsen, S. (1997). Aquaporin-1 water channel expression in human kidney. *J. Am. Soc. Nephrol.* **8,** 1–14.

Maurel, C. (1997). Aquaporins and water permeability of plant membranes. *Annu. Rev. Plant Physiol. Plant Mol. Biol.* **48,** 399–429.

Maurel, C., Reizer, J., Schroeder, J. I., and Chrispeels, M. J. (1993). The vacuolar membrane protein gamma-TIP creates water specific channels in Xenopus oocytes. *EMBO J.* **12,** 2241–2247.

Maurel, C., Reizer, J., Schroeder, J. I., Chrispeels, M. J., and Saier, M. H. J. (1994). Functional characterization of the *Escherichia coli* glycerol facilitator, GlpF, in *Xenopus* oocytes. *J. Biol. Chem.* **269,** 11869–11872.

Mitsuoka, K., Murata, K., Walz, T., Hirai, T., Agre, P., Heymann, B., Engel, A., and Fujiyoshi, Y. (1999). The structure of aquaporin-1 at 4.5 Å resolution reveals short α-helices in the center of the monomer. *J. Struct. Biol.* **128,** 34–43.

Moon, C., King, L. S., and Agre, P. (1997). AQP1 expression in erythroleukemia cells: Genetic Regulation of glucocorticoid and chemical induction. *Am. J. Physiol.* **273,** C1562–C1570.

Mulders, S. M., Preston, G. M., Deen, P. M. T., Guggino, W. B., van Os, C. H., and Agre, P. (1995). Water channel properties of major intrinsic protein of lens. *J. Biol. Chem.* **270,** 9010–9016.

Mulders, S. M., Bichet, D. G., Rijss, J. P., Kamsteeg, E. J., Arthus, M. F., Lonergan, M., Fujiwara, M., Morgan, K., Leijendekker, R., van der Sluijs, P., van Os, C. H., and Deen, P. M. T. (1998). An aquaporin-2 water channel mutant which causes autosomal dominant nephrogenic diabetes insipidus is retained in the Golgi complex. *J. Clin. Invest.* **102,** 57–66.

Murata, K., Mitsuoka, K., Hirai, T., Walz, T., Agre, P., Heymann, J. B., Engel, A., and Fujiyoshi, Y. (2000). Structural determinants of water permeation through aqua porin-1. *Nature* **407,** 599–605.

Nagelhus, E. A., Veruki, M. L., Torp, R., Haug, F. M., Nielsen, S., Agre, P., and Ottersen, O. P. (1998). Aquaporin-4 water channel protein in the rat retina and optic nerve: Polarized expression in Müller cells and fibrous astrocytes. *J. Neurosci.* **18,** 2506–2519.

Nakhoul, N. L., Davis, B. A., Romero, M. F., and Boron, W. F. (1998). Effect of expressing the water channel aquaporin-1 on the CO_2 permeability. *Am. J. Physiol.* **274,** C543–548.

Neely, J. D., Christensen, B. M., Nielsen, S., and Agre, P. (1999). Heterotetrameric composition of aquaporin-4 water channels. *Biochemistry* **38,** 11156–11163.

Nemeth-Cahalan, K. L., and Hall, J. E. (2000). pH and calcium regulate the water permeability of aquaporin 0. *J. Biol. Chem.* **275**, 6777–6782.

Nielsen, S., and Agre, P. (1995). The aquaporin family of water channels in kidney. *Kidney Int.* **48**, 1057–1068.

Nielsen, S., Smith, B. L., Christensen, E. I., Knepper, M., and Agre, P. (1993a). CHIP28 water channels are localized in constitutively water-permeable segments of the nephron. *J. Cell Biol.* **120**, 371–383.

Nielsen, S., Smith, B. L., Christensen, E. I., and Agre, P. (1993b). Distribution of aquaporin CHIP in secretory and resorptive epithelia and capillary endothelia. *Proc. Natl. Acad. Sci. USA* **90**, 7275–7279.

Nielsen, S., Chou, C. L., Marples, D., Christensen, E. I., Kishore, B. K., and Knepper, M. A. (1995a). Vasopressin increases water permeability of kidney collecting duct by inducing translocation of aquaporin-CD water channels to plasma membrane. *Proc. Natl. Acad. Sci. USA* **92**, 1013–1017.

Nielsen, S., Pallone, T., Smith, B. L., Christensen, E. I., Agre, P., and Maunsbach, A. (1995b). Aquaporin-1 water channels in short and long loop descending thin limbs and in descending vasa recta in rat kidney. *Am. J. Physiol. Renal* **37**, F1023–1037.

Nielsen, S., King, L. S., Mønster Christensen, B., and Agre, P. (1997a). Aquaporins in complex tissues: II. Subcellular distribution in respiratory and glandular tissues of rat. *Am. J. Physiol.* **273**, C1549–C1561.

Nielsen, S., Nagelhus, E. A., Amiry-Moghaddam, M., Bourque, C., Agre, P., and Ottersen, O. P. (1997b). Specialized membrane domains for water transport in glial cells: High resolution immunogold cytochemistry of aquaporin-4 in rat brain. *J. Neurosci.* **17**, 171–180.

Nielsen, S., Terris, J., and Andersen, D. *et al.* (1997c). Congestive heart failure in rats is associated with increased expression and targeting of aquaporin-2 water channel in collecting duct. *Proc. Natl. Acad. Sci. USA* **94**, 5450–5455.

Ohara, M., Martin, P. Y., Xu, D. L., St. John, J., Pattison, T. A., Kim, J. K., and Schrier, R. W. (1998). Upregulation of aquaporin 2 water channel expression in pregnant rats. *J. Clin. Invest.* **101**, 1076–1083.

Opperman, C. H., Taylor, C. G., and Conkling, M. A. (1994). Root-knot nematode-directed expression of a plant root-specific gene. *Science* **263**, 221–223.

Pallone, T., Kishore, B. K., Nielsen, S., Agre, P., and Knepper, M. A. (1997). A selective pathway mediates NaCl induced water flux across descending vasa recta. *Am. J. Physiol. Renal* **272**, F587–F596.

Preston, G. M., and Agre, P. (1991). Isolation of the cDNA for erythrocyte integral membrane protein of 28 kilodaltons: Member of an ancient channel family. *Proc. Natl. Acad. Sci. USA* **88**, 11110–11114.

Preston, G. M., Carroll, T., Guggino, W. B., and Agre, P. (1992). Appearance of water channels in Xenopus oocytes expressing red cell CHIP28 protein. *Science* **256**, 385–387.

Preston, G. M., Jung, J. S., Guggino, W. B., and Agre, P. (1993). The mercury-sensitive residue at cysteine-189 in the CHIP28 water channel. *J. Biol. Chem.* **268**, 17–20.

Preston, G. M., Jung, J. S., Guggino, W. B., and Agre, P. (1994a). Membrane topology of Aquaporin CHIP: Analysis of functional epitope-scanning mutants by vectorial proteolysis. *J. Biol. Chem.* **269**, 1668–1673.

Preston, G. M., Smith, B. L., Zeidel, M. L., Moulds, J. J., and Agre, P. (1994b). Mutations in aquaporin-1 in phenotypically normal humans without functional CHIP water channels. *Science* **265**, 1585–1587.

Raina, S., Preston, G. M., Guggino, W. B., and Agre, P. (1995). Molecular cloning and characterization of an aquaporin cDNA from salivary, lacrimal, and respiratory tissues. *J. Biol. Chem.* **270**, 1508–1512.

Ramón, Y., and Cajal, S. (1897). "Advice for a Young Investigator." Translated by N. Swanson and L. W. Swanson. MIT Press, Cambridge, MA, 1999.

Rao, Y., Jan, L. Y., and Jan, Y. N. (1990). Similarity of the product of the *Drosophila* neurogenic gene big brain and transmembrane channel proteins. *Nature* **345**, 163–167.

Rash, J., Yasumura, T., Hudson, C. S., Agre, P., and Nielsen, S. (1998). Direct immunogold labeling of aquaporin-4 in square arrays of astroglial and ependymal cell membranes from rat brain and spinal cord. *Proc. Natl. Acad. Sci. USA* **95**, 11981–11986.

Reizer, J., Reizer, A., and Saier, M. H. (1993). The MIP family of integral membrane channel proteins: Sequence comparisons, evolutionary relationships, reconstructed pathway of evolution, and proposed functional differentiation of the two repeated halves of the proteins. *Crit. Rev. Biochem. Mol. Biol.* **28**, 235–257.

Roberts, S. K., Yano, M., Ueno, Y., Pham, L., Alpini, G., Agre, P., and LaRusso, N. (1994). Cholangiocytes express aquaporin CHIP and transport water via a channel-mediated mechanism. *Proc. Natl. Acad. Sci. USA* **91**, 13009–13013.

Sabolic, I., Valenti, G., Verbavatz, J. M., van Hoek, A. N., Verkman, A. S., Ausiello, D. A., and Brown, D. (1992). Localization of the CHIP28 water channel in rat kidney. *Am. J. Physiol.* **263**, C1225–C1233.

Saboori, A. M., Smith, B. L., and Agre, P. (1988). Polymorphism in the M_r 32,000 Rh protein purified from Rh(D)-positive and -negative erythrocytes. *Proc. Natl. Acad. Sci. USA* **85**, 4042–4045.

Shiels, A., and Bassnett, S. (1996). Mutations in the founder of the MIP gene family underlie cataract development in the mouse. *Nat. Genet.* **12**, 212–215.

Skach, W. R., Shi, L. B., Calayag, M. C., Frigeri, A., Lingappa, V. R., and Verkman, A. S. (1994). Biogenesis and transmembrane topology of the CHIP28 water channel at the endoplasmic reticulum. *J. Cell Biol.* **125**, 803–815.

Smith, B. L., and Agre, P. (1991). Erythrocyte Mr 28,000 transmembrane protein exists as a multisubunit oligomer similar to channel proteins. *J. Biol. Chem.* **266**, 6407–6415.

Smith, B. L., Baumgarten, R., Nielsen, S., Raben, D. M., Zeidel, M. L., and Agre, P. (1993). Concurrent expression of erythroid and renal aquaporin CHIP and appearance of water channel activity in perinatal rats. *J. Clin. Invest.* **92**, 2035–2041.

Smith, B. L., Preston, G. M., Spring, F. A., Anstee, D. J., and Agre, P. (1994). Human red cell aquaporin CHIP: I. Molecular characterization of ABH and Colton blood group antigens. *J. Clin. Invest.* **94**, 1043–1049.

Smith, H. W. (1937). "The Physiology of the Kidney." Oxford University Press, New York.

Solomon, A. K. (1968). Characterization of biological membranes by equivalent pores. *J. Gen. Physiol.* **51**, S335–364.

Solomon, A. K. (1989). Water channels across the red blood cell and other biological membranes. *Methods Enzymol.* **173**, 192–222.

Solomon, A. K., Chasan, B., Dix, J. A., Lukacovic, M. F., Toon, M. R., and Verkman, A. S. (1983). The aqueous pore in the red cell membrane. *Ann. NY Acad. Sci.* **414**, 97–124.

Steinfeld, S., Cogan, E., King, L. S., Agre, P., Kiss, R., and Delporte, C. (2001). Defective cellular trafficking of aquaporin-5 water channel protein in Sjögren's syndrome lacrimal glands. *Lab Invest* (in press).

Terris, J., Ecelbarger, C. A., Marples, D., Knepper, M. A., and Nielsen, S. (1995). Distribution of aquaporin-4 water channel espression within rat kidney. *Am. J. Physiol.* **269**, F775–785.

Tsubota, K., Hirai, S. I., King, L. S., Agre, P., Ishida, N., and Mita, S. (2001). Defective cellular trafficking of aquaporin-5 water channel protein in Sjögren's syndrome lacrimal glands. *The Lancet* (in press).

Tsukaguchi, H., Shayakul, C., Berger, U. V., Mackenzie, B., Devidas, S., Guggino, W. M., van Hoek, A. N., and Hediger, M. A. (1998). Molecular characterization of a broad selectivity neutral solute channel. *J. Biol. Chem.* **273**, 24737–24743.

Umenishi, F., Carter, E. P., Yang, B., Oliver, B., Matthay, M. A., and Verkman, A. S. (1996). Sharp increase in rat lung water channel expression in the perinatal period. *Am. J. Resp. Cell Mol. Biol.* **15**, 673–679.

Ussing, H. H. (1965). Transport of electrolytes and water across epithelia. *Harvey Lect.* **59**, 1–30.

van Hoek, A. N., and Verkman, A. S. (1992). Functional reconstitution of the isolated erythrocyte water channel. *J. Biol. Chem.* **267**, 18267–18269.

van Hoek, A. N., Hom, M. L., Luthjens, L. H., de Jong, M. D., Dempster, J. A., and van Os, C. H. (1991). Functional unit of 30-kDa for proximal tubule water channels as revealed by radiation inactivation. *J. Biol. Chem.* **266**, 16633–16635.

Van Lieburg, A. F., Verkijk, M. A., Knoers, V. V., van Essen, A. J., Proesmans, W., Mallmann, R., Monnens, L. A., van Oost, B. A., van Os, C. H., and Deen, P. M. (1994). Patients with autosomal nephrogenic diabetes insipidus homozygous for mutations in the aquaporin-2 water-channel gene. *Am. J. Hum. Genet.* **55**, 648–652.

Verbavatz, J. M., Brown, D., Sabolic, I., Valenti, G., Ausiello, D. A., van Hoek, A. N., Ma, T., and Verkman, A. S. (1993). Tetrameric assembly of CHIP28 water channels in liposomes and cell membranes: A freeze-fracture study. *J. Cell Biol.* **123**, 605–618.

Verkman, A. S. (1992). Water channels in cell membranes. *Ann. Rev. Physiol.* **54**, 73–108.

Wade, J. B., Stetson, D. L., and Lewis, S. A. (1981). ADH action: Evidence for a membrane shuttle mechanism. *Ann. NY. Acad. Sci.* **372**, 106–117.

Walz, T., Smith, B. L., Zeidel, M. L., Engel, A., and Agre, P. (1994). Biologically active two-dimensional crystals of Aquaporin CHIP. *J. Biol. Chem.* **269**, 1583–1586.

Walz, T., Hirai, T., Murata, K., Heymann, J. B., Mitsuoka, K., Fujiyoshi, Y., Smith, B. L., Agre, P., and Engel, A. (1997). Three-dimensional structure of aquaporin-1. *Nature* **387**, 624–627.

Wistow, G. J., Pisano, M. M., and Chepelinsky, A. B. (1991). Tandem sequence repeats in transmembrane channel proteins. *Trends Biochem. Sci.* **165**, 170–171.

Xu, D. L., Martin, P. Y., and Ohara, M., *et al.* (1997). Upregulation of aquaporin-2 water channel expression in chronic heart failure rat. *J. Clin. Invest.* **99**, 1500–1505.

Yang, B. X., Brown, D., and Verkman, A. S. (1996). The mercurial insensitive water channel (AQP4) forms orthogonal arrays in stably transfected Chinese hamster ovary cells. *J. Biol. Chem.* **271**, 4577–4581.

Yang, B., Fukuda, N., van Hoek, A., Matthay, M. A., Ma, T., and Verkman, A. S. (2000). Carbon dioxide permeability of aquaporin-1 measured in erythrocytes and kidney of mice and in reconstituted proteoliposomes. *J. Biol. Chem.* **275**, 2686–2692.

Yasui, M., Kwon, T. H., Knepper, M. A., Nielsen, S., and Agre, P. (1999a). Aquaporin-6: An intracellular vesicle water channel protein. *Proc. Natl. Acad. Sci. USA* **96**, 5808–5813.

Yasui, M., Hazama, A., Kwon, T. H., Nielsen, S., Guggino, W. B., and Agre, P. (1999b). Rapid gating and anion permeability of an intracellular aquaporin. *Nature* **402**, 184–187.

Yool, A. J., Stamer, W. D., and Regan, J. W. (1996). Forskolin stimulation of water and cation permeability in aquaporin-1 water channels. *Science* **273**, 1216–1218.

Zeidel, M. L., Ambudkar, S. V., Smith, B. L., and Agre, P. (1992). Reconstitution of functional water channels in liposomes containing purified red cell CHIP28 protein. *Biochemistry* **31**, 7436–7440.

Zeidel, M. L., Nielsen, S., Smith, B. L., Ambudkar, S. V., Maunsbach, A. B., and Agre, P. (1994). Ultrastructure, pharmacologic inhibition, and transport-selectivity of aquaporin CHIP in proteoliposomes. *Biochemistry* **33**, 1606–1615.

Zelinkski, T., Kaita, H., Gibson, T., Coghlan, G., Philipps, S., and Lewis, M. (1990). Linkage between the Colton blood group locus and ASSP11 on chromosome 7. *Genomics* **6**, 623–625.

Zhang, R., Logee, K. A., and Verkman, A. S. (1990). Expression of mRNA coding for kidney and red cell water channels in Xenopus oocytes. *J. Biol. Chem.* **265**, 15375–15378.

CHAPTER 2

The Aquaporin Superfamily: Structure and Function

Henning Stahlberg*, Bernard Heymann*, Kaoru Mitsuoka†, Yoshinori Fuyijoshi† and Andreas Engel*

*M. E. Müller-Institute for Microscopy, Biozentrum, University of Basel, CH-4056 Basel, Switzerland;

†Department of Biophysics, Faculty of Science, Kyoto University, Kyoto 606-8502, Japan

I. INTRODUCTION

Aquaporins function as channels for nonionic compounds. They are found in most organisms, and they form a protein family that has been named after the first sequenced member, the major intrinsic protein (MIP; now known as AQP0) of lens fiber cells (Gorin *et al.,* 1984). However, using the MIP designation is problematic, because a search in any of the popular sequence and structure databases for the key word "MIP" also yields proteins not related to the aquaporins, such as the mitochondrial intermediate peptidase and the macrophage infectivity potentiator. Therefore, a renaming of the family to the "aquaporin family" has been suggested (Heymann and Engel, 1999). Because this protein family comprises channels not only for water but also for small hydrophilic solutes, we will use the term "aquaporin superfamily" in this review. This family comprises more than 160 channel proteins, of which the aquaporin from the human red blood cell, AQP1, was the first water channel discovered (Preston *et al.,* 1992). A recent phylogenetic analysis of this family (Heymann and Engel, 1999) showed subdivisions agreeing mostly with those given by Park and Saier (1996). While some subfamilies and groupings within subfamilies are overpopulated with highly similar members (such as the TIP and PIP subfamilies), others feature only single sequences (such as the single sequence for the Archaea, AQParc). Thus, phylogenetic distances were used to define 46 subtypes (Heymann and Engel, 1999) (Fig. 1) whose sequences have subsequently been submitted to an extensive analysis yielding important structural clues (Heymann and Engel, 2000).

Square arrays observed in freeze–fracture replicas of lens fiber cells (Fig. 2) identified as AQP0 (Kistler and Bullivant, 1980), square arrays of AQPcic in the membranes of the filter chamber of *Cicadella viridis* (Beuron *et al.,* 1995), and square arrays of AQP4 in astrocyte and ependymocyte plasma membranes (Rash *et al.,*

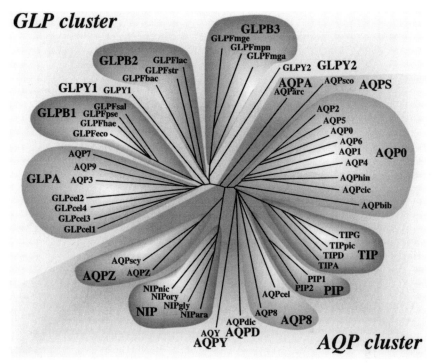

FIGURE 1 Phylogenetic analysis of the aquaporin superfamily suggests a classification into two clusters, AQP and GLP, 16 subfamilies, and 46 types. The types are considered to be representative of the whole family of more than 160 sequences obtained from Genbank, SWISS_PROT, EMBL, and the genome databases (Heymann and Engel, 1999). (See Color Plate.)

1998) document the propensity of aquaporins to assemble into two-dimensional (2D) crystals. This is related to the tetrameric nature of these proteins: AQP1 is a homotetramer containing four independent aqueous channels (Jung *et al.*, 1994; Shi *et al.*, 1994; Smith and Agre, 1991; Verbavatz *et al.*, 1993), and it sediments like a 190-kDa protein when solubilized in octylglucoside (Smith and Agre, 1991). AQP0 solubilized in decylmaltoside has a mass of 160 kDa (Hasler *et al.*, 1998b), while the *Escherichia coli* water channel, AqpZ, is a square-shaped particle having a 7-nm side length after solubilization in octylglucoside (Ringler *et al.*, 1999). The solubilized bacterial glycerol facilitator GlpF is a square-shaped particle having an 8-nm side length (Braun *et al.*, 2000), similar to the particles found in preparations of PM28, the water channel of spinach, suggesting tetrameric proteins as well (Fotiadis *et al.*, 2000).

The possibility of assembling 2D crystals from isolated aquaporin tetramers promoted structural analyses by electron crystallography. AQP1, the first water channel to be studied by this technique, was shown to assemble into highly ordered

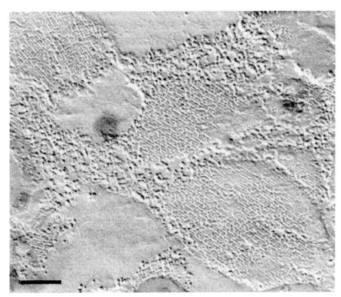

FIGURE 2   Tetragonal arrays in lens fiber cell membranes are revealed by freeze-fracture techniques. The arrays are assembled from the major itrinsic proteins, AQP0 (Kistler and Bullivant, 1980), the first member of the water channel protein family to be sequenced (Gorin *et al.*, 1984). Scale bar: 100 nm. (Courtesy of J. Kistler.)

arrays exhibiting a unit cell size of 9.6 nm and p42$_1$2 symmetry (Jap and Li, 1995; Mitra *et al.*, 1995; Walz *et al.*, 1995). But even the 0.35-nm resolution projection map (Jap and Li, 1995) did not allow the architecture of the molecule to be unraveled. Subsequent 3D maps at different resolution levels (Cheng *et al.*, 1997; Li *et al.*, 1997; Walz *et al.*, 1997), however, confirmed the hourglass model proposed by Jung *et al.* (1994). This model comprises six transmembrane α helices, which surround a central density formed by the two longest loops bearing the NPA sequence motif. A range of aquaporins have since been crystallized: AQP0 (Hasler *et al.*, 1998b), the bacterial water channel AqpZ (Ringler *et al.*, 1999), the bacterial glycerol facilitator GlpF (Braun *et al.*, 2000), a spinach plasma membrane water channel (Fotiadis *et al.*, 2000), and a plant vacuole water channel (Daniels *et al.*, 1999a).

Because sequence conservation is prominent in the six helical segments and the loops B and E (Heymann and Engel, 2000) that have been shown to fold back into the membrane, (Jung *et al.*, 1994), the largest structural variations among the aquaporins are expected to be found in the surface loops that connect the helices. Therefore, surface-sensitive techniques have been used to investigate the 2D crystals assembled from aquaporins. Surface reliefs reconstructed from freeze-dried, metal-shadowed crystals and topographs recorded with the atomic force microscope exhibit similar resolutions (in the best cases between 5 and 10 Å). However, the latter technique allows the surfaces to be studied in their native environment and function-related structural changes to be monitored (Müller and Engel, 1999).

X-ray crystallography is expected to provide the highest resolution structure among all techniques, but in spite of the availability of 3D crystals of AQP1 for several years (Wiener, 1999), and of AQP0 (Raimund Dutzler, personal communication, 1999), progress is hindered by either the fragility of the crystals or their limited order.

II. TWO-DIMENSIONAL CRYSTALLIZATION OF MEMBRANE PROTEINS

Membrane proteins are designed for the two-dimensional environment of the lipid bilayer. Therefore, an appropriate geometry for regularly packed membrane proteins is a 2D crystal assembled in the presence of lipids. Two-dimensional crystals are more easily produced than 3D crystals, because membrane proteins have a propensity to pack into a lipid bilayer. (For a recent review see Hasler *et al.*, 1998a). The reconstitution of solubilized membrane proteins into bilayers is achieved by mixing solubilized lipids and protein, and subsequent removal of the detergent. To this end, dialysis (Jap *et al.*, 1992), selective adsorption to Bio-Beads (Rigaud *et al.*, 1997), or dilution (Dolder *et al.*, 1996) can be used. The small micellar structures then coalesce into larger structures, leading to the formation of vesicles and sheets, both consisting of lipid bilayers with varying amounts of protein.

At high protein concentrations, the interaction between proteins in the membrane may lead to regular packing and 2D crystals (Fig. 3). With an excess of protein, crystals may still be obtained but a fraction of the protein ends up in amorphous aggregates. An important parameter is therefore the lipid:protein ratio (LPR), which should be low enough to promote crystal contacts between protein molecules, but not so low that the protein is lost to aggregation. The contents of both monodisperse protein and lipids are usually unknown, because protein assays do not indicate the amount of aggregates, and the amount of native lipids is difficult to assess. Therefore, the LPR must be optimized experimentally for every protein batch.

The lipid mixture used for reconstitution has an influence on the crystallization results. Crystallization requires the lipid bilayer to be in the fluid phase. While saturated lipids are chemically more stable and preferred, unsaturated lipids such as those from *E. coli* have been successfully used to produce highly ordered crystals. A good compromise is dimyristoylphosphatidylcholine (DMPC), a lipid frequently used with success, which has saturated fatty acids and a phase transition temperature close to room temperature (23°C). Native lipids are often ideal in terms of stability and transition temperatures, and they provide mixtures of headgroup

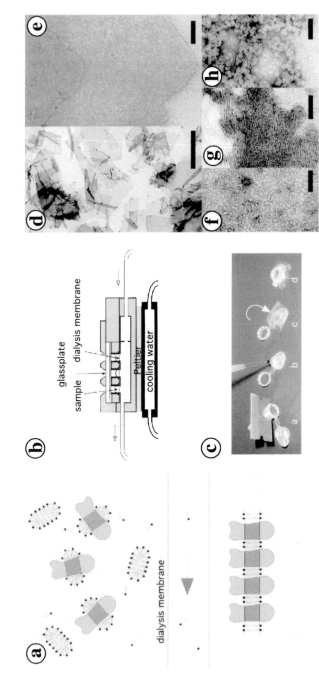

FIGURE 3 2D crystallization is achieved by mixing a purified solubilized membrane protein with solubilized lipids and removal of the detergent by dialysis (a). This can be done with a computer-controlled dialysis device (b) or with microdialysis devices produced from Eppendorf tubes. (c) Eventually well-ordered 2D crystals emerge (d, e), but frequently the dialysis is not complete, instead yielding wormlike structures (f) or stacked lamellar structures (g) after negative staining. Sometimes the protein does not integrate in the bilayer, leading to empty vesicles and protein aggregates (h). Scale bar represents 2 µm in (d) and 100 nm in (e–h).

charges and molecular geometries similar to those of the membranes from which the protein originated.

The detergents used for solubilization of protein and lipids have a major influence because they determine both the stability of the protein and the crystallization kinetics. Because protein–detergent interactions are not well understood, the selection of the detergent is usually accomplished experimentally, with the trend of solubilized proteins to aggregate being checked by electron microscopy. Critical are pH and counterions, both of which foster the protein–protein interactions that are ultimately responsible for crystallization. As in attempts to grow 3D crystals, such parameters are tested systematically in reconstitution experiments whose outcomes are inspected by electron microscopy.

Although it is difficult to attain atomic resolution with 2D membrane protein crystals, a few advantages should be kept in mind. First, the protein is reconstituted in a lipid bilayer and therefore in its native environment, allowing its functions to be assessed. Second, the protein is stabilized by the reconstitution process, allowing 2D crystals to be grown from fragile proteins that are not amenable to 3D crystallization. Third, electron microscopy and atomic force microscopy provide medium-resolution information very quickly: Once highly ordered crystals are available, images to better than 1-nm resolution can be acquired within a few days. Technical difficulties slow structural analyses that aim for resolving the 3D structure of a membrane protein at atomic resolution.

III. ELECTRON CRYSTALLOGRAPHY

Electron crystallography is ideally suited to the study of such 2D crystals, up to resolving atomic positions (see Walz and Grigorieff, 1998; Fujiyoshi, 1998). Samples in a trehalose solution are adsorbed to carbon films that are spanned flat over a metal grid (copper, molybdenum), blotted with a filter paper, and quickly frozen in liquid ethane (Dubochet et al., 1988). Frozen samples are then transferred into a cryo-electron microscope that allows specimens to be observed at 4 K (liquid helium stage) or 80 K (liquid nitrogen stage) (Fujiyoshi, 1998). Low temperature is important for reducing the damage introduced by the electron beam. Beam damage limits the dose that may be applied before the specimen structure is changed.

To achieve high resolution (better than 3 Å is possible), not only does the microscope need to be mechanically and electrically stable, but a coherent electron beam such as that produced by a field-emission tip is also required. Projections of the single-layer 2D crystals thus recorded contain the structural information of the unit cell sampled on a 2D array of lattice lines that span the 3D Fourier space (Fig. 4). According to the central section theorem, each projection represents a section through the origin of the Fourier space, intersecting the lattice lines in an array of points, the diffraction orders. Their amplitudes and phases are calculated from the

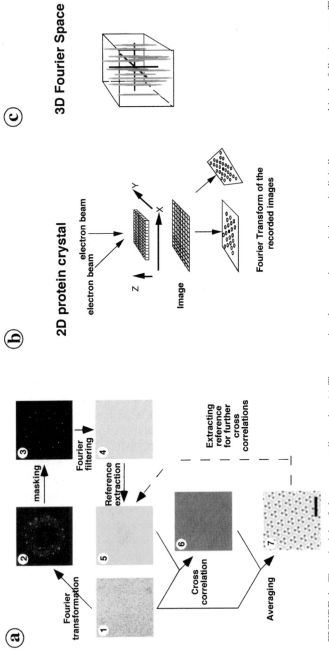

FIGURE 4 The principle of electron crystallography: (a) The raw noisy electron microscopic image is digitally processed in the following way: The image (1) is Fourier transformed (2), masked (3), and back-transformed (4) to yield a reference image (5). This reference is then cross-correlated with the original image to generate the cross-correlation map (6) that indicates the positions of the crystal unit cells. Using these positions, the unit cells in the raw image are averaged, resulting in an average image that shows the projection of the protein crystal with a much-reduced noise content (7). (b) Several images of differently tilted crystals are recorded in the microscope. The Fourier transformations of these images correspond to central sections at different angles through a 3D Fourier space (c). Once sufficient information about the data in this 3D Fourier space is obtained, a Fourier back-transformation reveals the 3D protein structure.

images (the projections) using a Fourier transform algorithm. Alternatively, the amplitudes can be directly measured when the electron microscope is operated in the diffraction mode. In the latter mode, the phase information cannot be acquired, analogous to X-ray diffraction. Thus, the advantage of electron crystallography lies in the possibility of directly measuring the phases, but this advantage is paid for with technical problems. Thus, electron crystallography is still a relatively difficult technique used throughout the world in a far smaller number of laboratories than X-ray crystallography.

IV. ATOMIC FORCE MICROSCOPY

2D protein crystals and their conformational changes can also be studied at high resolution by atomic force microscopy (AFM) (Müller *et al.*, 1995, 1999b; Müller and Engel, 1999). The atomic force microscope moves a sharp tip attached to a soft cantilever in a TV-raster-like pattern over a surface, and records deflections of the tip that correspond to the surface topography by a computer (Binnig *et al.*, 1986). Operated in physiological solutions the AFM allows biomolecules to be observed in their native environment (Engel *et al.*, 1999).

For biological applications, cantilevers are manufactured from silicon nitride, have a length of 100–200 μm, and are less than 1 μm thick. They have a force constant of approximately 0.1 N/m, and are thus very flexible. A pyramidal stylus at the cantilever end provides a tip sufficiently sharp to achieve a lateral resolution often better than 1 nm. Modern instruments monitor the deflection of the cantilever resulting from the tip–sample interaction with an optical detector at a resolution of 0.1 nm, allowing forces of 10–50 pN to be measured. To determine the surface topography, sample and stylus are brought into contact and the vertical force applied to the cantilever is adjusted to typically 50–100 pN. The sample is then raster-scanned laterally below the stylus, while a servo-system moves the sample vertically to maintain the cantilever force preset after the tip–sample approach. Lateral and vertical displacements are achieved by a precise, computer-controlled piezo-scanner. In this way, the sample moves up and down during the raster-scan while being contoured by the stylus. The contour is thus acquired line by line and assembled into an image (Fig. 5).

V. AQP1, THE ERYTHROCYTE WATER CHANNEL

A. Solubilized AQP1

Aquaporin 1 (AQP1), a protein comprising 269 amino acids (Preston and Agre, 1991), was the first water channel discovered (Preston *et al.*, 1992).

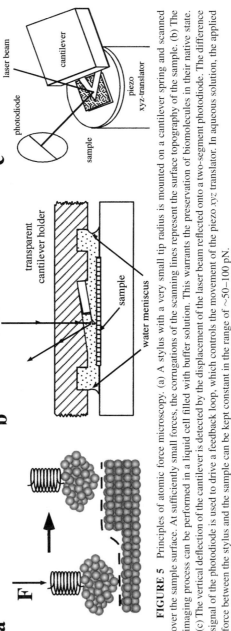

FIGURE 5 Principles of atomic force microscopy. (a) A stylus with a very small tip radius is mounted on a cantilever spring and scanned over the sample surface. At sufficiently small forces, the corrugations of the scanning lines represent the surface topography of the sample. (b) The imaging process can be performed in a liquid cell filled with buffer solution. This warrants the preservation of biomolecules in their native state. (c) The vertical deflection of the cantilever is detected by the displacement of the laser beam reflected onto a two-segment photodiode. The difference signal of the photodiode is used to drive a feedback loop, which controls the movement of the piezo *xyz* translator. In aqueous solution, the applied force between the stylus and the sample can be kept constant in the range of ~50–100 pN.

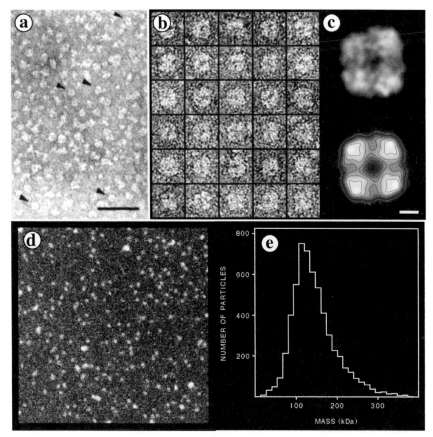

FIGURE 6 Electron microscopy and digital image processing of solubilized AQP1 oligomers. (a) A typical electron micrograph of negatively stained solubilized AQP1 particles, recorded at a magnification of 100,000×. The particles are rather homogenous in size, and many particles exhibit a square shape (*arrowheads*). (b) A gallery of particles after translational and angular alignment. The final average of 125 particles is shown in (c, *top*) before and (c, *bottom*) after 4-fold symmetrization and low-pass filtering to 1.6-nm resolution. The subframes of the gallery have a side length of 16.8 nm and the scale bars correspond to 50 nm in (a) and 2.5 nm in (c). Scanning transmission electron microscopy (STEM): (d) A low dose dark-field image of freeze-dried AQP1 oligomers. (e) Result of the STEM mass determination. The mass extrapolated to zero recording dose was 202 ± 3 kDa.

Negatively stained, octylglucoside-solubilized AQP1 shows a rather homogeneous size distribution (Fig. 6a) (Walz *et al.*, 1994a). Many particles exhibit a square shape and often a central, stain-filled pit. To enhance these features, approximately 350 particles were selected by eye, aligned with respect to a reference, and averaged. In addition to a central depression, aligned particles reveal a vertical or horizontal groove, in some cases both of them, dissecting the particle into four quadrants (Fig. 6b). An average of those 125 particles yielding the highest correlation coefficient with the final reference is shown in Fig. 6c; even without symmetrization the tetrameric structure of the particle is distinct. A possible asymmetry resulting from the glycosylation of one AQP1 per tetramer (Smith and Agre, 1991) could not be detected after uranyl formate staining. Therefore, the average displayed in Fig. 6c, top, was 4-fold symmetrized and low-pass-filtered to a resolution of 1.6 nm (Fig. 6c, bottom). The tetramer has a side length of 6.9 nm, and displays a square-shaped central depression with a side length of 2.5 nm. The four major stain-excluding units arranged about the central depression possess an elongated shape of 3 nm in length with one protrusion extending toward the 4-fold axis. The morphology and dimensions of these octylglucoside-solubilized AQP1 particles are similar to those of the intramembrane particles observed by freeze–fracture electron microscopy of liposomes reconstituted with AQP1, CHO cells transfected with AQP1 cDNA, or native kidney tubule plasma membranes (Verbavatz *et al.*, 1993).

Freeze-dried octylglucoside-solubilized AQP1 particles adsorbed to a thin carbon film and imaged in the scanning transmission electron microscope (STEM) exhibit uniform size and brightness (Fig. 6d). Mass analysis from such dark-field images recorded at doses between 200 and 1100 electron/nm^2 was performed as described (Müller *et al.*, 1992; Müller and Engel, 1998). Compensation for the beam-induced mass loss by extrapolation of the measured mass values to zero recording dose by linear regression yielded an average mass of 202 kDa with 99% confidence limits of ±3 kDa (Fig. 6e). This value is consistent with an AQP1 tetramer, including the glycosylation of one subunit and two octylglucoside micelles.

B. 2D Crystallization of AQP1

Purified solubilized AQP1 tetramers exhibited a marked propensity to assemble into square arrays when reconstituted at a low lipid:protein ratio (LPR < 2; w/w) (Walz *et al.*, 1994b). In the absence of divalent cations, closed unilamellar vesicles formed with an average diameter of 3 μm (Fig. 7a). Some vesicles flattened to planar, double-layer structures when adsorbed to carbon-coated formvar films rendered hydrophilic by glow discharge at low air pressure. Homogeneously stained flat areas revealed a faint regular pattern of lattice lines that gave rise to diffraction spots belonging to two square arrays with unit cell dimensions $a = b = 9.6 \pm 0.1$ nm ($n = 50$) (Fig. 7b). The presence of two distinct sets of spots rotated with respect to one another by an arbitrary angle evinced that both layers of the spread-flattened vesicles were crystalline, suggesting that the entire vesicle surface was covered by the square array of AQP1 oligomers. Correlation averages typically exhibited a resolution of 1.6 nm and a 4-fold rotational symmetry (root-mean-square deviation 6%) with a unit cell housing two tetramers (Fig. 7c).

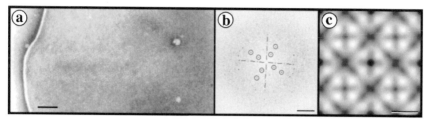

FIGURE 7 Electron microscopy and digital image processing of reconstituted 2D AQP1 crystals. (a) The micrograph shows a typical area of a negatively stained spread-flattened vesicle used for digital image processing. (b) The distinct square lattice ($a = b = 9.6$ nm) diffracts up to order (6,0), indicating a resolution of 1.6 nm. (c) The correlation averaged patch exhibits unit cells containing eight monomers organized as two tetramers (protein in bright shades and negative stain in dark shades). Scale bars represent 100 nm in (a), (2.5 nm)$^{-1}$ in (b), and 5 nm in (c).

One AQP1 tetramer had four stain-excluding elongated domains (bright shades in Fig. 7c) of approximately 2.7 nm in length and 1.6 nm in width surrounding a central stain-filled depression (dark shades). The tetramers were separated by rhomboid stained areas of approximately 7.3 nm in length and 4.7 nm in width that represent the lipid bilayer. By closer inspection of the computer-averaged projection displayed in Fig. 7c a distinct difference in stain penetration of the two types of AQP1 tetramers becomes evident.

Mass measurements performed with the STEM (Müller and Engel, 1998) yielded a mass per area (MPA) histogram with distinct peaks. The first peak at 2.7 ± 0.14 kDa/nm^2 ($n = 294$) represents lipid bilayers, whereas the peak at 4.1 ± 0.27 kDa/nm^2 ($n = 2160$) results from AQP1 square arrays. The mass of one unit cell is therefore 378 kDa, accommodating two AQP1 tetramers and phospholipids. Only one subunit of the AQP1 tetramer is likely to be glycosylated (Smith and Agre, 1991), yielding an estimated tetramer mass of 134 kDa, allowing for 110 kDa of interspersed lipid bilayer. Packed at 4 lipids/nm^2, i.e., 2.7 kDa/nm^2, the bilayer would thus cover 44% of the unit cell area, consistent with the area of the rhomboid structures of Fig. 7c, which contribute 40% of the unit cell area.

A single crystallographic packing arrangement was observed with AQP1. The best crystals were obtained at a LPR of 0.5, which could be related to the LPR found in the crystals themselves (Walz et al., 1994a). At lower LPR, the protein started to aggregate, whereas lattices were somewhat less ordered at higher LPR. Virtually identical crystals independent of the lipids used were produced by the dialysis method from octylglucoside-solubilized AQP1 from different sources (Jap and Li, 1995; Mitra et al., 1995; Walz et al., 1995). A subsequent phospholipase A$_2$ digestion step (Mannella, 1984) increased the amount of well-diffracting 2D crystals in one laboratory, but did not change the lattice parameters significantly (Jap and Li, 1995). This indicates that the lipid has little influence on the crystallinity or on the protein packing of AQP1 2D crystals, suggesting a single and dominant set of specific protein–protein interactions between the tetramers.

C. Water Activity of 2D AQP1 Crystals

The reconstituted protein–lipid vesicles containing AQP1 tightly packed in 2D crystals were evaluated for residual water channel activity. After the vesicles were loaded with carboxyfluorescein, the osmotic water permeability was measured by abruptly increasing the osmolality by 50% with a stopped flow device while monitoring the quenching of fluorescence. The 3-μm vesicles exhibited a high degree of water permeability and were fully shrunken in <20 ms (Fig. 8). Multiple recordings were averaged, and the coefficient of osmotic water permeability was calculated, $P_f = 0.472$ cm/s. The density of AQP1 within the 2D crystals in Fig. 7 [8 subunits/(9.6 nm)^2] permitted calculation of the unit water permeability, $p_f = P_f \times$ area/subunit $= 5.43 \times 10^{-14}$ cm^3/s/subunit. This value agrees remarkably with previous determinations of osmotic water permeability made on the small vesicles (approximately 0.14 μm in diameter), which were made by reconstitution of pure AQP1 at LPRs varying between 100 and 20, $p_f = 8.4 \pm 4.0 \times 10^{-14}$ cm^3/s/subunit (Zeidel *et al.*, 1992).

Other features of native water channels were also displayed by the 2D crystals: Incubation with submillimolar HgCl$_2$ reduced the osmotic water permeability to <10% of the original level (data not shown). In addition, water permeability measurements over a range of temperatures revealed a remarkably low Arrhenius activation energy, $E_a = 1.9$ kcal/mol (Fig. 8), which is similar to that of diffusion of water in bulk solution, indicating that AQP1 provides a pathway by which water may permeate the bilayer as a continuous unbroken stream. As AQP1 tetramers are incorporated in the p42$_1$2 lattices in both orientations (Fig. 9), the good agreement between water permeability of erythrocyte ghosts (Zeidel *et al.*, 1992) and 2D crystals indicates that the water channel is bidirectional.

D. 3D Structure of Negatively Stained AQP1 Crystals

The 3D density map calculated from tilt series of negatively stained crystals revealed two tetramers packed in opposite orientation per unit cell having a 9.6-nm width (Walz *et al.*, 1994a). Plots of either the standard deviation or the dynamic range of horizontal sections versus section number indicated that the reconstructed layer exhibited a nominal thickness of 4.5 nm (data not shown). Because flattening during dehydration and shrinkage resulting from the electron irradiation need to be taken into account, the effective thickness of a single layer is approximately 6 nm, consistent with estimates from freeze-dried, metal-shadowed crystals. The resulting horizontal sections of 0.3-nm thickness (Fig. 9a) show unit

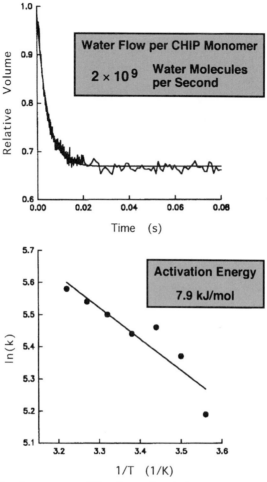

FIGURE 8 Osmotic water permeability of reconstituted AQP1 vesicles at 22°C (*top*) and Arrhenius plot of determinations at 8–39°C (*bottom*). AQP1 vesicles were incubated overnight at 4°C in a buffer [250 mM NaCl, 20 mM Tris-HCl (pH 7.55), 0.01% NaN$_3$] containing 20 mM carboxyfluorescein. The vesicles were pelleted, resuspended in the same buffer, and fluorescence at 490 nm was measured after exposure to a 50% increase in osmotic strength by stopped-flow analysis as described by Zeidel *et al.* (1992). Relative decrease in vesicle volumes was directly related to reduced fluorescence. Data from five to eight individual recordings served to determine the P_f.

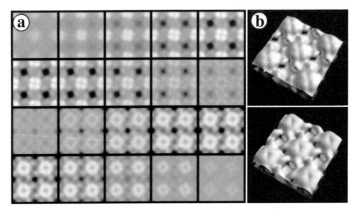

FIGURE 9 Horizontal sections and surface-rendered view of the 3D density map obtained from spread-flattened vesicle-type crystals. (a) Twenty horizontal sections of 0.3-nm thickness showing two tetramers of transmembrane AQP1 monomers per unit cell that are incorporated in opposite orientation. This results in the different surface topographies of adjacent tetramers revealed by the perspective view obtained by surface rendering the 3D map at a contour comprising 75% of the total mass (b). AQP1 monomers protrude by approximately 0.7 nm from the bilayer on one side and surround a 3-nm-diameter cavity (*top*). The AQP1 tetramer extends by approximately 1 nm on the other side (*bottom*).

cells housing two tetramers of membrane-spanning AQP1 monomers that are incorporated in the bilayer with opposite orientation. This yields different surface topographies of adjacent tetramers as displayed by the perspective view in Fig. 9b. Each tetramer exhibits stain-penetrated indentations about the 4-fold axis that extend from both membrane surfaces toward the inside.

E. Relief Reconstructions of Freeze-Dried, Metal-Shadowed AQP1 Crystals

A low magnification overview of a freeze-dried vesicular AQP1 2D crystal unidirectionally shadowed with Ta/W is shown in Fig. 10a. Micrographs of shadowed deglycosylated AQP1 2D crystals taken at a magnification of 96,000× with a CCD camera displayed no visible periodic structure (Fig. 10b), indicating that the surface corrugation is small. However, calculated power spectra revealed sharp diffraction spots up to the reciprocal lattice order 7,6 indicating a resolution of 1.04 nm (see Fig. 10c). After correlation averaging of more than 700 motifs from the area that yielded the diffraction pattern displayed in Fig. 10c, the tetramers can clearly be seen (Fig. 10d). Because of the unidirectional shadowing, the 4-fold symmetry is lost. Nevertheless, the windmill-like features of one AQP1 surface are evident and even the finest resolved features, with distances of 0.8–1.5 nm, show the directional nature of the shadowing process.

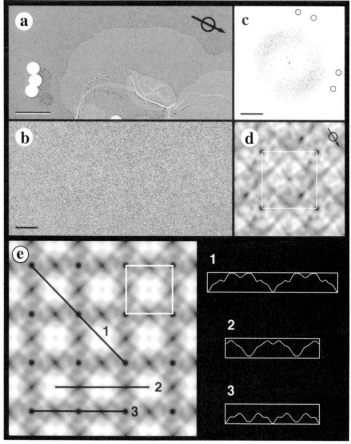

FIGURE 10 Electron microscopy and image processing of a unidirectionally metal-shadowed AQP1 2D crystal. (a) The low magnification overview shows a large vesicular 2D crystal after freeze drying and unidirectional metal shadowing. The shadowing direction is estimated from the shadow cast of the added latex spheres. (b) The high magnification view (96,000×) of the same vesicle does not exhibit visible periodic features. (c) The power spectrum of the same area, however, shows very sharp diffraction spots. Diffraction orders of 7, ±5, and 5, ±7 at a resolution of 1.1 nm are marked. (d) In the unsymmetrized correlation average, the two tetramers in the unit cell (outlined in white) display a clearly different structure. As a result of the unidirectional shadowing, the 4-fold symmetry of the AQP1 tetramer is lost. Scale bars represent 0.5 μm in (a), 25 nm in (b), and (2.5 nm)$^{-1}$ in (c). The side length of panel (d) corresponds to 19.2 nm. The arrows in (a) and (d) indicate the shadowing direction. (e) Surface reconstruction from a unidirectionally shadowed AQP1 2D crystal. The top view of the reconstruction is displayed in gray levels that are proportional to the height. The view demonstrates that the central tetramer in the unit cell outlined in white protrudes much more from the membrane than the adjacent tetramers. The height profiles along the lines marked 1, 2, and 3 reveal the height of the lipid surface, the major protrusion, and the central cavity, respectively.

Surface reliefs of AQP1 crystals were reconstructed from single micrographs exploiting the 4-fold symmetry (Smith and Kistler, 1977; Guckenberger, 1985). The relief reconstruction process revealed the difference in the vertical positions of two adjacent tetramer surfaces. In Fig. 10e the top view of a relief is presented with gray levels that are proportional to the height of the structure. Again, the square-shaped tetramers can clearly be seen. Neighboring tetramers are only slightly rotated with respect to each other, but exhibit a pronounced difference in brightness. This reflects a substantial difference in the protein mass protruding from the two sides of the lipid bilayer, consistent with the 3D map of negatively stained 2D crystals (Fig. 9). Height profiles along the lines marked 1, 2, and 3 further document this asymmetry. On one side, the four protrusions surround a small depression about the 4-fold axis. On the other side, a complex depression with a windmill-shaped disposition of small peripheral protrusions and a central cavity about the 4-fold axis is seen. The surface data are remarkably reproducible. Reconstructions from four different crystals with different shadowing azimuth (angle between lattice lines and shadowing direction) showed almost identical height profiles (data not shown).

F. Surface Topography of AQP1 Crystals as Determined by AFM

Surface topographs of native AQP1 2D arrays were recorded in buffer solution with the atomic force microscope (Walz *et al.*, 1996). Low-magnification recordings (Fig. 11a) reveal the flatness of the adsorbed AQP1 2D crystalline sheets. From lattice defects such as those circled in Fig. 10a, the height of mono-layered sheets was determined to be 5.8 ± 0.3 nm ($n = 66$). In addition, lattice fragments embedded in the bilayer allowed the height difference between protein and lipid membrane to be measured, 1.3 ± 0.2 nm ($n = 6$). At higher magnification (Fig. 11b), individual unit cells exhibiting quadruple protrusions can clearly be seen in the unprocessed image. The diffraction pattern (Fig. 11c) displays a homogeneous resolution of the recorded image along the fast scanning direction and perpendicular to it. Diffraction spots of the reciprocal lattice order 9,2 visible in the calculated power spectrum indicate a resolution of 1.04 nm. To further assess the lateral resolution, the radial correlation function (Saxton and Baumeister, 1982) as well as the SSNR profile (Unser *et al.*, 1987) were calculated from 134 unit cells to yield 0.9 and 1.1 nm, respectively. Finally, the reproducibility of the height values was estimated from the highest values in the standard deviation map to 0.2 nm (Schabert and Engel, 1994). This corresponds to 18% of the corrugation amplitude, 1.1 ± 0.2 nm ($n = 15$). The 4-fold symmetrized correlation average (Fig. 11d) documents the distinct difference between the surfaces of the two oppositely incorporated tetramers. This average is in excellent agreement with the surface relief reconstruction (Fig. 10e). The tetramer in the center of the unit

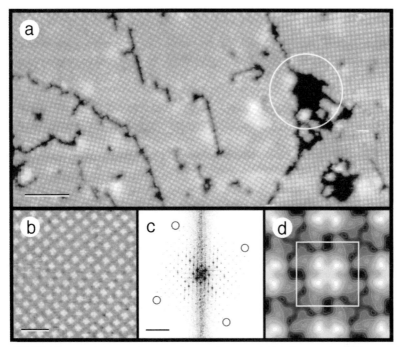

FIGURE 11 Atomic force microscopy image of a native AQP1 2D crystal. (a) The overview
was recorded in buffer solution at a loading force of about 0.2 nN, scan frequency 4.7 line/s (512
pixels). The AQP1 crystal is fragmented, exhibiting cracks and holes where the thickness of the layer
is determined. (b) At high magnification, the height signal displays the major tetrameric protrusions of
1 nm in height. (c) In the diffraction pattern calculated from the area shown in (a), diffraction spots up
to the reciprocal lattice order (9,2) can be discerned, corresponding to a resolution of 1.04 nm (d) The
correlation average reveals mainly one tetramer per unit cell, while the other tetramer is represented by
a pronounced depression with windmill-shaped peripheral protrusions of 0.5 nm in height. Scale bars
represent 100 nm in (a), 20 nm in (b), and $(2.5 \text{ nm})^{-1}$ in (c). The side length of panel (d) corresponds
to 19.2 nm.

cell is clearly resolved into four domains surrounding a small central depression.
Adjacent tetramers exhibit a less pronounced surface texture with small peripheral
protrusions arranged in a windmill-shaped pattern. The four maxima of the central
tetramer are exactly at the same positions as the corresponding maxima of the reconstructed surface relief shown in Fig. 10e. The height difference between major
and minor protrusions, 0.6 ± 0.2 nm ($n = 15$), allows the sections of the relief
reconstruction to be scaled (Fig. 10e). As a result of the finite tip size, however,
the central depressions of the adjacent tetramers are not as distinctly outlined as
in the surface relief reconstruction.

G. Projection Maps of Unstained AQP1 at Subnanometer Resolution

Glucose-embedded, highly ordered 2D AQP1 crystals yielded electron diffraction spots out to 0.5 nm (Fig. 12a). The calculated power spectra of images recorded at low temperature ($-180°$C) showed spots out to a resolution of 0.7 nm (Fig. 12b). After unbending and CTF correction of these images, some spots beyond 0.5 nm were still significantly above background, but in the merged data set of nine micrographs the phase error in the resolution band between 0.6 and 0.5 nm exhibited a value of $45.6°$. Therefore, the projection map was calculated from the merged amplitudes and phases to a resolution of 0.6 nm, enforcing the p42$_1$2 symmetry (Fig. 12c; Walz *et al.*, 1994a). Projection maps determined by different laboratories exhibited similar morphologies: Fig. 12d (Mitra *et al.*, 1995), Fig. 12e (Jap and Li, 1995), and Fig. 12f (Walz and Grigorieff, 1998). The maps displayed in Figs. 12e

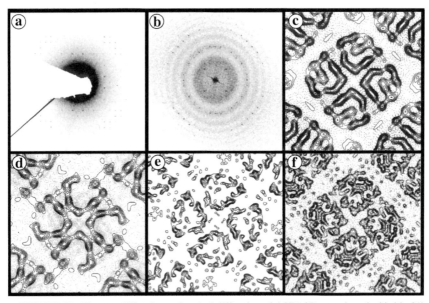

FIGURE 12   Electron diffraction and optical diffraction of AQP1 2D crystals embedded in 3% glucose. (a) The electron diffraction pattern displays a distinct 4-fold symmetry. Spots at a resolution of about 5 Å can be distinguished. (b) The optical diffraction displayed originates from an image recorded at 1.2 μm underfocus. Thus, many zero transitions of the CTF can be observed. It diffracts to spot $-9,9$, indicating a resolution of 7.5 Å. (c) The final projection map to a resolution of 6 Å was calculated by merging phases and amplitudes from nine individual images and imposing a p42$_1$2 symmetry. The tetramers reveal a central region of low density at the location of the 4-fold axis. The side length of the map is 96 Å (d) The 6-Å map produced by Mitra *et al.* (1995) exhibits slightly different features and was contoured at different levels. Even at 3.5-Å resolution, these projection maps cannot be interpreted in terms of predicted α helices: (e) from Jap and Li (1995). (f) from Walz and Grigorieff (1998).

and 12f include data to 3.5-Å resolution. Nevertheless, their interpretation in terms of assigning predicted helices to density maxima was not possible, because the density maxima did not exhibit the same features as those of bacteriorhodopsin at comparable resolution (Henderson *et al.*, 1986; see also Henderson *et al.*, 1990; Grigorieff *et al.*, 1996; Kimura *et al.*, 1997; Mitsuoka *et al.*, 1999).

H. Correlation between Surface Topography and Projection Map

To gain further insight into the molecular architecture of AQP1, the high-resolution projection map was overlaid with the surface topography. Two arrangements with superposed 4-fold centers are possible. The more likely one is based on two assumptions. First, the gaps dividing the projection of the AQP1 tetramer into four quadrants are taken as molecular boundaries. Second, loops connecting helices are thought to protrude out of the membrane mainly over the core of the protein rather than over the gaps between molecules (Fig. 13). With this arrangement, the strongest of all density peaks, peak 5, coincides with the highest protrusion of the surface topography, which partly covers peak 7 as well. If as predicted by the hour-glass model (Jung *et al.*, 1994) loops B and E fold back into the membrane, protruding domains could be loops A and C, but to resolve this question a high-resolution 3-D map is required.

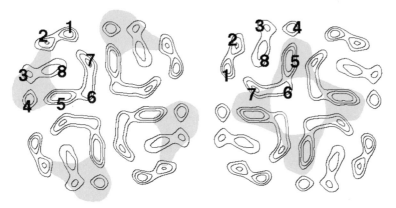

FIGURE 13 Projections of AQP1 tetramers from the cytosolic side (*left*) and the extracellular side (*right*). The contours represent the 6-Å projection of glucose-embedded crystals (Walz *et al.*, 1995), while shaded areas display the cytosolic protrusions (*left*) and the extracellular protrusions (*right*) revealed by surface relief reconstruction (Fig. 10) and AFM (Fig. 11). The alignment of the two graphs with respect to one another is based on the diamond-shaped lipid moieties most clearly seen in the average of negatively stained arrays (Fig. 7c). Extracellular protrusions are related to loops connecting putative α-helical spans.

Surface relief reconstructions calculated from freeze-dried, metal-shadowed 2D crystals prepared after carboxypeptidase Y treatment suggested a location of the C terminus close to the major protrusion of the surface topography, hence allowing the sidedness of the AQP1 molecule to be assigned (Walz *et al.*, 1996). However, attempts to apply the same technique to crystals of AQP0 (see below), or to reproduce the initial results achieved with AQP1, revealed a pronounced sensitivity of both crystals toward proteolytic digestion. The 2D crystals became disordered, preventing acquisition of high-resolution topographs suitable for unambiguous location of the digested protein mass. But when digestion was performed *on mica,* i.e., on 2D crystals firmly adsorbed to mica, digestion did not disrupt the crystals, allowing suitable topographies to be recorded. Preliminary experiments showed that mild digestion with carboxypeptidase Y of AQP1 did not change the major protrusion, but in contrast led to a loss of a small domain close to the 4-fold axis on the less corrugated surface (Fig. 14). Thus, the interpretation of (Walz *et al.*, 1996) had to be revised, giving a sidedness assignment as indicated in Fig. 13.

I. Secondary Structure Comparison of AQP1 and Bacteriorhodopsin by FTIR

Attenuated total reflection Fourier transform infrared (ATR-FTIR) spectroscopy has been used to compare reconstituted AQP1 2D crystals and bacteriorhodopsin (BR) in purple membranes. Samples either from an aqueous sample or from an aqueous sample flushed with N_2-saturated D_2O to allow H/D exchange were spread on a germanium plate (Cabiaux *et al.*, 1997). Fourier self-deconvolution of the deuterated spectra (Fig. 15a) shows that the main components of AQP1 (a) and bacteriorhodopsin (b) are located at 1658 and 1660 cm^{-1}, respectively, whereas the maximum at 1629 cm^{-1} results from the pure β-barrel protein OmpF. These maxima indicate the presence of a high percentage of α-helical structure in both AQP1 and BR. Their locations are not modified by deuteration. The 1636-cm^{-1} component observed in bacteriorhodopsin has been interpreted as a proof of some β sheet in the bacteriorhodopsin structure. This is compatible with the highest resolution structure obtained from electron crystallography that reveals the loop connecting helices B and C in a β-hairpin conformation (Kimura *et al.*, 1997; Mitsuoka *et al.*, 1999).

Fourier self-deconvolution and band shape analyses have been used for secondary structure determination. The AQP1 spectrum is characterized by an absorption maximum at 1658 cm^{-1}, very close to the absorption maximum of BR, both of which are indicative of a high α-helix content. The helix content found for AQP1 ranged from 42 to 48% depending on the algorithm used for secondary structure determination. As far as β structure was concerned, 0 to 20% β sheet were found depending on the algorithm used, but it must be kept in mind that

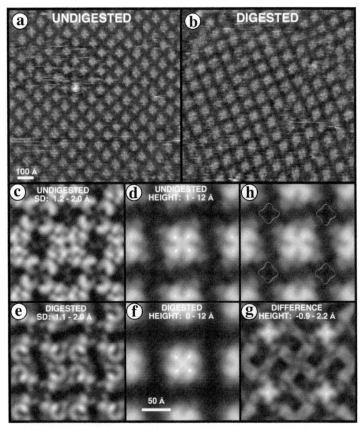

FIGURE 14 AQP1 digestion. (a) AFM image of the undigested AQP1 2D crystal. (b) AFM image of the carboxypeptidase Y digested AQP1 2D crystal (see text). Cross-correlational averaging of the undigested image in (a) gives the average image (d) and a standard deviation map (c). The same calculations for the digested image give the average image (f) and the standard deviation map (e). (g) The difference map between the undigested and digested topographies (d) and (f) exhibits a mass that is located about the 4-fold axis, and corresponds to a height difference of 2.2 Å between the two topographies. (h) Carboxypeptidase Y digestion resulted in a loss of protein mass at the AQP1 surface with smaller corrugations (Müller and Engel, unpublished).

the set of proteins used as a basis for the frequency assignments does not allow unambiguous discrimination between β sheets and 3_{10} helices. Thus, 20% should be considered a maximum value, and AQP1 is an all α-helical protein, with spectral characteristics being close to those of bacteriorhodopsin.

Information about the orientation of a given secondary structure can be obtained by recording ATR-FTIR protein spectra with polarized light, provided that

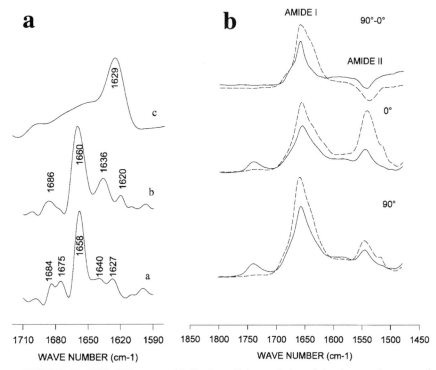

FIGURE 15 FTIR-ATR spectra (a) Fourier self-deconvolution of the deuterated spectra of (*a*) AQP1 reconstituted in proteoliposomes, (*b*) BR in the purple membrances, and (*c*) porin reconstituted in DMPC vesicles as described in Goormaghtigh *et al.* (1994). (b) Polarized deuterated spectra of AQP1 reconstituted in proteoliposomes (*solid line*) and of BR in purple membranes (*dashed line*). The subtraction coefficient was chosen to cancel the υ CH_2–CH_3 between 3000 and 2824 cm^{-1}. The difference spectra were enlarged 1.7 and 4 times as compared to the 90° spectrum for BR and AQP1, respectively.

the proteins are oriented with respect to the internal reflection element (the germanium plate). This prerequisite is met with films formed from membrane proteins reconstituted in lipid vesicles (Fringeli and Günthard, 1981). We have confirmed this by acquiring images of purple membrane films prepared on germanium plates for spectroscopy after checking their flatness by AFM. Figure 15b displays the 90° and 0° polarization spectra of AQP1 (solid line) and bacteriorhodopsin in purple membranes (dashed line) as well as the 90°–0° difference spectrum. Because of differences in the relative power of the 90° and 0° polarized evanescent fields, the subtraction coefficient (R_{iso}, see below) was chosen to zero the υ CH_2–CH_3 bands between 3000 and 2824 cm^{-1} instead of the lipid υ $C{=}O$ band (1763–1711 cm^{-1}).

A positive deviation in the polarization difference spectrum (Fig. 15b, bottom trace) indicates an orientation of the considered dipole parallel to the lipid acyl chains. For both proteins, a positive deviation was observed in the amide I ($C{=}O$) region of the difference spectrum, whereas a negative deviation was observed in the amide II (N–H) region. Both of these facts indicate that the α-helical component of AQP1 has a strong transmembrane orientation. Quantitative evaluation of the helix orientation can be obtained from the integrated dichroic ratio, assuming that α helices are the only strongly oriented components,

$$R_\alpha = \frac{R - \frac{R+2}{2R_{iso}+1}(1-x)}{1 - \frac{1}{R_{iso}}\frac{R+2}{2R_{iso}+1}(1-x)},$$

where R_α is the dichroic ratio of the α-helical component, R is the measured dichroic ratio (area amide I 90°/area amide I 0°), R_{iso} is the ratio of the intensities at 90° and 0° of a dipole with an isotropic orientation (the υ CH_2–CH_3 between 3000 and 2824 cm^{-1}) and x is the fraction of α-helical structure. A value of $x = 70\%$ was used for the percentage of α helix in bacteriorhodopsin. Because there is some variation in the determined percentages of α-helical structure of AQP1, the dichroic ratio was calculated using values of 40 and 50%. The dichroic ratio determined for the bacteriorhodopsin spectrum was 3.5, whereas four independent experiments with reconstituted AQP1 gave an average value of 2.0 ± 0.25 for both $x = 40$ and 50%.

The relationship between the dichroic ratio and the orientation of a given structure, α helical in this case, depends on the order parameter, which is in turn strongly dependent on the angle between the transition dipole being considered and the primary axis of the structure (Goormaghtigh et al., 1994). Another parameter influencing the angle determination is the refractive index (n) of the film. Values of 1.7 and 1.55 are currently used for protein and lipid films, respectively. A dichroic ratio of 3.5, as found for bacteriorhodopsin, can only be obtained with an order parameter of 1 and a value of 17° for the angle between the $C{=}O$ dipole and the helical axis. (For a discussion about this angle, see Goormaghtigh et al., 1994.) A dichroic ratio of 3.5 corresponds to a maximum average tilt of the α helices, with respect to the membrane surface normal, of 4° ($n = 1.7$) or 13° ($n = 1.55$). This result is in excellent agreement with the average angle of 13° found in the atomic models of bacteriorhodopsin (Grigorieff et al., 1996; Kimura et al., 1997; Mitsuoka et al., 1999). When the same $C{=}O$ dipole–helical axis angle and order parameter values are used to calculate the average orientation of the helices in AQP1, the maximum average angle to the membrane normal was found to be 21° $\pm4°$ ($n = 1.7$) or 27° $\pm 4°$ ($n = 1.55$) (four independent experiments). Thus, these data suggested that the helices in AQP1 are significantly more tilted than those in bacteriorhodopsin and explain the difficulty in interpreting the projection maps of AQP1 shown in Fig. 12.

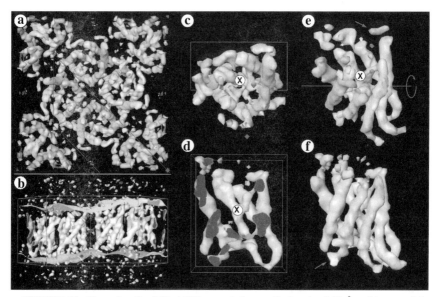

FIGURE 16 The unit cell of 2D AQP1 crystals has a side length of 96 Å and houses eight asymmetric units that form two tetramers integrated into the bilayer in opposite orientations. (a) The view along the 4-fold symmetry axis (♦) shows the cytoplasmic surface of the central tetramer, with one monomer colored in gold. Tetramers adjacent to the central tetramer are seen from the extracellular side, which exposes the connecting loop between monomers (*). Molecular boundaries are reflected by narrow gaps marked by arrows in the vertical slice displayed in (b). This slice with a width of 152 Å and a thickness of 60 Å contains four monomers and has been cut as outlined in (a). The overlaid surface reconstruction determined by metal shadowing and atomic force microscopy (Walz *et al.*, 1996) illustrates that AQP1 protrudes significantly from the membrane on the extracellular side. In addition, the surface extends down to the lipid bilayer between tetramers as indicated by a two-headed arrow. (c) The AQP1 monomer as seen in (a) from the cytosolic side and (d) cut open to expose the central density X. The slice shown in (d) is outlined in (c), and it is 16 Å thick. (e) The monomer after a clockwise rotation around the x axis by 45° and (f) after another 45° clockwise rotation around the x axis. (♦) marks the 4-fold axis in (a) and (c), while the horizontal line in (e) indicates the x axis. (See Color Plate.)

J. The 0.6-nm 3D Structure of AQP1

Figure 16a displays the unit cell calculated from 11,054 diffraction intensities and 13,734 phases measured over a tilt angle range of 60° (Walz *et al.*, 1997). The AQP1 tetramer in the center is viewed from the cytoplasmic side, whereas adjacent tetramers, which have the opposite orientation, are seen from the extracellular side. The sidedness has been determined by comparison of loop densities at 0.45-nm resolution (Mitsuoka *et al.*, 1999), and by atomic force microscopy of native and carboxypeptidase Y-digested membranes (Fig. 14; Müller and Engel, unpublished, 2000). The extracellular side of the tetramer protrudes further from the membrane than the cytoplasmic side, and the surface topography between tetramers defines the boundaries of the bilayer (marked by a two-headed arrow in Fig. 16b). Narrow vertical clefts between monomers span the membrane (marked by arrows in Fig. 16b). These gaps define the molecular boundaries within the tetramer, confirming the assumption made for aligning the surface topography with the 0.6-nm projection map (Fig. 13). Helix–helix contacts in the narrow cleft may contribute to the unusual stability of AQP1 tetramers. The square-shaped tetramers are rotated slightly around their 4-fold axes probably to provide the contacts between adjacent tetramers required for crystallization.

An AQP1 monomer viewed from the 4-fold axis (Fig. 16f) reveals six distinct tilted rods of varying lengths that form a right-handed bundle. Their shapes and dimensions suggest that these rods are membrane-spanning α helices, in agreement with the sequence-based structure prediction. The six helices surround a complex central density X, as illustrated by the molecule in Fig. 16d, where the innermost helices were cut away. This unusual structure is wide on the cytosolic side, has a side arm, and appears to end above a prominent loop folding back into the membrane. Therefore, density X is proposed to contain loop B with one of the NPA motifs and loop E with the other NPA motif. The hourglass model (Jung *et al.*, 1994) positions density X in proximity to the water channel.

The right-handed twist in the helix bundle is a feature commonly observed in soluble proteins, but has been observed in a membrane protein for the first time (Cheng *et al.*, 1997; Walz *et al.*, 1997). Compared to other helical membrane proteins, AQP1 helices exhibit a much larger tilt (over 20° in average), in excellent agreement with predictions from FTIR (Cabiaux *et al.*, 1997; Fig. 15). Meanwhile, two other membrane proteins have been solved that exhibit right-handed bundles of strongly tilted helices: the potassium channel from *Streptomyces lividans* (Chang *et al.*, 1998) and the mechanosensitive ion channel from *Mycobacterium tuberculosis* (Doyle *et al.*, 1998). The density X is likely to be involved in forming the channel that allows water but not ions to pass. To gain insight into this amazing specificity of AQP1 for water, however, higher resolution information is required.

K. The 0.45-nm 3D Structure of AQP1

As suggested by the electron diffraction patterns in Fig. 17, AQP1 crystals are sufficiently well ordered to provide atomic scale resolution (Mitsuoka *et al.*, 1999). The 3D map of AQP1 at 0.45-nm resolution is calculated from 85,251 diffraction intensities and 21,720 phases (Fig. 18). It contains six rodlike structures on which protrusions are clearly visible (marked by arrowheads in Fig. 18a). The 0.54-nm repeat of these protrusions along the axis of the rod and their helical arrangement is consistent with a canonical α helix of 3.6 residues per turn. Therefore, the protrusions were interpreted to represent side-chain densities of each amino acid

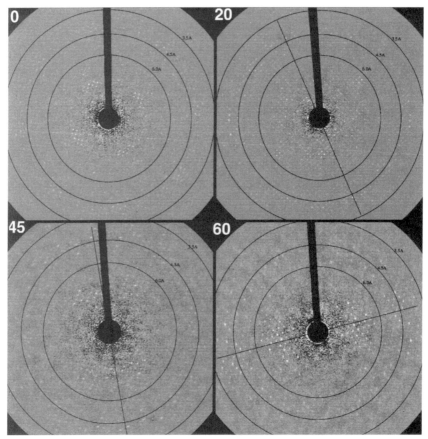

FIGURE 17   Typical electron diffraction patterns from 2D crystals of AQP1 at indicated tilting angles. The circles indicate the resolution, and the lines in the tilted diffraction patterns mark the position of the tilt axis. The black regions were excluded from the processing because of the beam stopper used and the lower and higher resolution cutoff.

of the helices. This allowed us to build poly(Ala) helices into the six transmembrane densities, confirming the right-handed bundle of highly tilted α helices observed in previous maps (Walz *et al.*, 1997).

In addition to the six transmembrane α helices, there are two other rodlike regions revealing a similar helical arrangement of protrusions. These densities are located in the center of the AQP1 monomer and are significantly shorter than the membrane-spanning helices (Figs. 18b and 18c). They are part of the X-shaped structure, which was already observed in the center of the AQP1 molecule in previous electron crystallographic studies. The 0.45-nm map indicates that one of the two branches to each membrane surface, which form the X-shaped density,

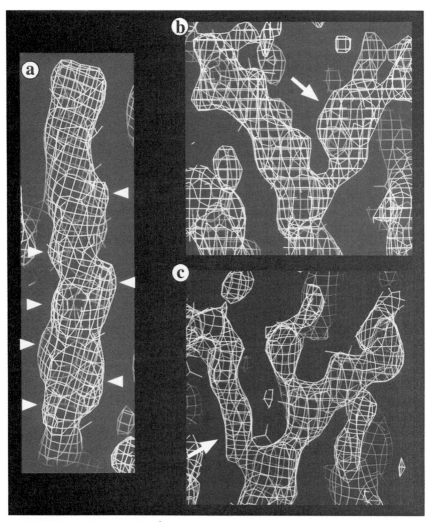

FIGURE 18   Views of the 4.5-Å resolution potential map contoured at 1.0 σ. (a) The potential map around a transmembrane rodlike structure shows protrusions corresponding to side chains of the transmembrane helix (*arrowheads*). Using the protrusions as markers, the poly-Ala helix (represented by the stick model) was manually built and subsequently refined. (b, c) The potential maps defining the two pore helix loop structures forming density X are shown in separate panels. The membrane surface is at the top of panels (b) and (c). Arrows indicate the densities of the pore loops while the stick models represent the pore helices. (See Color Plate.)

can be assigned to a short α-helix. One of the short α helices was noted in a previous study (Li *et al.*, 1997) and is definitely confirmed by the higher resolution map. In contrast, the other two branches of the X-shaped density do not display helical protrusions. Therefore, they may represent loops, which connect each short α-helix to a neighboring transmembrane helix. In Figs. 18b and 18c the densities of the loop regions are marked by white arrows while the densities for the short α helices contain a red stick model of the fitted poly(Ala) helices.

Previous electron crystallographic studies reported a noncrystallographic pseudo-two-fold symmetry within the AQP1 monomer (Cheng *et al.*, 1997; Li *et al.*, 1997). To determine whether our higher resolution map is in agreement with such a pseudo-twofold relationship, the six transmembrane helices and the two short helices were divided into two sets, grouping together the helices of the right and left half of the AQP1 monomer. These two sets were then compared to each other using a least-squares algorithm (Mitsuoka *et al.*, 1999). Although the twofold symmetry is not perfect, the least-squares comparison demonstrated that the pseudo-twofold symmetry within the AQP1 monomer applies not only to the membrane-spanning α helices but also to the short helices in the center of the monomer. Thus, the tandem-repeat structure found in the primary sequence of AQP1 (Wistow *et al.*, 1991; Heymann and Engel, 2000) is in fact reflected in its 3D structure.

Because of the pseudo-twofold symmetry in the structure, which results from the homologous repeats in the primary sequence, it is difficult to determine the orientation of the 3D structure in a cell from the potential map. Slight differences could be found for the potential maps around the two sets of the backbone poly(Ala) helices. The most prominent differences concern the connecting loops from the short α helices to the adjacent membrane-spanning helices. These regions are thought to correspond to loops B and E (Preston and Agre, 1991). It is plausible to assign the wider potential map of the connecting loop shown in Fig. 18b to loop B, which has more hydrophobic and bulky side chains than loop E. This assignment is opposite to that given by Walz *et al.* (1996), but compatible with recent carboxypeptidase Y digestion experiments *on mica* (Fig. 14; Müller and Engel, unpublished, 2000).

L. Clues from Sequence Analysis

1. The AQP Core Architecture

Multiple sequence alignments and phylogenetic analysis of 164 different sequences of the aquaporin superfamily revealed two distinct clusters, the AQP and GLP cluster, altogether composed of 46 subtypes (Fig. 1; Heymann and Engel, 1999). The core architecture derived from these aligned sequences consists of six transmembrane helices, two long functional loops, and three interlinking loops of

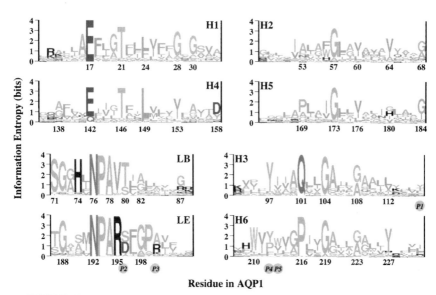

Residue in AQP1

FIGURE 19 Sequence logos reveal the conservation of residues at particular positions in the sequence. The core segments of the sequence alignments were converted to sequence logos (Schneider and Stephens, 1990) and shown with the residue numbers for AQP1. The five positions (P1–P5) that were found by Froger *et al.* (1998) to be different between the two clusters are shown in italics in circles. The scale gives the certainty of finding a particular amino acid type at each position, and is related to the entropy as $R_{seq} = \log_2 \{20 \cdot H \cdot c(m)\}$, where $c(m)$ is a correction factor for small sample size m. Colors: Gray, hydrophobic; light-blue, polar; green, amide; red, acidic; blue, basic. (See Color Plate.)

various lengths, constituting a minimum of 208 residues (Heymann and Engel, 1999). While the NPA motifs of the functional loops are considered to be the water channel signature, Fig. 19 documents many other equally conserved residues distributed over the entire core region. Helices 1 and 4 possess the remarkable pattern ExxxTxxL/F in their N-terminal half, whereas helices 3 and 6 show a distinct helical periodicity along their entire length. In addition to the NPA motifs, many other residues are conserved in the functional loops, indicating their structural and functional significance. The His 74 in loop B and the Arg 195 in loop E are the highly conserved, positively charged buried residues that may form ion pairs with the conserved buried Glu in helices 1 and 4. Moreover, conserved Gly/Ala indicate possible sites of helix–helix interaction.

2. Amphipathic Helices

The 3D structure determined by electron crystallography revealed a six α-helix bundle. Within each bundle, an additional density spans the membrane, inserted between the two highly tilted helices on the outer edge of the tetramer (Figs. 16 and 18). Two legs of this density facing the lipid environment are of a shape and size consistent with two short helices, together forming the equivalent of a transmembrane helix with a strong kink in the middle (Mitsuoka *et al.*, 1999). This density is thought to represent the two functional loops B and E, with the two short helices composed of the second hydrophobic halves of the loops of about 7–10 residues following the NPA motifs located at the center of the monomer.

The amphipathicity of especially the lipid-exposed helices should be evident from the sequence alignment, as analyzed for residue conservation and hydrophobicity patterns. This follows an extensive literature on amphipathic helix analysis to determine helix orientation in membrane proteins (Eisenberg *et al.*, 1984; Finer-Moore and Stroud, 1984, Donnelly *et al.*, 1994).

Examination of the average hydrophobicity of the aligned residues along the core segments suggested a periodicity of about four residues, consistent with helices with right-handed crossing angles. The tetramer of AQP1 shows five right-handed and two left-handed helix–helix crossing angles per monomer (Fig. 16), giving a predominance of right-handed crossing angles. Therefore, the periodicity analysis of the information entropy and the average hydrophobicity shown in Fig. 20 was performed with a frequency of four residues (Heymann and Engel, 2000).

The informational entropy periodicity amplitude smoothed with a 21-residue window shows six clear peaks associated with the six helices, ranked in decreasing peak amplitude as H3 > H6 > H1 > H4 > H5 > H2 (Fig. 20). The functional loops exhibit low entropy periodicity, in agreement with their strong conservation

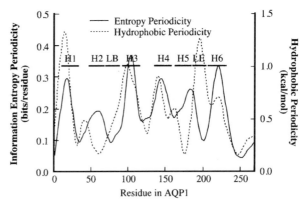

FIGURE 20 Periodicity of hydrophobicity and informational entropy. The amphipathicity is expressed as the amplitude of the Fourier transforms of entropy and hydrophobicity at the frequency associated with α helices with right-handed ($p = 4.0$) crossing angles. The solid bars indicate the conserved core segments. Curves are given with respect to the AQP1 sequence and were smoothed with a 21-residue window.

throughout the family. The hydrophobic periodicity reveals only five clear peaks associated with the core segments, with a ranking of H1 > LE > H3 > H4 > H6. The two remaining small peaks correspond to loops A and D, rather than to H2 and H5, respectively. The average phase differences between entropy and hydrophobic periodicity range from 133° to 178° for all helical segments, consistent with general opposition of residue conservation and hydrophobicity. The hydrophobic faces of the helices derived from this analysis are shown as residues with gray backgrounds in the topological model of Fig. 21.

3. Helix Assignment

The arrangement of densities in the tetramer can be described as three concentric rings of helices. The inner eight helices (labeled "i" in Fig. 21b) form a bundle with alternating right- and left-handed crossing angles. The middle ring has eight helices (labeled "m" in Fig. 21b), which are more tilted, each with one end exposed to the membrane lipid. The outer ring consists of eight highly tilted transmembrane helices (labeled "o" in Fig. 21b), and four pairs of short helices interpreted to be parts of loops B and E (labeled "o" in Fig. 21b). All the helices of the outer ring have one face exposed to the lipid environment.

Apart from the 24 transmembrane helices in the AQP1 3D map, only the kinked densities in the outer ring could be assigned to the helices in the functional loops. Loop E shows strong amphipathicity consistent with a short amphipathic helix located on the edge of the tetramer, while its high conservation means low entropy periodicity (Fig. 20). In contrast, the periodicity of loop B is rather low, but it is the more hydrophobic of the two short helices, suggesting that the second half may face the lipid environment.

The long tilted transmembrane helices in the outer ring are expected to show both hydrophobic and entropy periodicity. From the ranking of core segments in Fig. 20, the transmembrane helices exhibiting the highest entropy periodicity are H3 and H6, whereas H1 and H3 have the highest hydrophobic periodicity. The assignment to an "o" helix of H3 is therefore apparent, suggesting that its internal symmetry partner, H6, should be the other. The hydrophobic periodicity of H6 is marginal due to the prevalence of hydrophobic residues in the center of H6 on all faces, also leading to its higher hydrophobicity compared to H3 (Heymann and Engel, 2000). Thus, the hydrophobic periodicity is weak because the buried face of the helix also contains hydrophobic residues, and cannot be used to indicate exposure to the lipid environment. These considerations and the strong entropy periodicity of H6 suggest that it indeed is one of the outer helices. The striking similarity between H3 and H6 in the pattern of conservation evident from Fig. 19 further supports this conclusion.

The helices in the middle ring each have one end exposed to the lipid (Fig. 21b), suggesting periodicity in that part of the helix. After H3 and H6 are assigned to the outer ring, H1 and H4 are left as the transmembrane helices with the most

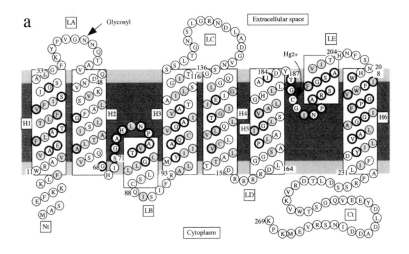

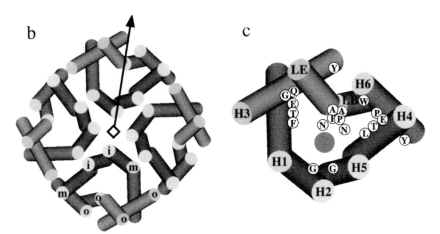

FIGURE 21 (a) A topological model of human AQP1 showing the six transmembrane helices and the two short helices in the functional loops B and E. Residues circled in bold are highly conserved (information entropy > 2 bitresidue) and those with a gray background are consistently hydrophobic (average hydrophobicity > 2 kcal/mol). Beginning and end of the core segments are indicated by the residue numbers. (b) The AQP1 tetramer (view from the extracellular side) consists of three rings of helices, labeled i (inner), m (middle) and o (outer). The helices within a monomer all show right-handed crossing angles, while the inner and middle helices form left-handed crossing angles with the helices from neighboring monomers. The pseudo-twofold axis is indicated in one monomer by the arrow. (c) Diagram of the monomer seen from the extracellular side. Helices are oriented so that highly conserved residues point toward the likely location of the water channel (*gray circle*). The two NPA motifs are shown in the middle, thought to cross over at the N termini of short helices in loops B and E. The hydrophobic residues on H1 (F) and H4 (L) are proposed to lie close to the water channel. The cylinders correspond to helical densities in the 4.5-Å structure of AQP1 (Mitsuoka *et al.*, 1999).

prominent periodicities, both entropy and hydrophobic (Fig. 20). In addition, the periodicity peaks are located at the N termini of these helices, associated with the ExxxT motifs (Fig. 19), implying that these ends are exposed to the lipid with the Glu and Thr buried.

The remaining two helices, H2 and H5, are associated with hydrophobic periodicity valleys, expected of buried helices. The entropy periodicity is significantly higher, presumably indicating packing constraints. These core segments are also the shortest of the transmembrane helices (Heymann and Engel, 2000), consistent with the shortest and least tilted helices of the inner ring in the 3D map (Fig. 21b).

Only a few arrangements of the helices in the 3D structure are consistent with the internal similarity and the pattern of hydrophobic and conserved residues. In summary, this assigns the outermost helices to H3 and H6, the innermost bundle to H2 and H5, and the intermediate ring to H1 and H4. The possible arrangements with these are thus only eight. Distinct differences in the densities of LB and LE in the 0.45-nm 3D map (Mitsuoka *et al.*, 1999) and most recent AFM experiments (Fig. 14) suggest a sidedness assignment as shown in Fig. 21c. The short interlinking loops between LB and H3, and LE and H6 (Heymann and Engel, 2000), distinguish between the two choices for the outer ring transmembrane helices. Similarly, the short interlinking loops between H1 and H2, and H4 and H5, suggest that these are pairs of interacting helices. This assignment is consistent with the internal similarity between the first and second halves, as well as with the pseudo-twofold axis in the membrane plane apparent in the 3D structure (Cheng *et al.*, 1997; Li *et al.*, 1997).

4. Differences between AQPs and GLPs: Residues Involved in Specificity

The major difference between AQPs and GLPs is that loop E is longer by ∼10–15 residues in the latter, but this difference still needs to be visualized and its functional implication unraveled. The variability of the other loops and the N and C termini is pronounced throughout the family. This leads to marked differences in the surface topography of aquaporins which are determined to subnanometer resolution by atomic force microscopy, as demonstrated by topographs of AQP1 (Walz *et al.*, 1996), AqpZ (Scheuring *et al.*, 1999b), and AQP0 (Fotiadis *et al.*, 2000) (see below).

Froger *et al.* (1998), have proposed to distinguish the AQPs and the GLPs based on five particular amino acid residues, called P1–P5 (Fig. 19). The position P1 is in general nonaromatic in the AQP cluster and aromatic in the GLP cluster. Positions P2 and P3 just follow the NPAR motif in loop E and form mostly an S–A pair in the AQP cluster and a D–R or D–K pair in the GLP cluster. The Asp at P2 is present in all aquaglyceroproteins sequenced so far. Positions P4 and P5 in helix 6 are mostly aromatic in the AQP cluster, with P4 usually a Pro in the GLP cluster. Unexpectedly, AQPcic carrying the S–A pair in positions P2 and P3 could be changed to a functional aquaglyceroporin by a double mutation in P4 and P5 (Lagree *et al.*, 1999).

Further clues are found in the more recent sequence analysis by Heymann and Engel (2000), who identified two critical conserved hydrophobic residues in the middle of helices 1 and 4. As shown in Fig. 19, these are either a Phe or a Leu in position 24, or mostly a Leu in position 149. In 17 of 21 GLP subtypes, these positions harbor the L–L pair, and in two subtypes it is an M–L pair. In contrast, 16 of 28 AQP subtypes exhibit an F–L, F–M, or L–F pair, while AQP0, AQP6, and BIB_DROME have a Tyr in position 24. However, seven AQPs possess the L–L pair characteristic for GLPs. Among them are three NIPs, two archaeal AQPs, and two fly AQPs, the latter including AQPcic. Because the two conserved hydrophobic residues are on the same helical face as the conserved Glu and Thr in helices 1 and 4, and because positions 24 and 149 are in the middle of these helices, the Leu and Phe may line the pore. In this case, these residues could be involved in determining the size and specificity of the pore.

5. Genetic Defects in AQP2

An interesting exercise is to compare the reported mutations of aquaporin-2 (AQP2) involved in nephrogenic diabetes insipidus (see http://www.uwcm.ac.uk/uwcm/mg/hgmd0.html) with the map of conserved residues (Fig. 19). AQP2 is essential for vasopressin (AVP)-regulated urine concentration (Deen *et al.*, 1994). AQP2 is predominantly expressed in intracellular vesicles under basal conditions and is redistributed to the apical membrane of collecting duct cells through activation of a cAMP signaling cascade initiated by binding of AVP to its V2 receptor (Marples *et al.*, 1995). Thus, dysfunction of AQP2 can be related to misfolding of the protein, to lack of transport of correctly folded protein, or to a mutation impairing water activity. As illustrated in Fig. 22 (in AQP2 notation), many of the point mutations correlate with highly conserved residues (Fig. 19, in AQP1 notation, hence a shift of sequence numbers). The mutant L22V alone is active when expressed in oocytes. However, it is found together with mutation C181W, which is unable to fold, as suggested by its endoplasmic reticulum-like intracellular distribution (Canfield *et al.*, 1997). Apparently, the replacement of the buried cystein in loop E (see Fig. 21a) with the much larger tryptophan prevents folding of the molecule. In G64R a bulky charged residue is placed at a site where loop B folds into the membrane; this mutant is unlikely to fold (van Lieburg *et al.*, 1994). Mutation E258K leads to much reduced plasma membrane expression, in spite of correct phosphorylation of S256 that is essential for plasma membrane integration. This mutant AQP2 is retarded in the Golgi, while the single-channel water permeability of this mutant is apparently not affected (Mulders *et al.*, 1998). Hence, the binding site for the transport machinery is disturbed, preventing shuttling to the plasma membrane. The ability of this latter mutant to heterotetramerize with wild-type AQP2, thereby impairing its further routing to the plasma membrane,

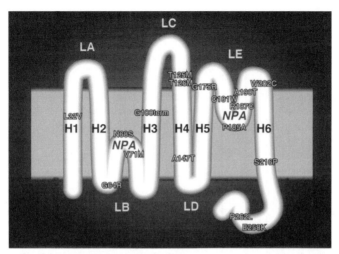

FIGURE 22 Patients suffering from hereditary nephrogenic diabetes insipidus (NDI) have defects in the vasopression 2 receptor (V2R) or in the waterchannel AQP2 of kidney collecting ducts. Mutations in AQP2 are generally located at highly conserved sites (compare to Fig. 19). In a few cases, these AQP2 mutations have been characterized at a cellular and molecular level. Three types of phenotypes have been described: (1) misfolding and retention in the ER, (2) lack of transport and retention in the Golgi, and (3) impaired water activity (see text).

provides an explanation for dominant NDI. Finally a number of AQP2 mutants have been described that show plasma membrane expression, but exhibit altered water transport activity. These concern A147T and T126M (Mulders *et al.,* 1997) and T125M and G175R (Goji *et al.,* 1998). Smaller residues are replaced with larger or even charged ones in these mutants, probably impairing the packing arrangement of helices 4 and 5 in a way that affects the pore, but not the folding of the mutant AQP2.

VI. AQP0: THE MAJOR INTRINSIC PROTEIN OF LENS FIBER CELLS

A. Purification of AQP0

The silver-stained SDS gel in Fig. 23a illustrates the different purification steps of AQP0 (Hasler *et al.,* 1998b). Lane 1 shows the proteins of urea/alkali stripped fiber cell membranes from the lens core, which contain as major integral proteins AQP0, MP20, and connexins 46 and 50 in the cleaved 38-kDa form. Lane 2 displays the protein bands after solubilization of the membrane with *n*-octyl-β-D-glucoside (OG). The solubilization efficiency was approximately 70% as estimated by the BCA protein assay. Because AQP0 tends to aggregate in OG (Koenig *et al.,* 1997),

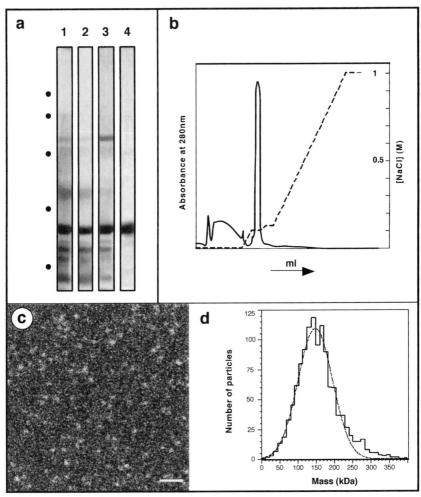

FIGURE 23 Purification of AQP0 from ovine lens fiber cell membranes. (a) SDS gel (12% w/v acrylamide with silver staining) showing the various steps in the purification of AQP0. Lane 1, urea/alkali stripped fiber cell membranes from the lens core; lane 2, supernatant after OG solubilization and centrifugation; lane 3, sucrose gradient fraction enriched in AQP0; lane 4, concentrated AQP0 after cation-exchange chromatography. Molecular mass standards marked by filled circles are (*from top to bottom*) 94, 66, 43, 31, and 21 kDa. (b) Cation-exchange chromotography of pooled sucrose gradient fractions enriched in AQP0. AQP0 eluted in a sharp peak at a NaCl concentration of approximately 110 mM NaCl in 10 mM HEPES (pH 7.2), 0.3% DM. Continuous line, elution profile ($\lambda = 280$ nm); broken line, NaCl gradient. (c, d) Mass analysis of soluble AQP0 oligomers in the STEM. (c) Dark-field image of freeze-dried, solubilized AQP0 particles recorded at 2000 electron/nm^2. (d) Mass histogram of 1154 particles recorded at an average does of 320 electron/nm^2. The Gauss peak was fitted at 147.5 ($\pm$47.0) kDa, omitting a tail of higher mass values resulting from some aggregates.

a sucrose density gradient was prepared with 0.3% decyl-β-D-maltoside (DM) in 10 mM HEPES (pH 7.2) to allow gentle detergent exchange. Sucrose gradient fractionation removed the 38-kDa connexin from (Fig. 23a, lane 3). A MiniS column was subsequently used to eliminate the MP20 protein and to concentrate AQP0. Figure 23b shows a sharp elution peak at a NaCl concentration of ∼100 mM. The flow-through fractions contained small amounts of all proteins detected by SDS–PAGE of pooled sucrose gradient fractions (Fig. 23a, lane 3), whereas the sharp peak was the result of highly purified AQP0 (Fig. 23a, lane 4). To yield a final protein concentration of about 1 mg/ml, AQP0 was further concentrated with Centricon 50.

B. Solubilized AQP0 is a Tetrameric Protein

To assess the homogeneity of the purified AQP0 oligomers and to determine their mass, DM-solubilized AQP0 particles were freeze-dried and observed with a scanning transmission electron microscope (STEM). Elastic dark-field images were recorded at doses of 320 electron/nm^2, 200,000× magnification, and 80-kV acceleration voltage (Fig. 23c). Figure 23d shows the mass peak of 1154 particles. The single peak at 147.5 ± 47.0 kDa (standard error, SE = 1.60 kDa) indicates that AQP0 is homogeneous and stable in DM. This mass value correlates favorably with recently published data, which document that solubilized single AQP0 particles have a tetrameric structure (Aerts *et al.,* 1990; Koenig *et al.,* 1997). The tetrameric nature of AQP0 is also consistent with earlier observations that lens fiber cell membranes contain square arrays which can be attributed to AQP0 (Fig. 2; Kistler and Bullivant, 1980; Zampighi *et al.,* 1982, 1989) and more recent reconstitution experiments yielding square arrays (Engel *et al.,* 1992).

C. 2D Crystallization of AQP0

Purified AQP0 complexes were reconstituted with solubilized *E. coli* lipids at low LPRs into 2D crystals. Highly ordered coherent areas of 0.5 μm in diameter were often observed (Fig. 24a). Their diffraction patterns (Fig. 24b) documented the presence of single-layer sheetlike 2D crystals. Diffraction spots indicated unit cell dimensions of $a = b = 6.4 ± 0.03$ nm ($n = 8$) and a resolution of 2.0 nm.

Projection maps of AQP0 crystals normally exhibited a pronounced 4-fold symmetry. One unit cell (Fig. 24c) housed a single tetramer with four stain-excluding regions (bright) and central stain-filled depression (dark), the tetramers being separated by strongly stained lipid bilayer domains. The protruding domains had a length and width of 2.7 and 1.8 nm, respectively. Disregarding the faint differences in staining, the unit cell exhibited 2-fold and 4-fold axes perpendicular to

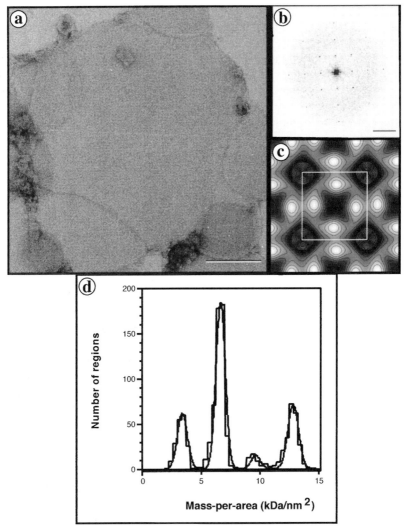

FIGURE 24 Negative stain electron microscopy and image analysis of 2D AQP0 crystals. (a) Negatively stained crystalline sheet of AQP0. Scale bar: 100 nm. (b) The power spectrum calculated from a selected area in (a) shows one set of diffraction spots corresponding to a single-layered crystal. The resolution is approximately 2.0 nm (diffraction order 3,0). Scale bar: $(5 \text{ nm})^{-1}$. (c) The averaged and 4-fold symmetrized unit cell calculated from AQP0 crystals exhibits four prominent elliptical domains. Protein is displayed in bright shades, the surrounding stain in dark shades. The outlined unit cell ($a = b = 6.4$ nm) includes one tetramer. The dark areas centered at the four corners of the unit cell represent the lipid bilayer. (d) Mass analysis of freeze-dried AQP0 crystals. The histogram shows peaks at 3.4 (± 0.3) kDa/nm^2 ($n = 175$), 6.63 (± 0.39) kDa/nm^2 ($n = 508$), 9.62 (± 0.44) kDa/nm^2 ($n = 49$), and 12.80 (± 0.46) kDa/nm^2 ($n = 224$). These peaks represent single sheets, collapsed vesicles, sheets and vesicles, and two vesicles, respectively.

the membrane plane, but no screw axes in the membrane plane. Therefore, the AQP0 crystal belongs to the plane group symmetry p4 (Fig. 24c).

Mass measurement of the AQP0 sheets with STEM yielded a mass per area histogram with distinct peaks (Fig. 24d). The first peak at 3.5 ± 0.3 kDa/nm^2 (standard error, SE $= 0.03$; $n = 175$) represents single-layer tetragonal AQP0 crystals. The mass per unit cell of 6.4 nm side length was calculated to 144 kDa. The second peak at 6.6 ± 0.32 kDa/nm^2 (SE $= 0.018$; $n = 508$) resulted from AQP0 square arrays of collapsed vesicles. The less populated peak at 9.8 ± 0.36 kDa/nm^2 (SE $= 0.025$; $n = 49$) represents the sum of the first and second peak. Finally, the fourth peak at 13.1 ± 0.38 kDa/nm^2 (SE $= 0.03$; $n = 224$) occurred when two crystalline vesicles were overlaid. The weighted average from all peaks yielded a mass per area of 3.35 kDa/nm^2, or 137 kDa per unit cell, disregarding a minor mass loss due to electron irradiation. Taking the latter value and considering the beam-induced mass loss (i.e., 2.4% at 320 electron/nm^2; Müller and Engel, 1998), the overall mass per unit cell is 140 kDa. Contouring the projection map of negatively stained tetragonal crystals at the steepest contrast gradient suggested that 37% of the unit cell surface be comprised of lipids. Assuming a lipid bilayer mass per area of 2.8 kDa/nm^2, the protein mass comes to approximately 98 kDa, compatible with one tetramer per unit cell.

D. Projection Maps of Freeze-Dried 2D AQP0 and AQP1 Crystals

The excellent crystallinity documented by negatively stained AQP0 crystals prompted a comparison of freeze-dried AQP0 and AQP1 lattices. Projection maps of such samples exhibited for both proteins a resolution of 0.8 nm (Fig. 25). The freeze drying compensates partly for the low contrast when working with glucose or vitrified ice. Therefore, projection maps revealed distinct and reproducicble features after averaging only 100 unit cells (Fig. 25). As demonstrated here for the first time, freeze-drying is a powerful method for analyzing the structure of 2D protein crystals to a resolution better than 1 nm. In Figs. 25g and 25h, freeze-dried projection maps of AQP0 and AQP1 are compared at a resolution of 0.9 nm.

Both projection maps show the tetramers arranged in square arrays with lattice vectors for AQP0 of $a = b = 6.4$ nm, $\gamma = 90°$ and for AQP1 $a = b = 9.6$ nm, $\gamma = 90°$. This is in agreement with the map of AQP1 at 0.6-nm resolution published by Walz et al. (1995) where the crystals were embedded in glucose. In addition, the lattice vectors are in agreement with the data from negatively stained preparations of both proteins (see above; Walz et al., 1994a). Both AQP0 and AQP1 show in their projection eight high-density maxima per monomer. However, AQP0 has a more pronounced internal symmetry than AQP1. This and the fact that AQP0 crystallizes into a p4 crystal with tetramers incorporated unidirectionally

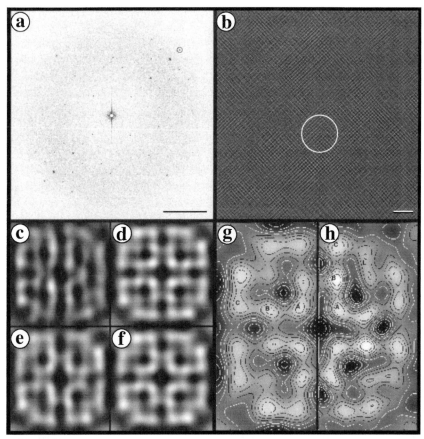

FIGURE 25 Correlation and single-particle averaging of freeze-dried 2D AQP0 crystals. (a) The diffraction pattern shows strong diffraction spots to about $(0.9 \text{ nm})^{-1}$ in one direction. The transfer function had its first zero value at a resolution of $(0.85 \text{ nm})^{-1}$ and does not indicate any major drift. Scale bar: $(2 \text{ nm})^{-1}$. (b) The cross-correlation function of the 2D crystal with a 4-fold symmetrized reference containing two tetramers exhibits sharp peaks as well as peaks blurred in one direction (area marked by a circle). The latter peaks suggest local defects, possibly resulting from charging. Scale bar: 20 nm. (c) The correlation average calculated from 395 fields centered at the sharp, unblurred peaks in (b) still exhibits some residual asymmetry, explaining the rms deviation of 0.29 for the 4-fold symmetrized tetramers in (d). (e) Single-particle averaging of 138 fields that exhibited a correlation coefficient >0.14 after angular and translational alignment with respect to a reference containing nine tetramers as shown in (d). The symmetry is improved as documented by the RMS deviation of 0.1 from the 4-fold symmetrized tetramer displayed in (f). Scale bar: 1 nm. Comparison of projection maps from freeze-dried AQP0 (g) and AQP1 seen from the cytosolic side (h). Pronounced similarities are recognized in the low-density regions as well as the three density maxima (A, B, and C) that line the interface between monomers. Differences are noticeable at the periphery of the tetramer as well as in the density L. The full width of the image corresponds to 6.4 nm.

in the bilayer rather than in both orientations are AQP1 could not be explained by these projection maps.

E. Visualization of AQP0 Domains Protruding from the Membrane

High-resolution AFM imaging was employed to visualize the peptide domains of AQP0 that protrude from the membrane (Fotiadis *et al.*, 2000). Crystalline sheets were adsorbed to freshly cleaved mica and imaged in aqueous solution. In addition to the atomically flat mica, three different surfaces were typically observed: the majority of the sheet surface appeared "smooth" and crystalline (Fig. 26a, 1), while smaller areas with a "rough" surface texture were located at the periphery of the sheets and appeared to be a step down from the "smooth" surface (Fig. 26a, 2). The texture of the "rough" surface ranged from unordered to crystalline, often with continuous transitions. The sheets were surrounded by a third type of surface that had no discernable structural features and represented lipid bilayer regions without protein (Fig. 26a, 3).

To image larger areas of the rough surface, the top layer was scraped away by scanning a selected area with a higher force applied to the tip (Hoh *et al.*, 1991; Schabert *et al.*, 1995). A typical experiment is shown in Fig. 26b. An area just

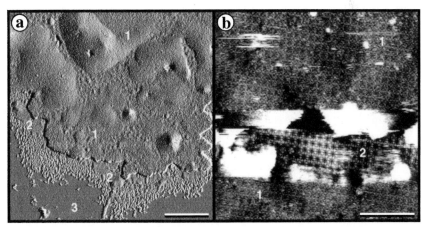

FIGURE 26   AFM of crystalline AQP0 double layers on mica. (a) Deflection AFM image of a collapsed vesicle showing three different regions. A "smooth" (1), a "rough" crystalline (2), and a structureless surface (3). (b) Height AFM image of both AQP0 surfaces after scraping away a part of the upper layer by increasing the applied force to the AFM tip. Imaging conditions: 10 mM Tris-HCl, 150 mM KCl, 50 mM MgCl$_2$, pH 8.8. Scan frequency 5.1 Hz (a and b). Scale bars represent 250 nm in (a) and 50 nm in (b). The vertical brightness range of the height topography in (b) is 1.8 nm.

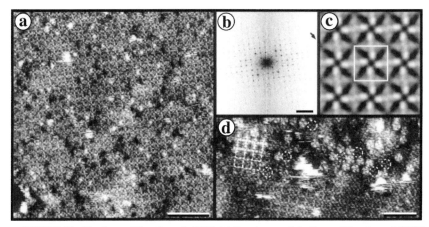

FIGURE 27 The "smooth" surface of AQP0. (a) Raw image of the "smooth" surface of AQP0. (b) The lateral resolution of the raw image (a) was estimated to be 0.61 nm according to its power spectrum. (c) The symmetrized correlation average (rms deviation of 18% from the 4-fold symmetry) of 542 unit cells revealed four major globular protrusions of 0.8 ± 0.1-nm height surrounding a distinct central cavity per tetramer. (d) The unit cell marked in the average (c) was defined according to single unordered tetramers in densely packed regions (white circles) and packing defects of "rough"-sided tetramers in the crystal (lattice). Imaging conditions: 10 mM Tris-HCl, 150 mM KCl, 50 mM MgCl$_2$, pH 8.8. Scan frequencies: 6.1 Hz in (a) and 5.5 Hz in (d). Scale bars represent 35 nm in (a), 2 nm^{-1} in (b) and 25 nm in (d). The frame size of the average (c) is that of three unit cells: 19.2 nm. Vertical brightness ranges: 1.0 nm in (a), 0.8 nm in (c), and 1.5 nm in (d).

below the middle of the image was first scanned with a force of 500–750 pN to remove the top layer, and then imaged at minimal force of 50–150 pN. The lower layer had a rough crystalline texture, which was identical to that seen in the peripheral regions of the sheets (Fig. 26a, 2). Examining Fig. 26b at glacing angles reveals that the lattices of the smooth crystalline top layer and the rough crystalline lower layer are stacked in register (Fotiadis *et al.,* 2000).

From a number of buffers tested, the best images of the smooth surface were obtained using 10 mM Tris–HCl, pH 8.8, 150 mM KCl, 50 mM MgCl$_2$ (buffer A; Fig. 27a). Diffraction spots extended to order (10,3) of the square lattice, representing a lateral resolution of 0.61 nm (Fig. 27b). The correlation average of the smooth side of the AQP0 tetramer revealed a major globular protrusion of 0.8 ± 0.1 nm ($n = 80$) height per monomer (Fig. 27c). "Crystal defects" resulting from the packing of tetramers with the rough side facing up (lattice; Fig. 27d) and the morphology of single smooth-sided tetramers (broken white circles; Fig. 27d) in disordered patches allowed the unit cell in the correlation-averaged image (Fig. 27c) to be defined. The resolution of the latter was 0.7 nm according to the Fourier ring correlation function (Saxton and Baumeister, 1982) and 0.8 nm according to the phase residual criterion (Frank *et al.,* 1981).

FIGURE 28 The "rough" surface of AQP0. In addition to crystalline areas (*top right*), densely packed regions containing single unordered AQP0 tetramers and even defective monomers (*arrows*) could be visualized by AFM. *Inset:* The correlation average of 445 unit cells exhibited a rms deviation of 7% from the 4-fold symmetry. Imaging conditions: 20 mM Na-acetate-HCl, 50 mM NaCl, pH 5. Scan frequency 4.7 Hz. Scale bar: 25 nm. The frame size of the inset is that of one unit cell: 6.4 nm. The vertical brightness range of the raw image is 1.6 nm. The raw image is displayed as relief tilted by 2°.

Large selected areas of the rough surface could be imaged following the removal of the top layer with the AFM tip (Fig. 27). Scraping away the upper layer was easier using a different buffer with reduced pH and salt (buffer B: 20 mM sodium acetate–HCl, pH 5.0, 50 mM NaCl). The rough surface exhibited crystalline areas, regions of densely packed tetramers, and regions with loosely packed tetramers with a distinct windmill appearance (Fig. 28). Scraping away the top layer often disturbed the crystalline order by introducing small cracks into the lower layer, and this limited the application of correlation averaging. Instead, we focused on areas with densely packed tetramers for structure determination using single particle analysis (Frank *et al.,* 1978; Schabert and Engel, 1994). The resolution was 0.6 nm according to both the Fourier ring correlation function (Saxton and Baumeister, 1982) and the phase residual criterion (Frank *et al.,* 1981), and 0.8 nm according to the spectral signal-to-noise ratio criterion (Unser *et al.,* 1987). The average of the rough side of the AQP0 tetramer showed two protrusions per monomer, a small globular and a larger elongated domain (Fig. 28, inset). The heights of these domains varied slightly, depending on whether measurements were taken from AQP0 tetramers in crystalline or densely packed areas, and on the buffer used for imaging. Average heights of the large and small domains were 1.4 ± 0.2 nm ($n = 120$) and 1.1 ± 0.2 nm ($n = 120$), respectively. Hence, the AQP0 monomer has two peptide domains protruding from the rough surface that are both significantly higher than the single domain visible on the smooth surface, meaning that the distribution of peptide mass with respect to the membrane plane is strongly asymmetric.

Lipid bilayers with or without AQP0 also showed different heights depending on the buffer used for imaging. When imaged in buffer A, the height of the double-layer AQP0 crystal was 13.1 ± 0.3 nm ($n = 80$) and the lipid alone exhibited a height of 4.5 ± 0.2 nm ($n = 120$). In buffer B the double-layer sheets had a height of 10.0 ± 0.4 nm ($n = 120$), while a single lipid bilayer had a height of 2.9 ± 0.2 nm ($n = 126$). These height differences measured in the different buffers are consistent with a change of the electrostatic tip–sample interaction (Müller and Engel, 1997).

F. Identification of the "Smooth" Surface as Cytoplasmic

The topology for AQP0 predicted from the amino acid sequence places the approximately 5-kDa carboxyl tail segment on the cytoplasmic side of the lens fiber cell plasma membrane (Gorin *et al.*, 1984). The same conclusion was reached by probing AQP0 in isolated membranes with site-specific proteolysis and antibodies (Keeling *et al.*, 1983). Removal of the carboxyl tail segment of AQP0 with carboxypeptides Y should result in a detectable loss of peptide mass in one of the topographs of AQP0, thereby identifying either the smooth or the rough surface as cytoplasmic. Sheets were digested *on mica* overnight and subsequently imaged by AFM in buffer A. Only the structure of the smooth surface was distinctly altered by the carboxypeptidase Y treatment. Its crystalline texture appeared coarser (Fig. 29a), and the averaged tetramer structure revealed the partial loss of the prominent four protrusions leaving a large cavity about the 4-fold symmetry axis (Fig. 29b). Compared to the situation before digestion (Fig. 29c), the remaining protruding mass was located more peripherally in the tetramer, and had a height of 0.6 ± 0.1 nm ($n = 80$). The difference map between the undigested and the digested cytoplasmic topographs (Fig. 29d) reveals that the carboxyl tail is located close to the center of the AQP0 tetramer similar to that of AQP1 (Fig. 14). In contrast to this readily detectable structure change in the smooth surface, the rough surface of the membranes did not appear to be affected by the carboxypeptidase Y treatment (data not shown).

The sidedness of the reconstituted sheets was also determined by immunogold labeling. HNPA1, an antibody against the cytoplasmic loop B (a kind gift of Dr. Ana B. Chepelinsky, National Eye Institute, NIH, Bethesda, Maryland), was bound to sheets adsorbed to mica and detected with a gold-labeled secondary antibody.

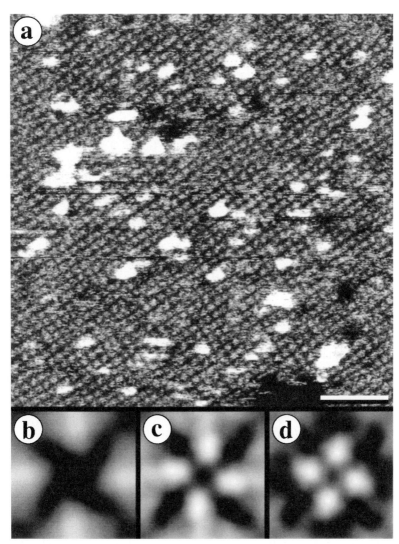

FIGURE 29 The "smooth" surface of AQP0 after digestion by carboxypeptidase Y. (a) The raw image shows clear structural differences compared to the undigested surface shown in Fig. 27a. A pronounced decrease in height from 0.8 ± 0.1 to 0.6 ± 0.1 nm and a wider cavity are characteristic for this new surface structure. (b) The calculated diffraction pattern of (a) reflects the partial loss in crystallinity after the carboxypeptidase Y treatment. (c) In the 4-fold symmetrized correlation average (rms deviation of 15% from the 4-fold symmetry) of 508 unit cells, the absence of the prominent initially existing protrusion (Fig. 27c) is clearly visible. This result identifies the "smooth" surface as the cytoplasmic side and highlights the position of the C terminus. Scale bar: 50 nm.

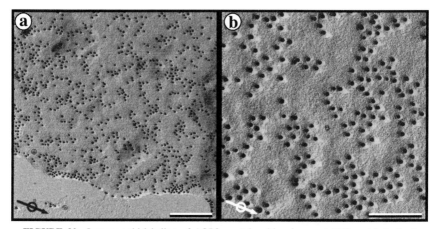

FIGURE 30 Immunogold labeling of AQP0 crystals with subsequent Pt/C metal shadowing. (a) An overview of a labeled vesicle is shown. Scale bar: 500 nm. (b) A region labeled with gold particles at higher magnification. Particles have a bright, sharp shadow and an elevated amount of Pt/C opposite to the metal shadowing direction. Scale bar: 100 nm.

A Pt/C replica was made and examined by electron microscopy (Fig. 30). The shadow casts of the gold beads had the same lengths irrespective of their location on sheets or mica, thus indicating that they were bound to the top surface of the sheets and not buried between the two crystalline layers. This result also identified the top surface as cytoplasmic in agreement with the AFM data.

G. Structural Interactions between the Extracellular Surfaces of AQP0

The survey recorded by AFM showed that the majority of reconstituted sheets consisted of two crystalline layers. To demonstrate that the lattices of the two membranes were in precise register, the cross-correlation function of Fig. 26b with the sum of the averages from both surface structures (Fig. 27c and the inset to Fig. 28) was calculated. Correlation peaks in Fig. 31a mark the position of all AQP0 tetramers. Peaks with a root-mean-square deviation of less than 10% from the fitted crystal lattice are highlighted by squares ($n = 92$) on the extracellular surface and by diamonds ($n = 278$) on the cytoplasmic surface. The large number of such peaks indicated that AQP0 tetramers were precisely superimposed on each other. Calculating the averages corresponding to the two sets of correlation peaks yielded the two insets to Fig. 31a. Using these averaged topographs the double-layer sheet was reconstructed, which demonstrated clearly that AQP0 tetramers from the opposing layers interact with each other in a "tongue-and-groove" fashion: The large protrusions in one membrane fit the small protrusions in the opposing

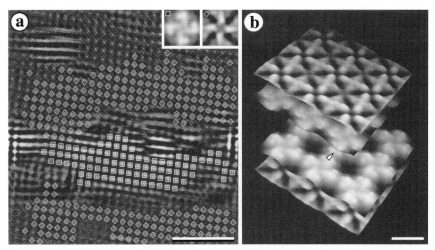

FIGURE 31 The in-register superposition of two crystalline layers in reconstituted AQP0 sheets. (a) The cross-correlation function of Fig. 26b with the sum of the extracellular (*left insert*) and cytoplasmic surface (*right insert*) topographs. The correlation peaks mark the positions of all AQP0 tetramers and show that they are in precise register. Peaks with a rms deviation of less than 10% from the fitted crystal lattice are highlighted by squares (extracellular surface) or diamonds (cytoplasmic surface). (b) Computer reconstruction of a double-layered AQP0 sheet showing the "tongue-and-groove" interaction of the extracellular surfaces (*white arrowhead*). Scale bars represent 50 nm in (a) and 5 nm in (b). The frame size of the insets in (a) is that of one unit cell: 6.4 nm. Vertical brightness ranges: 1.4 nm (b and inset: square) and 0.8 nm (inset: diamond).

membrane, and vice versa, as illustrated by the perspective view in Fig. 31b (white arrowhead).

Reconstituted sheets were further investigated by cryo-electron microscopy for two reasons: first, to verify that they were also double layered in solution and that this was not an artifact of specimen preparation for AFM; second, to verify the exact superposition of AQP0 tetramers by analyzing their projection structure at high resolution. Edge-on views of sheets that were suspended in a thin layer of vitrified negative stain solution (Adrian *et al.,* 1998) confirmed that they indeed existed mostly as double-layered structures in solution (white arrowheads in Fig. 32a).

Projection maps were calculated from sheets that were adsorbed to the carbon film and vitrified by rapid freezing. They fell into two categories: one exhibited eight density peaks arranged symmetrically in the monomer (Fig. 32b), whereas the other had eight densities arranged symmetrically about a mirror plane at 45° to the lattice (Fig. 32c). The former was only rarely observed: Its distinct handedness was similar to that of AQP1 (Walz *et al.,* 1995) and was likely to be derived from sheets with only one crystalline layer. In contrast, symmetrical projection maps

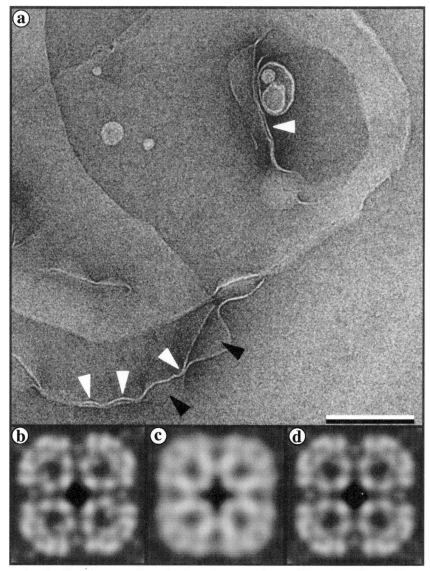

FIGURE 32 (a) Cryo-negative stain of reconstituted AQP0 sheets revealing their double membrane nature. Two layers can clearly be recognized at the borders of the sheets (*white arrowheads*). Apposing membranes are sometimes slightly opened at their edges and expose single layers (*black arrowheads*). Scale bar: 200 nm. (b) One type of projection map of AQP0 determined by cryo-electron microscopy shows eight density peaks per AQP0 monomer and exhibits a distinct handedness. A resolution of 0.57 nm was achieved. The seven electron micrographs used to calculate the map were most probably recorded from single-layered sheets. (c) The other type of projection map exhibits eight densities in the monomer, which are symmetrically arranged with respect to a mirror plane at 45° to the lattice.

were more frequently obtained and were consistent with the more abundant double-layer sheets, with tetramers of the two layers packed in register corresponding to space group p422. This assignment was supported by superpositioning the 0.57-nm asymmetric map with its own mirror image to simulate a double-layer sheet. The resulting projection map (Fig. 32d) was indeed similar to the 0.69-nm symmetrical map in Fig. 32c and also to the 0.90-nm map presented previously (Fig. 25). Hence, the data obtained by cryo-electron microscopy strongly support the precise stacking of two AQP0 tetramers by the "tongue-and-groove" interaction of extracellular surfaces.

These results obtained by atomic force and cryo-electron microscopy demonstrate that reconstituted square arrays of AQP0 can interact in a highly specific manner via their extracellular surfaces to form double-layer sheets. This supports the alleged ability of AQP0 to form membrane junctions (Bok *et al.,* 1982; Sas *et al.,* 1985). Figures 26 through 32 show that the interaction between AQP0 tetramers is highly specific insofar as they superimpose exactly and bind to each other with their extracellular surfaces in a "tongue-and-groove" fashion. This level of specificity is significant, and is more likely to represent an intrinsic property of AQP0 than merely an arbitrary aggregation of reconstituted membranes in the test tube.

Therefore AQP0 may function in the lens as a channel for water and selected small solutes, as well as serving for cell–cell adhesion, which supports the tight packing of fiber cell layers (Bond *et al.,* 1996). Immunolabeling experiments have shown that AQP0 is not restricted to membrane junctional domains but is also present throughout the nonjunctional membranes (Fitzgerald *et al.,* 1983; Paul and Goodenough, 1983; Varadaraj *et al.,* 1999). Hence, a sizable portion of AQP0 is available to serve as channels. However, an interesting novel option arises: If the precise superposition of AQP0 molecules as seen in reconstituted sheets also occurred between adjacent fiber cells in the lens, it is conceivable that they could form continuous channels connecting the cell cytoplasms. This would provide an alternative path for water and possibly small solutes that also flow through the cell–cell gap junction channels, known to be composed of connexins.

In conclusion, the crystallographic packing of two arrays of AQP0 implies a highly specific interaction between their extracellular surfaces. This lends further support to a dual function of AQP0, as a water channel, and as a cell–cell adhesion molecule. The future 3D reconstruction from tilted views of these sheets has the potential to provide full structural details for both of these functions.

The image has 0.69-nm resolution and was calculated from five micrographs most likely depicting the more frequent double-layered sheets. (d) A similar aspect is obtained when calculating the sum of the map shown in (b) with its mirrored projection, thus simulating a double-layered sheet. The frame size of (a), (b), and (c) is that of one unit cell: 6.4 nm.

VII. WATER CHANNEL OF *Escherichia coli*: AqpZ

A. Solubilized AqpZ Particles

Discovery of the *aqpZ* gene in wild-type *E. coli* was achieved by homology cloning using the sequence-related bacterial gene of the glycerol facilitator (*glpF*). The *aqpZ* DNA from *E. coli* contains a 693-base-pair open reading frame encoding a polypeptide whose sequence is 28–38% identical to other known aquaporins (Calamita *et al.*, 1995). Comparative transport analysis using injection of AqpZ cRNA in *Xenopus* oocytes and assays for water or glycerol permeability demonstrated the high water selectivity and negligible glycerol transport of AqpZ (Calamita *et al.*, 1995). Disruption of the chromosomal *aqpZ* gene is not lethal, but Calamita *et al.*, have shown that AqpZ knockout *E. coli* strains exhibited greatly reduced colony formation at maximum growth rate (39°C) or when grown at low osmolarity (Calamita *et al.*, 1998). These authors propose that the physiologic significance of AqpZ for the bacterium is to maintain cell turgor while facilitating volume expansion during cell division.

A construct of AqpZ with an additional 10-Histidine tag at the N terminus (molecular weight of 26.5 kDa) has been designed, overexpressed in *E. coli* and milligram amounts of pure and active protein were isolated by nickel-chelator-based chromatography (Borgnia *et al.*, 1999). Solubilized AqpZ migrated at a relative molecular weight of 70 kDa in a denaturing SDS–polyacrylamide gel electrophoresis (SDS–PAGE 12%), suggesting a tetrameric form (Fig. 33a).

Micrographs of negatively stained AqpZ particles solubilized in octyl-β-D-glucopyranoside (OG) recorded at a magnification of 70,000× showed a homogenous size distribution (Fig. 33b). The protein particles appear white on a dark background (negative staining) and many particles exhibit a square shape. To enhance these features, 639 particles were selected by eye on a single micrograph, aligned with respect to a reference, and classified into clusters by multivariate statistical analysis (MSA; van Heel and Frank, 1981). An average of 181 aligned particles from five clusters exhibited a squarelike shape with 8-nm side length and showed the tetrameric structure without symmetrization (Fig. 33b, inset). Although other cluster averages had similar dimensions, they did not show a pronounced 4-fold symmetry, most likely because these tetramers were tilted with respect to the carbon film.

Previous studies of water channel proteins by freeze-fracture electron micro–scopy showed tetragonal arrays of the lens major intrinsic protein AQP0 (Kistler and Bullivant, 1980; Zampighi, *et al.*, 1982), of the insect aquaporin AQPcic (Beuron *et al.*, 1995), and tetramers of the human aquaporin AQP1 (Verbavatz *et al.*, 1993). These are all present in native membranes, suggesting a tetrameric assembly of native water channels. Both AQP0 and AQP1 have been solubilized by OG as tetramers (Hasler *et al.*, 1998b; Smith and Agre, 1991, Walz *et al.*, 1994a;

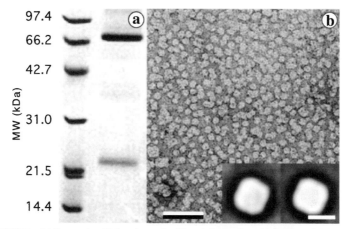

FIGURE 33 (a) Denaturing SDS–polyacrylamide gel electrophoresis. *Left lane:* Molecular weight markers. *Right lane:* Solubilized AqpZ. (b) Electron micrograph of negatively stained AqpZ particles solubilized in OG and recorded at 70,000× magnification. The particles are rather homogenous in size and shape. *Inset:* Out of 639 particles, 181 particles with the highest 4-fold symmetry were selected to calculate the final average. It has a clear squarelike shape and is shown before (*at left*) and after 4-fold symmetrization (*at right*). Scale bars represent 50 nm in (b) and 10 nm in the inset.

Koenig *et al.,* 1997). Our finding that solubilized AqpZ remains associated as a tetramer corroborates the idea that the functional oligomer *in vivo* is tetrameric.

B. Reconstitution and Crystallization

The crystallization procedures were derived from the conditions yielding highly ordered 2D crystals of AQP1 (Walz *et al.,* 1994a) or AQP0 (Hasler *et al.,* 1998b). Solubilized and purified AqpZ was mixed with various phospholipids in the presence of OG and reconstituted at various LPRs (w/w) by dialysis against different buffers (Jap *et al.,* 1992; Hasler *et al.,* 1998a). The crystallization trials were checked for 2D crystals by transmission electron microscopy of negatively stained samples. Whenever the AqpZ protein was reconstituted into lipid bilayers, it exhibited a marked propensity to self-assemble into square arrays. At LPR-2 flattened vesicles were observed with densely packed square-shaped particles corresponding to AqpZ tetramers. They were similar to the single particle averages in Fig. 33b, and self-assembled into mosaic tetragonal lattices (Fig. 34a).

The main feature observed in a large variety of crystallization conditions was that the 2D crystals of AqpZ had a high propensity to pile up into 3D stacks (Figs. 34b–d). Some rare side views allowed the thickness of a single-layer 2D crystal to be estimated to ~6.1 nm (Fig. 34d), close to the thickness of AQP1

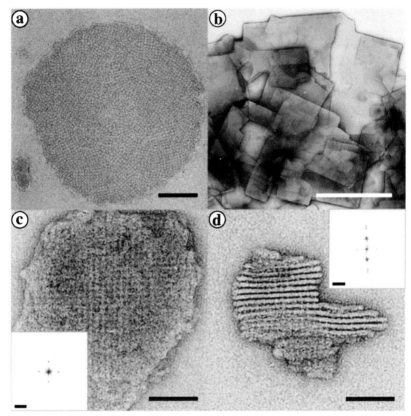

FIGURE 34 Negative stain electron microscopy of AqpZ tetramers reconstituted into lipid bilayers. (a) Flattened vesicles with densely packed square-shaped particles corresponding to AqpZ tetramers self-assembled into mosaic tetragonal lattices at a LPR of ~2. (b) Large piled-up AqpZ 2D crystal sheets with typical straight edges. (c) Small 3D stack of AqpZ 2D crystals with the corresponding optical diffraction pattern (*bottom left corner*). (d) Side view of a 3D stack and the corresponding optical diffraction pattern (*top right corner*) with a layer spacing of ~6.1 nm. Scale bars represent 100 nm in (a), 1 μm in (b), 50 nm in (c) and (d), and 5 nm^{-1} in the power spectrum inserts.

crystals, 5.8 nm, as measured with by AFM. (Engel *et al.*, 1999). To test whether the stacking was related to N-terminal poly-histidine, this charged tail was digested with trypsin before crystallization. Under the conditions used, no difference in the crystallization behavior could be detected.

Crystallization was optimized when purified AqpZ protein was mixed in presence of OG at a 0.5 mg/ml final protein concentration with synthetic lipids (POPC and DMPC in the ratio 1:1, w/w) at an LPR between 0.3 and 0.4, and dialyzed against 20 mM Citrate buffer (pH 6.0) containing 200 mM NaCl, 100 mM $MgCl_2$, 3 mM NaN_3, and 10% glycerol. Under these conditions, a mixture of micron-sized unilamellar vesicles and single-layer sheets with straight edges assembled that were all found to have their entire surface covered with ordered arrays of the AqpZ oligomer.

Due to the stability of purified, OG-solubilized AqpZ preparations, the 2D crystallization trials could be performed at room temperature over several days using dialysis. Tetragonal arrays of AqpZ tetramers were obtained under various conditions, at LPR between 0.3 and 1, but were often stacked in multilayers (5–10 layers) indicating the possibility to initiate the growth of 3D crystals suitable for X-ray crystallography. Interestingly the stacking was not affected by the presence or absence of the N-terminal poly-histidine tail. The use of both a dialysis buffer containing citrate, and a POPC/DMPC lipid mixture was found to be critical to obtain large vesicles that had their entire surface packed with square arrays of AqpZ. At an optimized LPR of 0.35, micrometer large single 2D crystal sheets were found together with vesicle-like 2D crystals, without any detectable aggregation.

C. Analysis of Negatively Stained 2D Crystals

Transmission electron microscopy of negatively stained vesicles and sheets showed AqpZ 2D crystals with $p42_12$ symmetry (Fig. 35) and unit cell dimensions of $a = b = 95.0 \pm 0.1$ Å ($n = 30$). Single-layer 2D crystals exhibited characteristic straight edges (Fig. 35a). Power spectra of such crystals often showed diffraction spots up to the sixth order in negative stain (Fig. 35a, inset). However, in most cases, power spectra of single-layer 2D crystals had a prominent first diffraction order that is expected to vanish for $p42_12$ crystals. Folded sheets (Fig. 35b) had power spectra showing two distinct sets of diffraction spots rotated with respect to one another by an arbitrary angle. The negative staining of one layer was usually more uniform than that of the other layer as judged from the diffraction pattern of untilted AqpZ 2D crystals. However, the expected systematic absences for ($2n + 1,0$) reflections indicated an appropriate negative staining for both layers, as is demonstrated in Fig. 35b, inset.

To study this in more detail, tilt series were acquired from the two crystals in Fig. 35 using automated tomography (Dierksen *et al.*, 1992). The 3D maps are displayed either as horizontal sections of 3 Å thickness (Figs. 35c, left, and 35d, right) or as surface rendered views (Figs. 35c, right, and 35d, left). The 3D map calculated from the negatively stained single-layer crystal (Fig. 35c) is consistent with an up and down packing arrangement of the AqpZ tetramers. Adjacent tetramers expose different surfaces to one side of the layer, one tetramer being bold and the other being thinner, having a cross shape. Similar features were seen on the other side of the layer, except that the surface of the bold tetramer showed less detail and was flatter. The flattening is likely to result from the interaction of this tetramer surface with the carbon support, preventing appropriate embedding in the negative stain (Baumeister *et al.*, 1986).

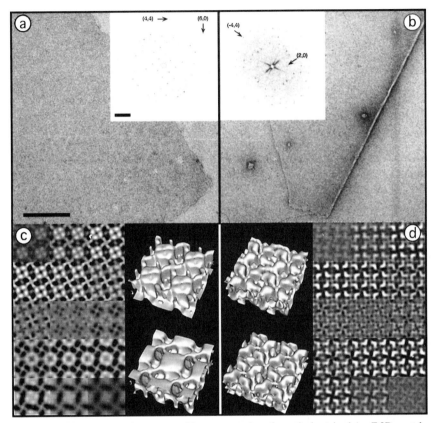

FIGURE 35 Electron microscopy and image processing of negatively stained AqpZ 2D crystals.
(a) A single-layer AqpZ 2D crystal. The crystalline sheet has straight edges, and a regular rectangular
pattern can be seen when observed at glancing angle. The power spectrum exhibits diffraction spots up
to the sixth order (*see arrow*). The presence of the (1,0) reflection is characteristic for different staining
of the two surfaces of the crystal. (b) Folded AqpZ 2D crystal sheet. Two sets of diffraction spots are
visible rotated with respect to one another by ~12° in the power spectrum. (c, d) 3D maps of negatively
stained AqpZ 2D crystals calculated from a tilt series comprising 25 projections distributed over −60°
to +60°. In (c) the left side shows fifteen 0.3-m-thick, horizontal sections of the single–layer crystal of
(a). The perspective view on the right illustrates the flattening of the surface that was in contact with the
carbon film. In (d) 15 horizontal sections of one layer of the folded crystal of (b) are shown on the right.
Both surfaces of this layer are nearly identical, considering the shift of half a unit cell, as illustrated by
the perspective view on the left. The unit cells house two adjacent tetramers, incorporated in opposite
orientations. Horizontal sections in the middle of the membrane show only low contrast because this
region is not accessible to the negative stain. Perspective views of the 3D maps were isocontoured at
the steepest gradient by a tool kindly provided by Dr. B. Heymann. Scale bars represent 300 nm in (a)
and 5 nm^{-1} in the diffraction pattern. The side length of the horizontal section is 19 nm.

The 3D map obtained from one layer of the negatively stained double-layer crystal also showed two membrane-spanning AqpZ tetramers incorporated in the bilayer with opposite orientation (Fig. 35d). However, this 2D crystal is more symmetrically stained, resulting in a map that is very similar at the top and bottom. The AqpZ tetramer on one surface has a windmill shape with a distinct vorticity, whereas the other surface exposes a smaller, cross-shaped structure.

Because the aqueous staining solution cannot penetrate into the lipid bilayer, the horizontal sections in the 3D map corresponding to the inner part of the bilayer show only low contrast compared to upper and lower surfaces (Figs. 35c and 35d). The remaining contrast is reflecting the limited resolution along the normal to the membrane plane.

D. Mass Analysis

Scanning transmission electron microscopy mass analysis (Müller and Engel, 1998) of single-layer 2D crystals yielded a mass loss corrected mass/area histogram with one single peak at 3.19 ± 0.11 kDa/nm^2 (Fig. 36). Therefore, each unit cell comprising an area of 90.25 nm^2 has a mass of 288 kDa. Considering two tetramers

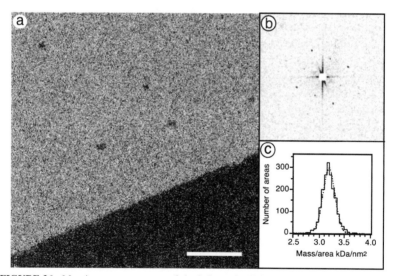

FIGURE 36 Mass/area measurements of single-layer AqpZ 2D crystal using STEM. (a) Elastic dark-field image of unstained sample recorded at a dose of 330 ± 23 electron/nm^2 reveals the typical straight edges of AqpZ single-layer 2D crystals and a faint regular pattern. (b) The power spectrum of the dark-field image of (a) shows reflections extending to the second order, allowing the unit cell dimensions of the freeze-dried tetragonal lattice to be estimated to $a = b = 95.0 \pm 0.1$ Å. (c) The mass loss corrected mass/area histogram shows one single peak of 3.19 ± 0.11 kDa/nm^2 indicating that each unit cell comprising an area of 90.25 nm^2 has a mass of 288 kDa. Scale bars represent 100 nm in (a) and 5 nm^{-1} in (b).

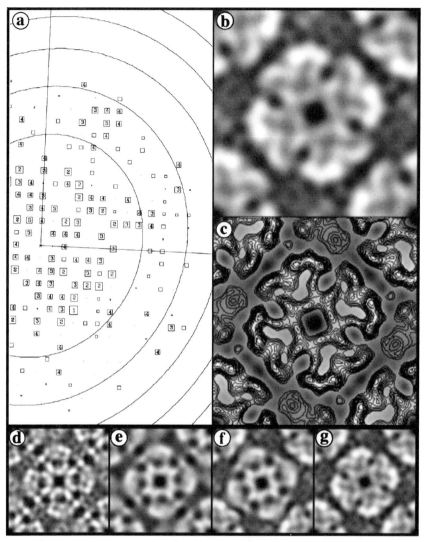

FIGURE 37   Cryo-electron microscopy power spectrum of an untilted AqpZ 2D crystal and the
8 Å projection map. (a) The power spectrum is shown with boxed numbers of the calculated diffraction
pattern of one single crystalline area representing the calculated IQ values for the measured diffraction
spots as defined by Henderson *et al.* (1986). (b) Averaged projection map calculated to a resolution
of 8 Å obtained by merging the data from six highly ordered crystalline areas (with p42$_1$2 symmetry
imposed). In the displayed unit cell, the central AqpZ tetramer resembles a four-leaf clover where each
leaf is proposed to be the monomeric AqpZ. (c) The projection map from (b) displayed with contour
levels showing that the monomer of AqpZ is formed by seven high-density regions separated by about
8 Å. The side length of the projection maps of (b) and (c) is 95 Å. (d–g) Comparison between AqpZ
and AQP1 untilted cryo-electron projection maps at similar resolution. (d) Projection map of a single
unit cell of AQP1 at 4 Å resolution, showing the central tetramer viewed from the extracellular side.

of AqpZ (8 monomers estimated at 212 kDa total mass according to their amino acid sequence), this unit cell mass allows for 76 kDa of surrounding phospholipids, giving an LPR of 0.35. Taking the value of 717 Da for the mean molecular mass of the POPC/DMPC 1:1 w/w lipid mixture used for crystallization, and a packing density of 4 lipid/nm^2, the interspersed lipid bilayer should cover 30% of the unit cell area (mean value of 106 lipid molecules per unit cell).

E. Cryo-Electron Microscopy of Unstained AqpZ 2D Crystals

Power spectra of micrographs of unstained crystalline patches showed weak diffraction spots up to 0.8 nm, indicating a significant improvement in resolution compared to negative staining (Fig. 37a). The six best images (corresponding to ~11,000 unit cells) were unbent, CTF corrected, and an averaged projection map was calculated from the merged amplitudes and phases to a resolution of 0.8 nm, enforcing the p42$_1$2 symmetry (Figs. 37b and 37c).

The overall shape of the projection of the untilted AqpZ tetramer resembles a four-leaf clover. The monomeric AqpZ is proposed to correspond to one leaf and is formed by seven high-density regions separated by ~0.8 nm (Fig. 37c).

AqpZ packs with p42$_1$2 crystal symmetry, as does AQP1 (Walz et al., 1994a), but not AQP0 (Hasler et al., 1998b). From the mass/area value (3.19 kDa/nm^2) measured by STEM on single-layer 2D crystals the LPR can be estimated, and the surface covered by the phospholipids predicted to 30%. When contouring the cryo-electron projection map of unstained 2D crystals at the steepest contrast gradient, the area covered by the lipid bilayer was found to be 31%, in excellent agreement with the STEM analysis. Mass analysis of AQP1 crystals yielded 4.1 kDa/nm^2 (Walz et al., 1994b), explained by the additional mass of the glycosyl, about 80 kDa per unit cell.

The contrast in projections of these vitrified 2D crystals is expected to result mainly from transmembrane domains rather than from flexible external loops. Closer analysis of the amino acid sequence homology between AqpZ and AQP1 reveals that the identical residues are mostly located either in the predicted six transmembrane-spanning α helices or in the highly conserved B and E loops that fold back into the membrane (Calamita et al., 1995). Therefore, it is not surprising that the unstained projection maps of AQP1 and AqpZ display such a strong

---

(e) The same projection map of AQP1 as in (d) is rendered at a resolution of around 8 Å. (f) The central AQP1 tetramer of (e) is rotated counter-clockwise by an angle of 15° and pasted into the AqpZ projection map. (g) The AqpZ projection map at 8 Å can be compared to the AQP1 projection map of (f). Similar localization of the density maxima and minima within the tetramers viewed from the cytoplasmic side suggests a similar helix packing arrangement between AqpZ and AQP1. The sizes of the unit cells are 96 Å (AQP1) and 95 Å (AqpZ).

similarity (Figs. 12c and 37b). The similar disposition of the density maxima and minima within the tetramers suggests a helix packing arrangement in AqpZ similar to that of AQP1, which is a right-handed bundle of six tilted transmembrane α helices that surround a central density (Cheng *et al.*, 1997; Mitsuoka *et al.*, 1999; Walz *et al.*, 1997). This central density was proposed to contain the extended loops B and E (Mitsuoka *et al.*, 1999; Walz *et al.*, 1997) and is also visible in the AqpZ projection map (Figs. 37b and 37c).

Orientational cross-correlation of the central AqpZ tetramer from a unit cell with that of AQP1 yielded a rotation angle of 15° (counter clockwise) that had to be applied to the AQP1 tetramer to fit it onto the AqpZ tetramer (Figs. 37d–g). Therefore, the contacts between aquaporin tetramers, which drive the assembly of 2D crystals, may be different between AqpZ and AQP1. Comparing the projection maps of AqpZ and AQP1 tetramers after rotational alignment, correlation analysis indicated that the central tetramer of AqpZ displayed in Fig. 37b is the view from the cytosolic side as defined for AQP1 (Fig. 13).

F. AqpZ Surface Topography as Assessed by AFM

We have used the atomic force microscope (AFM) (Binnig *et al.*, 1986) to measure the surface topography of AqpZ crystals in a native environment. As previously reported, this method allows protein surfaces to be imaged at subnanometer resolution (Mou *et al.*, 1996; Müller *et al.*, 1995, 1999b; Schabert *et al.*, 1995; Scheuring *et al.*, 1999a).

In the presence of 25 mM MgCl$_2$ the 2D crystals of AqpZ adsorbed to mica without wrinkles and were thus suitable for high-resolution imaging. To this end, the adsorption buffer was exchanged with the recording buffer that was adjusted to electrostatically balance the van der Waals forces (Müller *et al.*, 1999a). The overview in Fig. 38a demonstrates the flatness of the 2D crystals whose thickness was found to be 5.7 ± 0.4 nm ($n = 45$; Fig. 38b).

At higher magnification the square lattice became distinct (Fig. 38c, main frame, top). The unit cell dimensions were $a = b = 9.5 ± 0.2$ nm, in excellent agreement with results from electron microscopy (Ringler *et al.*, 1999). Correlation averaging revealed a unit cell housing two tetramers in opposite orientations with respect to the membrane plane (p42$_1$2 packing; Fig. 38c, top right). The high signal-to-noise ratio of the AFM also allowed high-resolution imaging on densely packed vesicles, which only exhibited small crystalline areas comprising about 30 tetramers arranged with p4 symmetry (Fig. 38c, main frame, bottom). The p4 crystals had unit cell dimensions of $a = b = 7.2 ± 0.2$ nm and housed a single tetramer (Fig. 38c, bottom right).

The recombinant AqpZ crystallized has an N-terminal fragment of 26 amino acids, containing a trypsin cleavage site at Arg 26 and a 10 his tag at amino acid

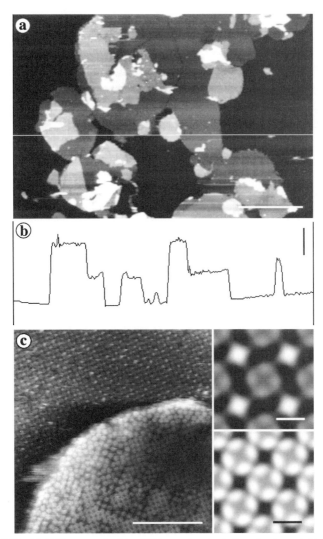

FIGURE 38 (a) AFM topograph of AqpZ 2D crystals adsorbed to mica (recorded in buffer solution: 17 mM Tris–HCl, pH 7.2, 150 mM KCl). Double- and multilayer areas can clearly be distinguished from single-layer crystals by their higher appearance (scale bar: 2 μm; full grayscale: 30 nm). (b) Section analysis along the white line in image (a). The 2D crystals show a uniform height of 57 ± 4 Å ($n = 45$). Double-layer areas appear as plateaus twice as high as the single-layer crystal sheets (vertical scale bar: 50 Å). (c) AFM topograph of reconstituted AqpZ (recorded in buffer solution: 17 mM Tris–HCl, pH 7.2, 150 mM KCl). A densely packed vesicle containing crystalline areas with p4 symmetry adsorbed onto a crystal sheet with p42$_1$2 symmetry (scale bar: 100 nm; full grayscale: 20 nm). *Top inset:* average of the sheet with p42$_1$2 symmetry (scale bar: 50 Å). *Bottom inset:* average of the crystalline areas with p4 symmetry within the densely packed vesicle (scale bar: 50 Å).

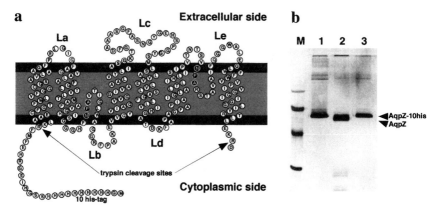

FIGURE 39 (a) Amino acid sequence model of AqpZ showing the six membrane-spanning helices derived from hydropathy analysis. Trypsin cleavage sites are located on Arg26 and Arg230. (b) Silver stained SDS–polyacrylamide 10% (w/v) gel. Lane M, marker: 97.4, 66.2, 42.7, and 31.0 kDa; lane 1, AqpZ solubilized in 2% OG; lane 2, AqpZ crystals after overnight trypsin treatment. The AqpZ band is broadened indicating that a minor part of the protein remained undigested. The two diffuse bands in the low molecular weight region (below 30 kDa) correspond to trypsin and the cleaved N-terminal fragments; lane 3, AqpZ-10his crystals. The faint bands at high molecular weight (∼200 kDa) in all three lanes arise from specific aggregates.

positions 2 to 12 (Fig. 39a; Borgnia *et al.,* 1999). Consequently, a total of 104 amino acids, including 40 histidines, protruded from the cytoplasmic side of each tetramer. These peptides produced a strong signal in the AFM, resulting in a 2.0 ± 0.2 nm high peak (Figs. 38c and 40a). The exact position and appearance of the protrusion depended critically on the force applied to the stylus, the scan speed, and the direction of the scan. This extreme flexibility prevented the resolution of substructure. To prove that the large protrusion observed indeed arose from the N-terminal domain, crystals were treated with trypsin to cleave off this flexible end domain as documented in Fig. 39b.

The digested crystals exhibited a striking change; instead of an ill-defined protrusion of 2.0-nm height, the cytoplasmic side now showed four distinct protrusions each with a height of 0.35 ± 0.04 nm above the lipid bilayer (Fig. 40b). In contrast, the extracellular side was neither changed in shape nor height by the trypsin treatment (Fig. 40c).

The extracellular side was not sensitive to trypsin digestion; however, it underwent a reversible conformational change when the force applied to the tip was increased during imaging. At minimal force (∼80 pN) each AqpZ subunit showed three major protrusions probably related to the loops connecting the membrane spanning helices on the extracellular side (Figs. 41a and 41c). On recording a second image of the same areas with a force increased by +80 pN, a drastic

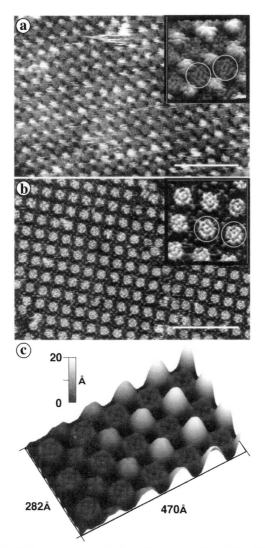

FIGURE 40 (a) AFM topograph (recorded in buffer solution: 17 m*M* Tris–HCl, pH 7.2, 150 m*M* KCl) of an AqpZ-10his 2D crystal with p42₁2 symmetry recorded using minimal force (scale bar: 500 Å, full grayscale: 30 Å). *Inset:* relief view (tilt: 85°) of a 300-Å square, raw data; extracellular surfaces are marked by circles. (b) AFM topograph (recorded in buffer solution: 17 m*M* Tris–HCl, pH 7.2, 150 m*M* KCl) of an AqpZ 2D crystal with p42₁2 symmetry after trypsin treatment recorded using minimal force (scale bar: 500 Å; full grayscale: 10 Å). *Inset:* relief view (tilt: 85°) of a 300 Å square, raw data; extracellular surfaces are marked by circles. (c) 3D reconstruction of the trypsin cleavage process (see Fig. 39a) observed on the cytoplasmic surface. On digestion, this surface changes shape and height drastically in the location of the N-terninal his tags (*right top,* undigested state; *left bottom,* digested state).

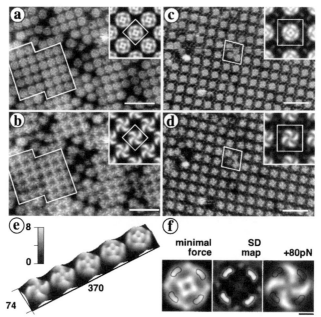

FIGURE 41 (a) AFM topograph of AqpZ-10His densely packed in a vesicle, scanned in buffer solution (17 m*M* Tris–HCl, pH 7.2, 150 m*M* KCl) at minimal force (∼80 pN) (scale bar: 250 Å; full grayscale: 18 Å). A small crystalline area with p4 symmetry is outlined. *Inset:* average of (a). The white square indicates the unit cell ($a = b = 72 \pm 2$ Å), which houses one tetramer. (b) AFM topograph of the same area as (a) recorded in buffer solution (17 m*M* Tris–HCl, pH 7.2, 150 m*M* KCl) applying an additional force of +80 pN to the tip during scanning. The outlined area corresponds to the area marked in (a) (scale bar: 250 Å; full grayscale: 18 Å). *Inset:* average of (c). The white square indicates the unit cell ($a = b = 72 \pm 2$ Å), which houses one tetramer. (c) AFM topograph of trypsin treated AqpZ 2D crystals with p42₁2 symmetry recorded in buffer solution (Tris–HCl, pH 7.5, 150 m*M* KCl) using minimal force (∼80 pN) (scale bar: 250 Å; full grayscale: 20 Å). The outlined area shows a pronounced lattice distortion. *Inset:* average of (c). The white square indicates the unit cell ($a = b = 95 \pm 2$ Å), which houses two tetramers. (d) AFM topograph of the same area as (c) recorded in buffer solution (10 m*M* Tris–HCl, pH 7.5, 150 m*M* KCl) applying an additional force of +80 pN to the tip during scanning. Note the same lattice irregularity as in (c) (scale bar: 250 Å; full grayscale: 20 Å). *Inset:* average of (d). The white square indicates the unit cell ($a = b = 95 \pm 2$ Å), which houses two tetramers. (e) 3D reconstruction illustrating the effect observed on the extracellular surface when the imaging force is increased by +80 pN during scanning (*right top*, minimal force; *left bottom*, +80 pN). (f) Comparison of the averages of the extracellular surface at minimal force (*left*), the standard deviation (SD) map (*middle*), and the extracellular surface average at an additional force of +80 pN (*right*) (full image sizes: 72 Å). The outlined regions in the middle image represent a SD ∼0.7 Å. These regions correspond to the four elongated peripheral protrusions in the minimal force average, which are strongly displaced in the average gained from images recorded at +80 pN.

conformational change was observed (compare Figs. 41a and 41b, and Figs. 41c and 41d). The extracellular AqpZ surface reversibly changed its rather circular appearance into a left-handed windmill, which still protruded out of the membrane by 0.7 nm (Fig. 41e). This force-induced conformational change was not influenced by the trypsin treatment, as illustrated by comparing the topographs of digested (Figs. 41c and 41d) and undigested AqpZ crystals (Figs. 41a and 41b). The digested cytoplasmic surface did not show the same force dependence; the minor force induced conformational change was barely noticeable (Figs. 41c and 41d, insets). The standard deviation (SD) map of 289 densely packed single tetramers (such as that shown in Fig. 41a) recorded at minimal force was calculated to identify the flexible regions of the extracellular surface (Müller *et al.,* 1998). As displayed in Fig. 41f, one region exhibited a pronounced SD, while the rest yielded highly reproducible heights (SD $\leq$ 0.02 nm). Interestingly, this flexible region also exhibited the major force-induced conformational change (compare left and right tetramers in Fig. 41f).

Sequence-based structure prediction postulates AqpZ to be a protein consisting of six transmembrane helices connected by three loops on the extracellular side and two loops on the cytoplasmic side, as well as two cytoplasmic termini (Fig. 39a). In agreement with this, on imaging at minimal force, three protrusions were found on the extracellular surface of the AqpZ monomer. One was close to the 4-fold symmetry center, and one small and one elongated protrusion at the periphery (Fig. 42a). Volume calculations on the protrusion close to the 4-fold symmetry center resulted in 1.278 $\pm$ 0.150 nm^3. The small peripheral protrusion

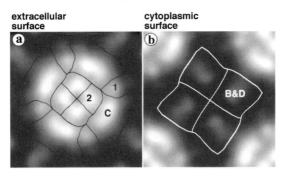

FIGURE 42 Proposed assignment and borderlines between adjacent loops of the AqpZ tetramer. The *x, y,* and *z* scaling used for 3D integration of the protruding volumes is derived from high-resolution AFM topographs (full image sizes: 95 Å). (a) Extracellular surface exposing protrusions 1, 2, and C with volumes of 984 $\pm$ 134, 1278 $\pm$ 150, and 3187 $\pm$ 528 Å^3, respectively. Protrusions 1 and 2 correspond to loops A and E. The similarity of their volumes prevents an unambiguous assignment. (b) Cytoplasmic surface having only one defined protrusion housing loops B and D with a volume of 1222 $\pm$ 144 Å^3.

yielded a volume of 0.984 ± 0.134 nm^3, while the elongated protrusion had a volume of 3.187 ± 0.528 nm^3, the larger SD reflecting the flexibility of this region. Nevertheless, the assignment of this protrusion to loop C is consistent with all data presently available. The single protrusion observed per monomer on the digested cytoplasmic surface had a volume of 1.222 ± 0.144 nm^3 (Fig. 42b). Since the termini are removed (Figs. 39a and 39b), this protrusion is expected to house loops B and D.

VIII. GLYCEROL CHANNEL OF *Escherichia coli:* GlpF

The glycerol uptake facilitator in *E. coli* (GlpF; see Fig. 1; Boos *et al.*, 1990) is the first identified member of the GLP cluster and is one of the few known diffusion facilitators in the *E. coli* plasma membrane. Its biological role is to mediate the diffusion of glycerol into the cell where it is phosphorylated by the glycerol kinase (GlpK) to prevent back-diffussion out of the cell. Besides transport of glycerol, diffusion of polyols and urea derivatives were reported (Maurel *et al.*, 1994), but none of these substrates is transported in a phosphorylated state. In contrast, the diffusion of water through GlpF remains unclear.

A sequence comparison of GlpF with AQP1 reveals additional amino acids in loop E of GlpF. Forger *et al.* (1998) identified other conserved residues to discriminate GLPs from AQPs (see Fig. 19). In addition, Heymann and Engel (2000) have identified the pair of residues F24/L149 in AQPs and L24/L149 in GLPs (see Fig. 19) that may determine the size of the pore.

The tetrameric organization of all aquaporins seems to be a general structural feature (Hasler *et al.*, 1998b; Ringler *et al.*, 1999; Smith and Agre, 1991; Walz *et al.*, 1994a). However, the French group at the University of Rennes determined a monomeric organization for GlpF by analytical ultracentrifugation (Lagree *et al.*, 1998), and later supported this finding by freeze–fracture electron microscopy (Bron *et al.*, 1999).

A. Solubilized GlpF Particles

GlpF carrying 10 C-terminal histidine residues was overexpressed in *E. coli,* solubilized, and purified in OG as described by Borgnia *et al.* (1999). Analysis of purified and solubilized GlpF in SDS–PAGE revealed a prominent band at 30-kDa apparent molecular weight. Negatively stained preparations of the solubilized and purified GlpF were imaged with a transmission electron microscope (TEM) (Braun *et al.*, 2000). Figure 43a shows the homogenous size distribution of the purified protein, exhibiting square-shaped particles. Occasional larger complexes can be seen (arrows). Single particles were windowed and subjected to

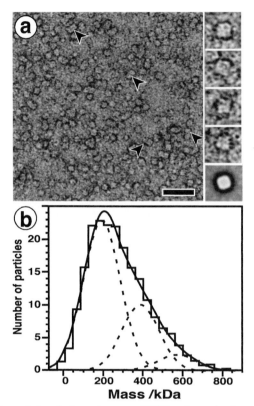

FIGURE 43 Transmission electron microscopy (TEM) and scanning transmission electron microscopy (STEM) of solubilized GlpF. (a) Overview of solubilized GlpF in TEM. Octamers are marked with arrows. Note the homogenous size distribution. *Insets:* Four selected particles and a 4-fold symmetrized average after reference free single particle analysis. (b) Results of STEM measurements. The histogram was fitted with three Gauss peaks at 194, 388, and 575 kDa.

reference-free alignment, resulting in an average projection with square structure and a edge length of about 9 nm (Fig. 43b). Four weak densities were distinguished within the square.

Purified GlpF particles were also freeze-dried and recorded at low dose by STEM for mass measurement (Müller and Engel, 1998). The mass histogram could be fitted with three peaks centered at 194 kDa (total error of ± 12.3 kDa), 388 $\pm$ 27.1 kDa, and 575 $\pm$ 32.6 kDa after mass-loss correction (Müller *et al.*, 1992). A value of 194 kDa is compatible with a GlpF tetramer and four His tags including spacers (12 kDa), plus a number of OG molecules equivalent to two micelles (60 kDa) (Walz *et al.*, 1994a). A value of 388 kDa would correspond to an octamer with approximately four OG micelles and 575 kDa to a dodecamer.

B. Two-Dimensional Crystallization

Solubilized GlpF was reproducibly crystallized using a continuous-flow dialyzing device (Jap *et al.*, 1992) exploring a variety of conditions (Braun *et al.*, 2000). Analysis by TEM showed polycrystalline vesicles with diameters up to 40 μm and mostly mono crystalline vesicles with rectangular shapes and diameters up to 8 μm (Fig. 44a).

C. Electron Microscopy and Image Treatment

Images of negatively stained crystals were recorded in a TEM and analyzed by correlation averaging (Saxton and Baumeister, 1982). The unit cell containing two tetrameric structures exhibited a p4 symmetry and dimensions of $a = b = 104$ Å. Adjacent tetramers were differently stained (data not shown) similar to the AQP1 crystals (Walz *et al.*, 1994b), indicating an up–down orientation of adjacent tetramers.

Images of frozen hydrated crystals recorded at low dose revealed sharp spots when examined by optical diffraction out to a resolution of 7 Å. The 12 best images were digitized and processed by the MRC program package (Henderson *et al.*, 1986, 1990). The calculated power spectrum with boxed IQ numbers of one image is reproduced in Fig. 44b. The phase residuals of the merged map obtained after lattice unbending and transfer function correction indicated significant information up to a resolution of 3.7 Å (Braun *et al.*, 2000). The resulting projection map is shown in Fig. 45.

D. Tetrameric Organization of GlpF

GlpF showed a tetrameric organization throughout the purification and after reconstitution into two-dimensional crystals. This was monitored by electron microscopy of negatively stained samples taken at various steps during the purification of the 2D crystals and was corroborated by mass measurements of freeze-dried, unstained preparations in the STEM. Thus, GlpF exhibits the same tetrameric structure as other aquaporins previously studied: AQP1 (Smith and Agre, 1991; Walz *et al.*, 1994a), AQPZ (Ringler *et al.*, 1999), and AQP0 (Hasler *et al.*, 1998b). This result is in stark contrast to reports of monomeric GlpF (Lagree *et al.*, 1998), and challenges the hypothesis that glycerol facilitators are monomeric whereas water channels require tetramerization (Lagree *et al.*, 1999). The tetrameric nature of GlpF supports the model proposed by Voegele *et al.*, (1993) that GlpF interacts with the tetrameric glycerol kinase GlpK to stimulate glycerol phosphorylation. Octamer formation of GlpF did not occur in the presence of a reducing

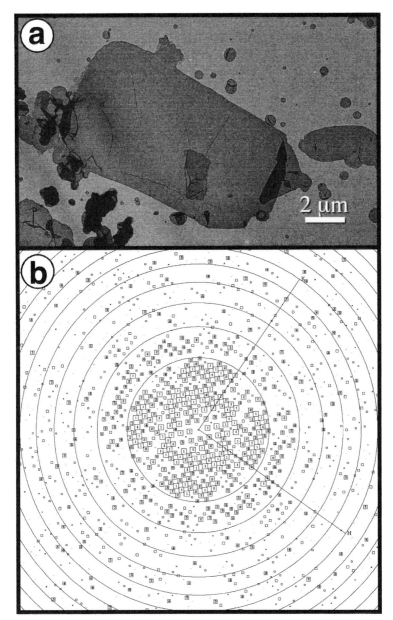

FIGURE 44   Two-dimensional crystal of GlpF. (a) This smaller rectangular type of crystal was usually monocrystalline and well ordered. (b) The cryo-electron microscopy power spectrum is shown with boxed numbers of the calculated diffraction pattern of one single crystalline area, representing the calculated IQ values for the measured diffraction spots as defined by Henderson *et al.* (1986). The zero crossings of the microscope's contrast transfer function are indicated by the circles.

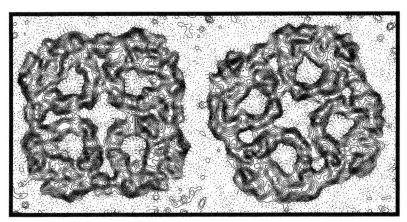

FIGURE 45   Projection structure of GlpF. The map was calculated at 3.7 Å resolution, applying a negative temperature factor of -20 Å^2.

agent such as mercaptoethanol or dithiothretiol (DTT), suggesting that there are surface-exposed sulfhydrils.

GlpF crystallized under similar conditions as AQP1 (Walz *et al.*, 1994a, 1995), but at pH 8.5 instead of pH 6 and at much lower MgCl$_2$ concentration. The unit cell length of 104 Å is slightly larger than that of AQP1 (96 Å) (Walz *et al.*, 1994a), indicating a different packing arrangement. Indeed, while the up–down orientation of neighboring tetramers is maintained as visible in their mirrored handedness (Fig. 45), GlpF crystallizes in a p4 symmetry and not p42$_1$2 as is true for AQP1. This is manifest in the different rotations relative to the lattice vectors of adjacent tetramers packed in the up or down orientations.

IX. COMPARISON OF THE HIGH-RESOLUTION PROJECTION STRUCTURES OF GlpF AND AQP1

It is interesting to compare the high-resolution projection structure of the archetypal member of the GLP subcluster with that of AQP1, the first aquaporin structurally analyzed to high resolution (Cheng *et al.*, 1997; Li *et al.*, 1997; Mitsuoka *et al.*, 1999; Walz *et al.*, 1994a, 1997) (Fig. 46). The density maxima marked in the AQP1 map appear to be slightly shifted in the GlpF map. These density maxima represent the projection of highly tilted helices that surround a central density produced by loops B and E, marked by an X in Fig. 46 (Mitsuoka *et al.*, 1999). According to the hourglass model (Jung *et al.*, 1994), this is the site of the channel, and differences between GlpF and AQP1 in this region are of particular

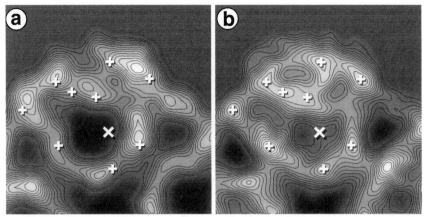

FIGURE 46 Comparison of GlpF and AQP1. The tetramers of GlpF and AQP1 were resolution limited to 4.0 Å and rotationally aligned using the SEMPER program (Saxton *et al.,* 1979). The monomers are reproduced at the same scale, and the image side lengths correspond to 4.1 nm. (a) The projection of the monomer of GlpF is marked with the positions of the maxima of mass densities in the projection of the monomer of AQP1 in (b). GlpF shows a displacement of the helices in the top right area, resulting in a narrowing of the monomer. The density in AQP1 in the channel-associated central hole (marked with a X) is missing in GlpF, and the central hole appears larger in GlpF.

interest. While the projection map of AQP1 shows a weak density at this position, GlpF seems to have a much larger hole with no inner structure discernable in the projection map.

In summary, the projections of GlpF and AQP1 are comparable in their overall architecture but present interesting differences, which may be related to their different biological functions. It is compelling to speculate that the larger depression seen in the GlpF monomer represents the pore where glycerol and similar solutes pass (Maurel *et al.,* 1994). This in turn would support the hourglass model. However, to understand the molecular mechanisms and specificity of water and solute transport in aquaporins, the atomic structures of both AQP1 and GlpF are required.

X. CONCLUSION AND PERSPECTIVES

Structural data are at the present state available for five members of the aquaporin superfamily: AQP0, AQP1, AQPZ, α-TIP (Daniels *et al.,* 199b), and GlpF. A sequence comparison among these five reveals the major differences in the loop regions (Fig. 47). These five channels differ however in their biological role and their function. While the members of the AQP cluster (APQ0, AQP1, AQPZ, and α-TIP) are permeable for water, it remains unclear whether GlpF allows the passage

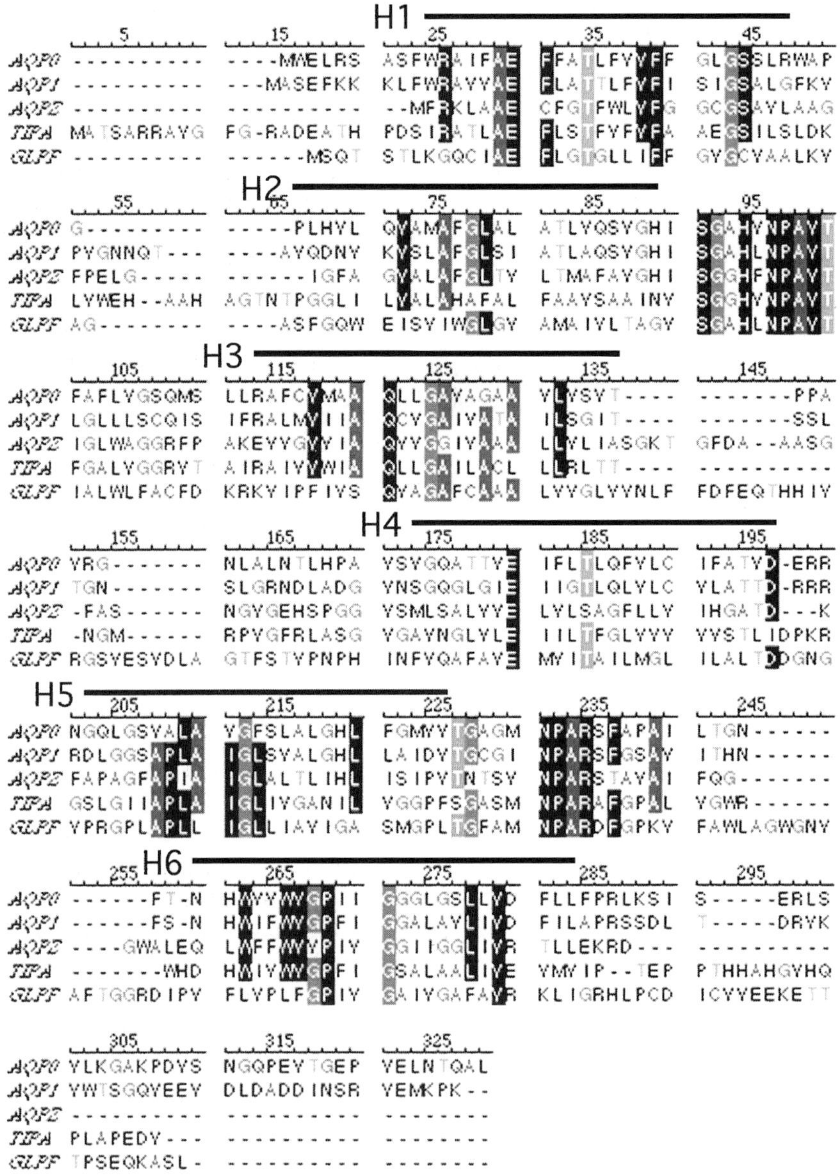

FIGURE 47 Comparison of five aquaporin sequences aligned according to Heymann and Engel (2000). Helices are indicated by solid lines. The highlighted residues are similar to those highlighted in Fig. 19. Remarkable variations are observed in the surface loop regions, explaining the differences of surface topographies (Fig. 49). Functional loop B is equal in length in all proteins and highly conserved in the first half, but loop E is by 10–12 residues longer in GlpF than in the others. (See Color Plate.)

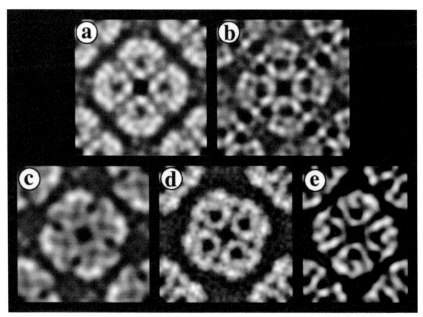

FIGURE 48 Comparison of the 2D projection structures of (a) AQP0 at 5.5 Å, (b) AQP1 at 3.5 Å, (c) AQPZ at 8 Å, (d) GlpF at 3.7 Å, and (e) α-TIP at 7.7-Å resolution (We are thankful to Mark Daniels for providing the α-TIP image; see also Daniels *et al.*, 1999b.) The projections resemble each other in their gross features, with the exception of the larger central hole in each of the four monomers of the GlpF tetramers. An important common feature is the remarkably low density region about the 4-fold axis. The meaning of this cavity remains to be unveiled.

of water at all. However, the later allows the passage of glycerol at a high rate. Individual differences in permeability rates for the different channels demonstrate their adaptation to their different environments (AQP0 and AQP1 are mammalian, α-TIP was isolated from beans, and AQPZ and GlpF are from *E. coli*).

These highly selective channels allow an energy-free traversal of small solutes. At the same time, they are blocking the passage of ions, therefore preventing the dissipation of the membrane potential. To fully understand the underlying mechanisms, the structural studies presented here are the most direct approach. 2D crystallization and cryoelectron microscopy revealed the projection structure (Fig. 48) and in one case the 3D structure at 4.5 Å of these proteins (Fig. 18). All channels structurally investigated so far revealed a tetrameric organization of four monomers that, according to the hourglass model (Jung *et al.*, 1994), have the channel located in their centers. While the rough structural features of these channels resemble each other in their projection, small differences become apparent at higher resolution. GlpF in particular appears to have a larger pore diameter.

a # AQP0

b # AQP1

c # AQPZ

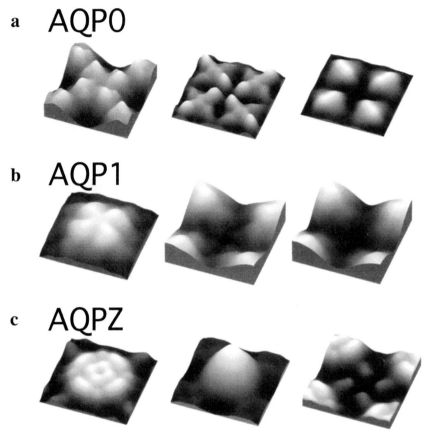

FIGURE 49 Surface topographies of water channel proteins are distinctly different, reflecting the differences of their sequences that are mainly found in the helix-connecting loops. Surface reliefs are acquired in buffer solution by atomic force microscopy. (a) AQP0. *Left:* the extracellular surface; *center:* the cytoplasmic surface before; *right;* after carboxypeptidase Y treatment *on mica* (Fotiadis *et al.*, 2000). (b) AQP1. *Left:* the extracellular surface; *center:* the cytosolic surface with four elongated peripheral and four small central domains produce a windmill-shaped structure with a height of 6 Å; *right:* preliminary data from carboxypeptidase Y treatment *on mica* indicates that the carboxyl terminus is located at a similar position as with AQP0 (see Fig. 14). (c) AqpZ. *Left:* the extracellular surface; *center:* the poly-His tag used for purification produces a massive protrusion on the cytosolic surface; *right:* after trypsin cleavage the cytosolic surface exhibits four small protrusions. The elongated extracellular protrusion in AqpZ has been identified as loop C based on its volume and flexibility (Scheuring *et al.*, 1999b). Correlation averages displayed comprise two tetramers and have a side length of approximately 95 Å. Horizontal and vertical scaling are identical, but the gray-level range is adapted to go from black to full white within each topograph.

Major differences between the aligned sequences are found in the loop regions. Their functions have yet to be elucidated, whereas their structure can be revealed by AFM (Fig. 49). Future work will concentrate on the identification of function-related conformations of these loops.

Based on the 4.5 Å density map of AQP1 (Mitsuoka *et al.*, 1999), data derived from sequence analyses (Heymann and Engel, 2000), and measurements of water mobility by NMR spectroscopy, we dare to speculate as to how the channel might work. Size exclusion is likely to be an important aspect of specificity, which could be provided by the hydrophobic F–L pairs in AQPs and the L–L pairs in GLPs. The question arises of whether the pore exhibits altogether a hydrophobic surface. From NMR measurements taken at different temperatures the activation energy for water displacement in nonpolar surface pockets of ribonuclease A (Denisov and Halle, 1998) and in the minor groove of a B-DNA duplex (Denisov *et al.*, 1997) have been estimated as 10 and 13 kcal/mol, respectively. This is compatible with the activation energy $E_a > 10$ kcal/mol for the diffusion of water through pure lipid bilayers, but incompatible with the osmotically driven water flux that can amount to two water molecules per channel and nanosecond. The activation energy for movement of water molecules in bulk water, 5 kcal/mol, is close to the activation energy measured for passage of water in AQPs. This suggests that some polar residues or main chain carbonyls in the vicinity of the pore facilitate a water hydrogen-bonding network. Such polar sites are also required to interrupt hydrogen bonds in a single file of water molecules to prevent the creation of a "proton wire" (Pomès and Roux, 1996). Water molecules may thus tumble from one set of tetrahedral hydrogen bonds to the next and pass through the channel without noticing it.

These studies underscore the need to determine the structure of aquaporins and aquaglyceroporins at high resolution. It is essential that structural landmarks be established in aquaporin proteins, since these may ultimately be used to develop therapeutic agents for important clinical disorders. 3D maps from electron crystallographic analyses now approach a sufficient resolution to build an atomic model. This and 3D crystals of AQP1 (Wiener, 1999) and AQP0 (Raimund Dutzler, personal communication, 1999) promise that this goal should be reached soon to eventually answer the question of how these proteins establish their amazing selectivity.

Acknowledgments

The work was supported by Swiss National Foundation for Scientific Research (NF grant no. 4036 P44062), the M. E. Müeller Foundation of Switzerland, and the EU Biotech Program (grant no. B104-CT98-00024).

References

Adrian, M., Dubochet, J., Fuller, S.D., and Harris, J. R. (1998). Cryo-negative staining. *Micron* **29**, 145–160.

Aerts, T., Xia, J. Z., Slegers, H., de Bock, J., and Clauwaert, J. (1990). Hydrodynamic characterization of the major intrinsic protein from the bovine lens fiber membranes. *J. Biol. Chem.* **265**, 8675–8680.

Baumeister, W., Barth, M., Hegerl, R., Guckenberger, R., Hahn, M., and Saxton, W. O. (1986). Three-dimensional structure of the regular surface layer (HPI layer) of Deinococcus radiodurans. *J. Mol. Biol.* **187,** 241–253.

Beuron, F., Le Cahérec, F., Guillam, M.-T., Cavalier, A., Garret, A., Tassan, J.-P., Delamarche, C., Schultz, P., Mallouh, V., Rolland, J.-P., Hubert, J.-F., Gouranton, J., and Thomas, D. (1995). Structural analysis of a MIP family protein from the digestive tract of *Cicadella viridis*. *J. Biol. Chem.* **270**(29), 17414–17422.

Binnig, G., Quate, C. F., and Gerber, C. (1986). Atomic force microscope. *Phys. Rev. Lett.* **56**(9), 930–933.

Bok, D., Dockstader, J., and Horwitz, J. (1982). Immunocytochemical localization of the lens main intrinsic polypeptide (MIP26) in communicating junctions. *J. Cell. Biol.* **92,** 213–220.

Bond, J., Green, C., Donaldson, P., and Kistler, J. (1996). Liquefaction of cortical tissue in diabetic and galactosemic rat lenses defined by confocal laser scanning microscopy. *Invest. Ophthalmol. Vis. Sci.* **37,** 1557–1565.

Boos, W., Ehmann, U., Forkl, H., Klein, W., Rimmele, M., and Postma, P. (1990). Trehalose transport and metabolism in *Escherichia coli*. *J. Bacteriol.* **172**(6), 3450–3461.

Borgnia, M., Kozono, D., Calamita, G., Nielsen, S., Maloney, P. C., and Agre, P. (1999). Functional reconstitution and characterization of *E. coli* aquaporin-Z. *J. Mol. Biol.* **291,** 1169–1179.

Braun, T., Philippsen, A., Wirtz, S., Borgnia, M. J., Agre, P., Kühlbrand, W., Engel, A., and Stahlberg, H. (2000). The glycerol facilitator (GlpF) at 3.5 Å resolution: Structural variant of the aquaporin tetramer. *EMBO J.* **2,** 183–189.

Bron, P., Lagrée, V., Froger, A., Rolland, R., Hubert, J. F., Delamarche, C., Deschamps, S., Pellerin, I., Thomas, D., and Haase, W. (1999). Oligomerization state of MIP proteins expressed in *Xenopus* oocytes as revealed by freeze-fracture electron-microscopy analysis. *J. Struct. Biol.* **128**(3), 287–296.

Cabiaux, V., Oberg, K. A., Pancoska, P., Walz, T., Agre, P., and Engel, A. (1997). Secondary structures comparison of aquaporin-1 and bacteriorhodopsin: A Fourier transform infrared spectroscopy study of two-dimensional membrane crystals. *Biophys. J.* **73**(1), 406–417.

Calamita, G., Bishai, W., Preston, G., Guggino, W., and Agre, P. (1995). Molecular cloning and characterization of AqpZ, a water channel from *Escherichia coli*. *J. Biol. Chem.* **270**(49), 29063–29066.

Calamita, G., Kempf, B., Bonhivers, M., Bishai, W., Bremer, E., and Agre, P. (1998). Regulation of the *Escherichia coli* water channel gene *aqpZ*. *Proc. Nat. Acad. Sci. USA* **95**(7), 3627–3631.

Canfield, M. C., Tamarappoo, B. K., Moses, A. M., Verkman, A. S., and Holtzman, E. J. (1997). Identification and characterization of aquaporin-2 water channel mutations causing nephrogenic diabetes insipidus with partial vasopressin response. *Human Mol. Gen.* **6**(11), 1865–1871.

Chang, G., Spencer, R. H., Lee, A. T., Barclay, M. T., and Rees, D. C. (1998). Structure of the MscL homolog from Mycobacterium tuberculosis: A gated mechanosensitive ion channel. *Science* **282**(5397), 2220–2226.

Cheng, A., van Hoek, A. N., Yeager, M., Verkman, A. S., and Mitra, A. K. (1997). Three-dimensional organization of a human water channel. *Nature* **387,** 627–630.

Daniels, M. J., Chrispeels, M. J., and Yeager, M. (1999a). 2D crystallization of a plant vacuole membrane aquaporin and determination of its projection structure by electron crystallography. *Biophys. J.* **76,** A456.

Daniels, M. J., Chrispeels, M. J., and Yeager, M. (1999b). Projection structure of a plant vacuole membrane aquaporin by electron cryo-crystallography. *J. Mol. Biol.* **294**(5), 1337–1349.

Deen, P., Verdijk, M., Knoers, N., Wieringa, B., Monnens, L., van Os, C., and van Oost, B. (1994). Requirement of human renal water channel aquaporin-2 for vasopressin-dependent concentration of urine. *Science* **264,** 92–95.

Denisov, V. B., and Halle, B. (1998). Thermal denaturation of ribonuclease A characterized by water ^{17}O and ^2H magnetic relaxation dispersion. *Biochem,* **37,** 9595–9604.

Denisov, V. P., Carlström, G., Venu, K., and Halle, B. (1997). Kinetics of DNA hydration. *J. Mol. Biol.* **268,** 118–136.

Dierksen, K., Typke, D., Hegerl, R., Koster, A. J., and Baumeister, W. (1992). Towards automatic electron tomography. *Ultramicroscopy* **40,** 71–87.

Dolder, M., Engel, A., and Zulauf, M. (1996). The micelle to vesicle transition of lipids and detergents in the presence of a membrane protein: Towards a rationale for 2D crystallization. *FEBS Letters* **382,** 203–208.

Donnelly, D., Overington, J. P., and Blundell, T. L. (1994). The prediction and orientation of helices from sequence alignments: The combined use of environment-dependent substitution tables, Fourier transform methods and helix capping rules. *Protein Eng.* **7**(5), 645–653.

Doyle, D. A., Cabral, J. M., Pfuetzner, R. A., Kuo, A., Gulbis, J. M., Cohen, S. L., Chait, B. T., and MacKinnon, R. (1998). The structure of the potassium channel: Molecular basis of K^+ conduction and selectivity. *Science* **280**(5360), 69–77.

Dubochet, J., Adrian, M., Chang, J.-J., Homo, J.-C., Lepault, J., McDowall, A. W., and Schultz, P. (1988). Cryo-electron microscopy of vitrified specimens. *Quart. Rev. Biophys.* **21**(2), 129–228.

Eisenberg, D., Schwarz, E., Komaromy, M., and Wall, R. (1984). Analysis of membrane and surface protein sequences with the hydrophobic moment plot. *J. Mol. Biol.* **179**(1), 125–142.

Engel, A., Hoenger, A., Hefti, A., Henn, C., Ford, R. C., Kistler, J., and Zulauf, M. (1992). Assembly of 2-D membrane protein crystals—dynamics, crystal order, and fidelity of structure analysis by electron microscopy. *J. Struct. Biol.* **109**(3), 219–234.

Engel, A., Lyubchenkov, Y. L., and Müller, D. J. (1999). Atomic force microscopy: A powerful tool to observe biomolecules at work. *Trends Cell Biol.* **9,** 77–80.

Finer-Moore, J., and Stroud, R. M. (1984). Amphipathic analysis and possible formation of the ion channel in an acetylcholine receptor. *Proc. Nat. Acad. Sci. USA* **81**(1), 155–159.

Fitzgerald, P. G., Bok, D., and Horwitz, J. (1983). Immunocytochemical localization of the main intrinsic polypeptide (MIP) in ultrathin frozen sections of rat lens. *J. Cell Biol.* **97,** 1491–1499.

Fotiadis, D., Hasler, L., Müller, D. J., Stahlberg, H., Kistler, J., and Engel, A. (2000). Surface tongue-and-groove contours on lens MIP facilitate cell-to-cell adherence. *J. Mol. Biol.* **300,** 779–789.

Frank, J., Goldfarb, W., Eisenberg, D., and Baker, T. S. (1978). Reconstruction of glutamine synthetase using computer averaging. *Ultramicroscopy* **3,** 283–290.

Frank, J., Verschoor, A., and Boublik, M. (1981). Computer averaging of electron micrographs of 40S ribosomal subunits. *Science* **214,** 1353–1355.

Fringeli, U. P., and Günthard, H. H. (1981). Infrared membrane spectroscopy. *In* "Membrane Spectroscopy" (E. Grell, ed.). Springer Verlag, Berlin.

Froger, A., Tallur, B., Thomas, D., and Delamarche, C. (1998). Prediction of functional residues in water channels and related proteins. *Prot. Sci.* **7**(6), 1458–1468.

Fujiyoshi, Y. (1998). The structural study of membrane proteins by electron crystallography. *Adv. Biophys.* **35,** 25–80.

Goji, K., Kuwahara, M., Gu, Y., Matsuo, M., Marumo, F., and Sasaki, S. (1998). Novel mutations in aquaporin-2 gene in female siblings with nephrogenic diabetes insipidus: evidence of disrupted water channel function. *J. Clin. Endocrinol. Metab.* **83**(9), 3205–3209.

Goormaghtigh, E., Cabiaux, V., and Ruysschaert, J. M. (1994). Determination of soluble and membrane protein structure by Fourier transform infrared spectroscopy. *Subcell. Biochem.* **23,** 329–450.

Gorin, M. B., Yancey, S. B., Cline, J., Revel, J.-P., and Horwitz, J. (1984). The major intrinsic protein (MIP) of the bovine lens fiber membrane: Characterization and structure based on cDNA cloning. *Cell* **39,** 49–59.

Grigorieff, N., Ceska, T. A., Downing, K. H., Baldwin, J. M., and Henderson, R. (1996). Electron-crystallographic refinement of the structure of bacteriorhodopsin. *J. Mol. Biol.* **259**(3), 393–421.

Guckenberger, R. (1985). Surface reliefs derived from heavy-metal-shadowed specimens—Fourier space techniques applied to periodic objects. *Ultramicroscopy* **16**, 357–370.

Hasler, L., Heymann, J. B., Engel, A., Kistler, J., and Walz, T. (1998a). 2D crystallization of membrane proteins: Rationales and examples. *J. Struct. Biol.* **121**(2), 162–171.

Hasler, L., Walz, T., Tittmann, P., Gross, H., Kistler, J., and Engel, A. (1998b). Purified lens major intrinsic protein (MIP) forms highly ordered tetragonal two-dimensional arrays by reconstitution. *J. Mol. Biol.* **279**, 855–864.

Henderson, R., Baldwin, J. M., and Ceska, T. A. (1990). Model for the structure of bacteriorhodopsin based on high-resolution electron cryo-microscopy. *J. Mol. Biol.* **213**, 899–929.

Henderson, R., Baldwin, J. M., Downing, K. H., Lepault, J., and Zemlin, F. (1986). Structure of purple membrane from *Halobacterium halobium:* Recording, measurement and evaluation of electron micrographs at 3.5 Å resolution. *Ultramicroscopy* **19**, 147–178.

Heymann, B., and Engel, A. (2000). Structural clues in the sequences of the aquaporins. *J. Mol. Biol.* **295**(4), 1039–1053.

Heymann, J., and Engel, A. (1999). Aquaporins: Phylogeny, structure and physiology of water channels. *News Physiolog. Sci.* **14**, 187–193.

Hoh, J. H., Lal, R., John, S. A., Revel, J. P., and Arnsdorf, M. F. (1991). Atomic force microscopy and dissection of gap junctions. *Science* **253**, 1405–1408.

Jap, B. K., and Li, H. (1995). Structure of the osmo-regulated H_2O-channel, AQP-CHIP, in projection at 3.5 Å resolution. *J. Mol. Biol.* **251**, 413–420.

Jap, B. K., Zulauf, M., Scheybani, T., Hefti, A., Baumeister, W., and Aebi, U. (1992). 2D crystallization: From art to science. *Ultramicroscopy* **46**, 45–84.

Jung, J., Preston, G., Smith, B., Guggino, W., and Agre, P. (1994). Molecular structure of the water channel through aquaporin CHIP. The hourglass model. *J. Biol. Chem.* **269**(20), 14648–14654.

Keeling, P., Johnson, K., Sas, D., Klukas, K., Donahue, P., and Johnson, R. (1983). Arrangement of MP26 in lens junctional membranes: Analysis with proteases and antibodies. *J. Membr. Biol.* **74**, 217–228.

Kimura, Y., Vassylev, D. G., Miyazawa, A., Kidera, A., Matsushima, M., Mitsuoka, K., Murata, K., Hirai, T., and Fujiyoshi, Y. (1997). Surface structure of bacteriorhodopsin revealed by high resolution electron crystallography. *Nature* **389**, 206–211.

Kistler, J., and Bullivant, S. (1980). Lens gap junctions and orthogonal arrays are unrelated. *Febs Lett.* **111**(1), 73–78.

Koenig, N., Zampighi, G. A., and Butler, P. J. G. (1997). Characterization of the major intrinsic protein (MIP) from bovine fibre membranes by electron microscopy and hydrodynamics. *J. Mol. Biol.* **265**, 590–602.

Lagree, V., Froger, A., Deschamps, S., Pellerin, I., Delamarche, C., Bonnec, G., Gouranton, J., Thomas, D., and Hubert, J. F. (1998). Oligomerization state of water channels and glycerol facilitators. Involvement of loop E. *J. Biol. Chem.* **273**(51), 33949–33953.

Lagree, V., Froger, A., Deschamps, S., Hubert, J. F., Delamarche, C., Bonnec, G., Thomas, D., Gouranton, J., and Pellerin, I. (1999). Switch from an aquaporin to a glycerol channel by two amino acids substitution. *J. Biol. Chem.* **274**(11), 6817–6819.

Li, H., Lee, S., and Jap, B. K. (1997). Molecular design of aquaporin-1 water channel as revealed by electron crystallography. *Nat. Struct. Biol.* **4**(4), 263–265.

Mannella, C. A. (1984). Phospholipase-induced crystallization of channels in mitochondrial outer membranes. *Science* **224**, 165–166.

Marples, D., Knepper, M. A., Christensen, E. I., and Nielsen, S. (1995). Redistribution of aquaporin-2 water channels induced by vasopressin in rat kidney inner medullary collecting duct. *Am. J. Physiol.* **269**(3 Pt 1), C655–C664.

Maurel, C., Reizer, J., Schroeder, J. I., Chrispeels, M. J., and Saier, M. H., Jr. (1994). Functional characterization of the *Escherichia coli* glycerol facilitator, GlpF, in *Xenopus* oocytes. *J. Biol. Chem.* **269**(16), 11869–11872.

Mitra, A. K., van Hoek, A. N., Wiener, M. C., Verkman, A. S., and Yeager, M. (1995). The CHIP28 water channel visualized in ice by electron crystallography. *Nat. Struct. Biol.* **2**, 726–729.

Mitsuoka, K., Murata, K., Walz, T., Hirai, T., Agre, P., Heymann, J. B., Engel, A., and Fujiyoshi, Y. (1999). The structure of aquaporin-1 at 4.5 Å resolution reveals short alpha-helices in the center of the monomer. *J. Struct. Biol.* **128**(1), 34–43.

Mou, J., Czajkowsky, D. M., Sheng, S., Ho, R., and Shao, Z. (1996). High resolution surface structure of *E. coli* GroES oligomer by atomic force microscopy. *FEBS Lett.* **381**, 161–164.

Mulders, S., Knoers, N., Van Lieburg, A., Monnens, L., Leumann, E., Wuhl, E., Schober, E., Rijss, J., Van Os, C., and Deen, P. (1997). New mutations in the AQP2 gene in nephrogenic diabetes insipidus resulting in functional but misrouted water channels. *J. Am. Soc. Nephrol.* **8**(2), 242–248.

Mulders, S. M., Bichet, D. G., Rijss, J. P., Kamsteeg, E. J., Arthus, M. F., Lonergan, M., Fujiwara, M., Morgan, K., Leijendekker, R., van der Sluijs, P., van Os, C. H., and Deen, P. M. (1998). An aquaporin-2 water channel mutant which causes autosomal dominant nephrogenic diabetes insipidus is retained in the Golgi complex. *J. Clin. Invest.* **102**(1), 57–66.

Müller, D. J., and Engel, A. (1997). The height of biomolecules measured with the atomic force microscope depends on electrostatic interactions. *Biophys. J.* **73**, 1633–1644.

Müller, D. J., and Engel, A. (1999). Voltage and pH-induced channel closure of porin OmpF visualized by atomic force microscopy. *J. Mol. Biol.* **285**(4), 1347–1351.

Müller, D. J., Büldt, G., and Engel, A. (1995). Force-induced conformational change of bacteriorhodopsin. *J. Mol. Biol.* **249**, 239–243.

Müller, D. J., Fotiadis, D., Scheuring, S., Müller, S. A., and Engel, A. (1999a). Electrostatically balanced subnanometer imaging of biological specimens by atomic force microscope. *Biophys. J.* **76**, 1101–1111.

Müller, D. J., Sass, H.-J., Müller, S., Büldt, G., and Engel, A. (1999b). Surface structures of native bacteriorhodopsin depend on the molecular packing arrangement in the membrane. *J. Mol. Biol.* **285**(5), 1903–1909.

Müller, S., Goldie, K. N., Bürki, R., Häring, R., and Engel, A. (1992). Factors influencing the precision of quantitative scanning transmission electron microscopy. *Ultramicroscopy* **46**, 317–334.

Müller, S. A., and Engel, A. (1998). Mass measurement in the scanning transmission electron microscope: A powerful tool for studying membrane proteins. *J. Struct. Biol.* **121**(2), 219–230.

Park, J. H., and Saier, M. H. (1996). Phylogenetic characterization of the MIP family of transmembrane channel proteins. *J. Membr. Biol.* **153**(3), 171–180.

Paul, D. L., and Goodenough, D. A. (1983). Preparation, characterization, and localization of antisera against bovine MIP26, an integral protein from lens fiber plasma membrane. *J. Cell. Biol.* **96**, 625–632.

Pomès, R., and Roux, B. (1996). Structure and dynamics of a proton wire: A theoretical study of H^+ translocation along the single-file water chain in the gramicidin A channel. *Biophys. J.* **71**(1), 19–39.

Preston, G. M., and Agre, P. (1991). Isolation of the cDNA for erythrocyte integral membrane protein of 28 kilodaltons: Member of an ancient channel family. *Proc. Nat. Acad. Sci. USA* **88**, 11110–11114.

Preston, G. M., Carroll, T. P., Guggino, W. B., and Agre, P. (1992). Appearance of water channels in *Xenopus* oocytes expressing red cell CHIP28 protein. *Science* **256**, 385–387.

Rash, J. E., Yasumura, T., Hudson, C. S., Agre, P., and Nielsen, S. (1998). Direct immunogold labeling of aquaporin-4 in square arrays of astrocyte and ependymocyte plasma membranes in rat brain and spinal cord. *Proc. Nat. Acad. Sci. USA* **95**(20), 11981–11986.

Rigaud, J.-L., Mosser, G., Lacapere, J.-J., Olofsson, A., Levy, D., and Ranck, J.-L. (1997). Bio-Beads: An efficient strategy for two-dimensional crystallization of membrane proteins. *J. Struct. Biol.* **118**(3), 226–235.

Ringler, P., Borgnia, M. J., Stahlberg, H., Agre, P., and Engel, A. (1999). Structure of the water channel AqpZ from *Escherichia coli* revealed by electron crystallography. *J. Mol. Biol.* **291**, 1181–1190.

Sas, D. F., Sas, J., Johnson, K. R., Menko, A. S., and Johnson, R. G. (1985). Junctions between lens fiber cells are labeled with a monoclonal antibody shown to be specific for MP26. *J. Cell. Biol.* **100**, 216–225.

Saxton, W. O., and Baumeister, W. (1982). The correlation averaging of a regularly arranged bacterial cell envelope protein. *J. Microsc.* **127**, 127–138.

Saxton, W. O., Pitt, T. J., and Horner, M. (1979). Digital image processing: The Semper system. *Ultramicroscopy* **4**, 343–354.

Schabert, F. A., and Engel, A. (1994). Reproducible acquisition of *Escherichia coli* porin surface topographs by atomic force microscopy. *Biophys. J.* **67**, 2394–2403.

Schabert, F. A., Henn, C., and Engel, A. (1995). Native *Escherichia coli* OmpF porin surfaces probed by atomic force microscopy. *Science* **268**, 92–94.

Scheuring, S., Müller, D. J., Ringler, P., Heymann, J. B., and Engel, A. (1999a). Imaging streptavidin 2D crystals on biotinylated lipid monolayers at high resolution with the atomic force microscopy. *J. Microsc.* **193**, 28–35.

Scheuring, S., Ringler, P., Brognia, M., Stahlberg, H., Müller, D., Agre, P., and Engel, A. (1999b). High resolution AFM topographs of the *Escherichia coli* waterchannel aquaporin Z. *EMBO J.* **18**, 4981–4987.

Shi, L., Skach, W., and Verkman, A. (1994). Functional independence of monomeric CHIP28 water channels revealed by expression of wild-type mutant heterodimers. *J. Biol. Chem.* **269**(14), 10417–10422.

Smith, B. L., and Agre, P. (1991). Erythrocyte M_r 28,000 transmembrane protein exists as a multisubunit oligomer similar to channel proteins. *J. Biol. Chem.* **266**(10), 6407–6415.

Smith, P. R., and Kistler, J. (1977). Surface reliefs computed from micrographs of heavy metal-shadowed specimens. *J. Ultrastruct. Res.* **61**, 124–133.

Schneider, T. D., and Stephens, R. M. (1990). Sequence logos: A new way to display consensus sequences. *Nucl. Acids Res.* **18**, 6097–6100.

Unser, M., Trus, D. L., and Steven, A. C. (1987). A new resolution criterion based on spectral signal-to-noise ratios. *Ultramicroscopy* **23**, 39–52.

van Heel, M., and Frank, J. (1981). Use of multivariate statistics in analysing the images of biological macromolecules. *Ultramicroscopy* **6**, 187–194.

van Lieburg, A. F., Verdijk, M. A., Knoers, V. V., van Essen, A. J., Proesmans, W., Mallmann, R., Monnens, L. A., van Oost, B. A., van Os, C. H., and Deen, P. M. (1994). Patients with autosomal nephrogenic diabetes insipidus homozygous for mutations in the aquaporin 2 water-channel gene. *Am. J. Hum. Gen.* **55**(4), 648–652.

Varadaraj, K., Kushmerick, C., Baldo, G. J., Bassnett, S., Shiels, A., and Mathias, R. T. (1999). The role of MIP in lens fiber cell membrane transport. *J. Membr. Biol.* **170**, 191–203.

Verbavatz, J.-M., Brown, D., Sabolic, I., Valenti, G., Ausiello, D. A., Van Hoek, A. N., Ma, T., and Verkman, A. S. (1993). Tetrameric assembly of CHIP28 water channels in liposomes and cell membranes: A freeze-fracture study. *J. Cell Biol.* **123**, 605–618.

Voegele, R. T., Sweet, G. D., and Boos, W. (1993). Glycerol kinase of *Escherichia coli* is activated by interaction with the glycerol facilitator. *J. Bacteriol.* **175**(4), 1087–1094.

Walz, T., and Grigorieff, N. (1998). Electron crystallography of two-dimensional crystals of membrane proteins. *J. Struct. Biol.* **121**(2), 142–161.

Walz, T., Smith, B., Agre, P., and Engel, A (1994a). The three-dimensional structure of human erythrocyte aquaporin CHIP. *EMBO J.* **13**(13), 2985–2993.

Walz, T., Smith, B., Zeidel, M., Engel, A., and Agre, P. (1994b). Biologically active two-dimensional crystals of aquaporin CHIP. *J. Biol. Chem.* **269**(3), 1583–1586.

Walz, T., Typke, D., Smith, B. L., Agre, P., and Engel, A (1995). Projection map of aquaporin-1 determined by electron crystallography. *Nat. Struct. Biol.* **2**(9), 730–732.

Walz, T., Tittmann, P., Fuchs, K. H., Müller, D. J., Smith, B. L., Agre, P., Gross, H., and Engel, A. (1996). Surface topographies at subnanometer resolution reveal asymmetry and sidedness of aquaporin-1. *J. Mol. Biol.* **264**(5), 907–918.

Walz, T., Hirai, T., Murata, K., Heymann, J. B., Mitsuoka, K., Fujiyoshi, Y., Smith, B. L., Agre, P., and Engel, A. (1997). The 6Å three-dimensional structure of aquaporin-1. *Nature* **387**, 624–627.

Wiener, M. C. (1999). Three-dimensional crystallization and preliminary characterization of human aquaporin-1. *Biophys. J.* **76**, A277.

Wistow, G., Pisano, M., and Chepelinsky, A. (1991). Tandem sequence repeats in transmembrane channel proteins. *Trends Biochem. Sci.* **16**(5), 170–171.

Zampighi, G., Simon, S. A., Robertson, J. D., McIntosh, T. J., and Costello, M. J. (1982). On the structural organization of isolated bovine lens fiber junctions. *J. Cell Biol.* **93**, 175–189.

Zampighi, G. A., Hall, J. E., Ehring, G. R., and Simon, S. A. (1989). The structural organization and protein composition of lens fiber junctions. *J. Cell. Biol.* **108**, 2255–2275.

Zeidel, M. L., Ambudkar, S. V., Smith, B. L., and Agre, P. (1992). Reconstitution of functional water channels in liposomes containing purified red cell CHIP28 protein. *Biochemistry* **31**, 7436–7440.

CHAPTER 3

Physiological Roles of Aquaporins in the Kidney

Mark A. Knepper, * **Søren Nielsen,**[†] **and Chung-Lin Chou***

*National Heart, Lung and Blood Institute, National Institutes of Health, Bethesda, Maryland 20892;
[†]Department of Cell Biology, University of Aarhus, DK-8000 Aarhus, Denmark

Current Topics in Membranes, Volume 51
121

I. INTRODUCTION

Lipid bilayers are permeable to water even in the absence of integral membrane proteins (Finkelstein, 1976). Mammalian cells that do not express water channels can equilibrate osmotically in response to changes in extracellular fluid osmolality in a matter of seconds to minutes (Fig. 1). Thus, at first glance, it might be suspected that molecular water channels such as the aquaporins are superfluous. It could be argued that if the lipid bilayers that make up the plasma membranes of cells are permeable to water, there is no physiological need for water channels. The answer to this argument is that certain physiological functions require extraordinarily rapid water transport, which can only occur with proteinaceous water pores. The discovery of the aquaporins (Agre *et al.*, 1987; Denker *et al.*, 1988; Preston and Agre, 1991) and the subsequent flurry of investigation concerning physiological roles of aquaporins in various tissues have made it clear that these water channels do play important roles in cellular and systemic homeostasis. Nowhere is this clearer than in the kidney, where rapid water transport is critical to the mechanisms that regulate the tonicity and volume of the extracellular fluid. An examination of the physiological roles of aquaporins in the kidney points to specific cellular function of aquaporins which may be shared by other tissues. The purpose of this focused review is to highlight three such roles of rapid aquaporin-mediated water transport in the kidney with a view toward generalizing about principles of water transport in epithelial tissues. The strategy is to begin by reviewing fundamental knowledge regarding renal epithelial water transport and the cellular localization of the aquaporins in the kidney, and then to discuss the three known functional roles of aquaporin-mediated water transport in kidney: (1) facilitation of isosmotic fluid transport in proximal tubule; (2) mediation of rapid osmotic water transport in the descending part of Henle's loop and in the descending vasa recta, necessary for concentration of solutes in the medullary interstitium: and (3) provision of molecular targets for regulation of water transport in the collecting ducts.

II. WATER TRANSPORT ALONG THE RENAL TUBULE

The rate of excretion of water by the kidney is determined by the rate of ultra-filtration by the glomerulus minus the sum of the rates of water reabsorption of each of the renal tubule segments. Micropuncture studies have defined the sites of water transport along the nephron (Jamison, 1970; Lassiter *et al.*, 1961; Pennell *et al.*, 1974) (Fig. 2). In terms of the percent of water filtered by the glomerulus, one-third reaches the end of the proximal convoluted tubule, and approximately 14% reaches the bend of Henle's loop or the beginning of the distal tubule. In the renal collecting duct, water absorption is variable and dependent on the neurohypophyseal hormone, vasopressin. In the presence of high circulating levels

Osmotic Equilibration in Animal Cells

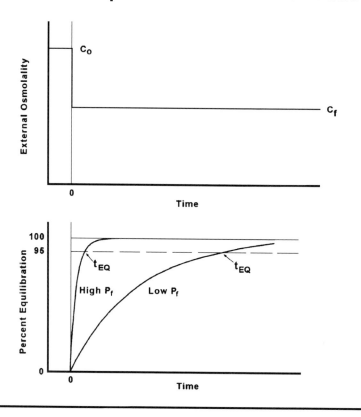

Calculation of Equilibration Times for Mammalian Cells
(Assumes 5 μm radius)

	P_f (μm/s)	t_{EQ} (s)
No water channels	20	31
With water channels		
Erythrocyte	370	1.6
PCT Cell	2,500	0.13

FIGURE 1 Results of mathematical modeling of osmotic equilibration process in mammalian cells. Upper plot shows the assumed reduction in extracellular osmolality; lower plot shows time course of osmotic equilibration of a hypothetical spherical cell (radius 5 μm) assuming that osmotic equilibration occurs only by water flux. (Equation is a first-order differential equation with time as an independent variable representing mass balance for water.) Equilibration time is a function of the osmotic water permeability of the plasma membrane of the cell. t_{EQ} represents the time that it takes for volume of cell to change to 95% of equilibrium value. Bottom of figure shows t_{EQ} values for a hypothetical cell with no water channels with a P_f value typical of an artificial lipid bilayer (31 s) as well as for P_f values typical of an erythrocyte (1.6 s) and a proximal convoluted tubule cell (0.13 s). (Note that much larger cells such as *Xenopus* oocytes require significantly longer periods to swell.)

Mark A. Knepper *et al.*

Water Absorption along the Renal Tubule

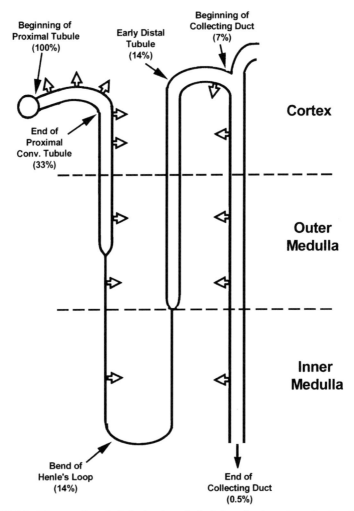

FIGURE 2 Diagram of renal tubule showing principal sites of water reabsorption in antidiuretic rats (arrows) as demonstrated by micropuncture studies (Jamison, 1970; Lassiter *et al.,* 1961; Pennell *et al.,* 1974). Numbers represent the fraction of filtered water reaching various sites along the renal tubule.

of vasopressin, most of the water reaching the beginning of the collecting duct is reabsorbed, whereas water absorption is markedly attenuated in the absence of vasopressin. Thus, it was clear from the micropuncture literature that the most important sites of water absorption are the proximal tubule, the descending limb of Henle's loop, and the collecting duct.

III. WATER PERMEABILITY ALONG THE RENAL TUBULE

Studies in isolated perfused tubules of rabbits have defined the osmotic water permeabilities of the epithelia that line the different segments of the renal tubule (Andreoli *et al.,* 1978; Burg and Green, 1973; Grantham and Burg, 1966; Gross *et al.,* 1975; Imai, 1979; Imai and Kokko, 1976; Kokko, 1970; Rocha and Kokko, 1973, 1974b; Schafer and Andreoli, 1972; Schafer *et al.,* 1978) (Fig. 3). Epithelial

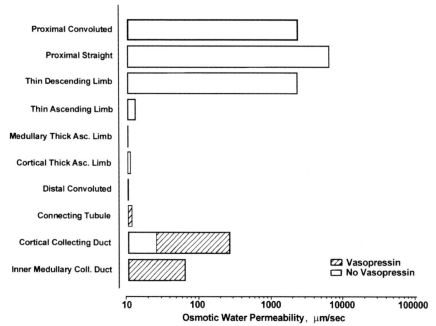

FIGURE 3   Osmotic water permeability coefficients measured in isolated perfused renal tubule segments from rabbits (Andreoli *et al.,* 1978; Burg and Green, 1973; Grantham and Burg, 1966; Gross *et al.,* 1975; Imai, 1979; Imai and Kokko, 1976; Kokko, 1970; Rocha and Kokko, 1973, 1974b; Schafer and Andreoli, 1972; Schafer *et al.,* 1978). Note the logarithmic scale. A similar pattern has been obtained in perfused tubules from rats (not shown), except that higher osmotic water permeability values have been noted in rat collecting ducts (Nonoguchi *et al.,* 1988; Reif *et al.,* 1984; Sands *et al.,* 1987) than are shown here. (Adapted from Knepper and Burg, 1983.)

water permeabilities are extremely high in the proximal tubule and descending limb of Henle's loop, which are sites of aquaporin-1 expression in the kidney (Denker *et al.*, 1988; Nielsen *et al.*, 1993c; Sabolic *et al.*, 1992). Water permeabilities are extremely low in the ascending portions of the loop of Henle (Burg and Green, 1973; Rocha and Kokko, 1973). No water channels have been identified in these segments. Finally, water permeability is either high or low in collecting duct segments depending on whether or not vasopressin is present (Grantham and Burg, 1966; Morgan *et al.*, 1968; Reif *et al.*, 1984; Rocha and Kokko, 1974a; Sands *et al.*, 1987). The collecting duct principal cell is the site of expression of three water channels: aquaporin-2 (Fushimi *et al.*, 1993; Nielsen *et al.*, 1993a), aquaporin-3 (Ecelbarger *et al.*, 1995; Ishibashi *et al.*, 1994; Ma *et al.*, 1994a) and aquaporin-4 (Frigeri *et al.*, 1995; Terris *et al.*, 1995). Thus the localization of aquaporins at sites of high water permeability along the renal tubule provided early evidence that the aquaporins provide the molecular pathway for osmotic water transport across renal tubular epithelia.

IV. AQUAPORIN-1 FACILITATES ISOSMOTIC FLUID TRANSPORT IN PROXIMAL TUBULE

A. Principle of Isosmotic Fluid Absorption

Figure 4 illustrates the principle of isosmotic or near-isosmotic fluid transport across so-called leaky epithelia such as the proximal tubule of the kidney. These epithelia carry out active transport of NaCl, which tends to dilute the fluid bathing the apical aspect of the cells relative to the fluid bathing the basolateral aspect (Fig. 4A). Because the proximal tubule has an extraordinarily high water permeability (Andreoli *et al.*, 1978; Green and Giebisch, 1989; Schafer *et al.*, 1978), water rapidly follows the actively transported NaCl. The water permeability is normally so high that the water flow can occur at rates proportional to active NaCl absorption in response to an extremely small transepithelial osmotic gradient. Direct measurements in rat proximal tubules perfused *in vivo* by Green and Giebisch (1984) have demonstrated that the rapid water absorption seen there is driven by a transepithelial osmotic gradient of only about 4 mOsm/kg H_2O. As a result, the fluid absorbed has an osmolality very close to the osmolality of the luminal fluid.

Figure 5 illustrates the hypothetical effect of lowering the epithelial osmotic water permeability (P_f) of the proximal tubule as would be the case in the absence of aquaporin-1. If the rate of net NaCl transport is unchanged, it is predicted that net fluid absorption would be unchanged since the transepithelial osmolality gradient would simply increase to compensate for the lower water permeability (Fig. 5B). Thus, in the proximal tubule and other epithelia that exhibit near-isosmotic

Near Isosmotic Fluid Transport
in Leaky Epithelia

A. Transepithelial Osmolality Difference

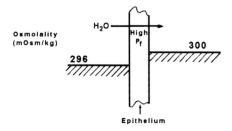

B. Local Osmolality Difference

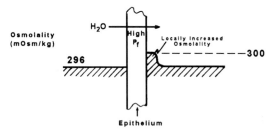

FIGURE 4 Principle of near-isosmotic fluid transport in leaky epithelia like the proximal tubule. (A) A small transepithelial osmolality difference generated by active transepithelial NaCl transport is sufficient to drive transepithelial water transport at a rapid rate. The osmolality of the reabsorbed fluid is nearly the same as that of the fluids on either side of the epithelium. (B) The active NaCl transport may cause a local elevation of extracellular osmolality in vicinity of cell, responsible for the driving force for transepithelial water flux. In most formulations, the intercellular space between epithelial cells is considered the region of high local osmolality. See text. [Modified from Whittembury, G., and Reuss, L. (1992). Mechanisms of coupling of solute and solvent transport in epithelia. Chap. 13 in "The Kidney: Physiology and Pathophysiology" D. W. Seldin and G. Giebisch, eds., 2nd ed.]

transport, net NaCl transport is rate limiting for transepithelial fluid transport, and water absorption is theoretically independent of water permeability. *What then is the value of having a high osmotic water permeability in the proximal tubule?* The answer appears to lie with the leaky nature of these epithelia, that is, the high permeability of the paracellular pathway for small ions. It has long been recognized that the net rate of NaCl absorption across the proximal tubule is not solely determined by the rate of active transcellular NaCl transport but is also dependent on the rate of paracellular NaCl backleak across the epithelium. Thus, as the transepithelial NaCl gradient across the epithelium increases (as would occur in

A. With Aquaporin-1 (P_f = 1200 μm/s)

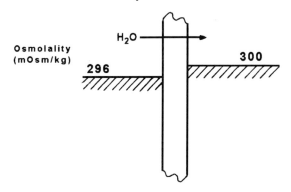

B. Without Aquaporin-1 (P_f = ~240 μm/s)

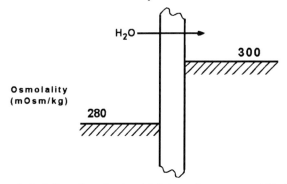

FIGURE 5   The hypothetical effect of lowering the epithelial osmotic water permeability (P_f) of the proximal tubule as would be the case in the absence of aquaporin-1. (A) Osmotic profile across proximal tubule with a normal transepithelial osmotic water permeability ($P_f = 1200$ μm/s). (B) Osmotic profile across the proximal tubule with a transepithelial osmotic water permeability hypothetically reduced to 20% of normal ($P_f = \sim240$ μm/s). If the rate of net NaCl transport is unchanged, it is predicted that net fluid absorption would be unchanged since the transepithelial osmolality gradient would simply increase to compensate for the lower water permeability.

the absence of aquaporins, Fig. 5B), the rate of NaCl backleak would increase and the rate of net NaCl and fluid absorption would decrease. Therefore, it appears that the high osmotic water permeability of the proximal tubule is important because it allows fluid absorption to occur with the smallest possible transepithelial osmotic and NaCl gradients, limiting backleak, and maximizing the efficiency of NaCl and fluid absorption. This concept seems to have been bourne out by studies in aquaporin-1 knockout mice (see below).

B. Evidence That Aquaporin-1 Is Responsible for the High Water Permeability of the Proximal Tubule

Immunocytochemistry and immunoelectron microscopy with polyclonal antibodies raised to aquaporin-1 revealed heavy labeling of the apical and basolateral plasma membranes of proximal tubules (Nielsen *et al.*, 1993c; Sabolic *et al.*, 1992), suggesting that aquaporin-1 is very abundant in the proximal tubule cells. Direct measurements of aquaporin-1 protein abundance in microdissected rat proximal tubules using a fluorescence-based ELISA method indicated that S-1 proximal convoluted tubules have approximately 20 million copies of the aquaporin-1 protein per cell or 11 femtomoles per millimeter of tubule length (Maeda *et al.*, 1995). The unit conductance of aquaporin-1 derived from studies involving reconstitution of purified or semipurified aquaporin-1 in lipid vesicles (van Hoek and Verkman, 1992; Zeidel *et al.*, 1992) is approximately 10^{-13} cm^3/(s-molecule). Calculation of the expected epithelial permeability calculated from the absolute aquaporin-1 abundance in the proximal S-1 segment and the unit conductances measurements gives a value of $\sim$2000 μm/s (Maeda *et al.*, 1995). Measured osmotic water permeability values in rat proximal tubules *in vivo* were in the range of 1000–1500 μm/s (Green and Giebisch, 1989; Preisig and Barry, 1985). Thus there is more than enough aquaporin-1 expressed in rat proximal tubule to account for the water permeability.

Verkman and his associates (Ma *et al.*, 1998) have developed an aquaporin-1 knockout mouse that has permitted investigations of the role of aquaporin-1 in proximal tubule transport. Schnermann *et al.* (1998) have reported that the epithelial osmotic water permeability of S-2 proximal tubules from aquaporin-1 knockout mice is approximately 20% of the value in wild-type mice, suggesting that aquaporin-1 accounts for about 80% of the osmotic water permeability. The apical plasma membrane of the proximal tubule is highly amplified in area because of the presence of microvilli, and the basolateral plasma membrane is similarly amplified because of complex interdigitations and basal infoldings. Hence, it has been proposed that the residual water permeability could have been largely due to water permeation through the lipid bilayer (Schnermann *et al.*, 1998). Nevertheless, it seems possible that a part of the residual osmotic water permeability in proximal tubules of aquaporin-1 mice could be due to water permeation through proteins in the tight junctions or through nonaquaporin proteins in the plasma membranes. Recently, it has been demonstrated that the aquaporin-4 water channel is expressed in the basolateral plasma membrane of S-3 proximal straight tubules in mouse (van Hoek *et al.*, 2000). Thus, it appears possible that part of the residual osmotic water permeability in S-2 proximal segments in aquaporin-1 knockout mice could be due to aquaporin-4, although aquaporin-4 is not strongly expressed in the S-2 segment.

C. Fluid Transport in Proximal Tubules of Aquaporin-1 Knockout Mice

As discussed earlier, for models of isosmotic fluid transport, the rate of fluid absorption is theoretically independent of proximal tubule water permeability, unless there is significant paracellular backleak of NaCl. Measurements of spontaneous fluid absorption in isolated perfused S-2 proximal tubule from knockout mice and wild-type controls demonstrated that fluid absorption is decreased by an average of 52% in the absence of aquaporin-1 (Schnermann *et al.*, 1998). A similar decrease was seen by *in vivo* micropuncture measurements (Schnermann *et al.*, 1998). As emphasized in Fig. 5, it can be predicted that the transepithelial osmolality difference increases in proximal tubules from aquaporin-1 knockout mice to compensate in part for the decrease in osmotic water permeabilty.

This prediction was recently validated by micropuncture measurements (Vallon *et al.*, 2000) showing an approximately 3-fold increase in lumen-to-blood osmolality difference in proximal convoluted tubules of aquaporin-1 knockout mice relative to wild-type controls. The fact that the fluid absorption rate was decreased relative to controls demonstrates that there was a decrease in net NaCl absorption in the proximal tubules of aquaporin-1 knockout mice, consistent with the possibility of significant NaCl backleak. Thus, although water channels are theoretically unnecessary for fluid absorption to occur in the renal proximal tubule, they permit fluid absorption to occur efficiently by limiting NaCl backleak, thus reducing the ATP cost of reabsorbing NaCl and cotransported solutes.

V. AQUAPORIN-1 ALLOWS OSMOTIC EQUILIBRATION IN THE THIN DESCENDING LIMB OF HENLE DESPITE RAPID FLOW OF TUBULE FLUID

A. Thin Descending Limb Function and the Principle of Countercurrent Multiplication

Another important function of water channels is to allow rapid osmotic equilibration of tubule fluid of the thin descending limb of Henle's loop with the surrounding interstitium. Osmotic equilibration in the thin descending limb is crucial to the countercurrent multiplication mechanism (Fig. 6) that concentrates solutes in the renal medullary interstitium, resulting in an interstitial osmolality that is much greater than that of systemic blood plasma. The generation of a hypertonic medullary interstitium is important because it provides the osmotic driving force for absorption of water from the medullary collecting ducts, resulting in efficient water conservation. When water absorption from the medullary collecting duct reaches osmotic equilibrium with the hypertonic medullary interstitium, the amount of water excreted can be reduced to as low as 0.2% of the amount of water

Countercurrent Multiplier Mechanism

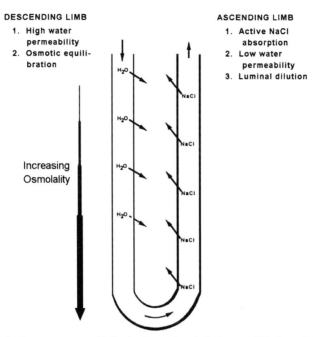

DESCENDING LIMB
1. High water
   permeability
2. Osmotic equili-
   bration

Increasing
Osmolality

ASCENDING LIMB
1. Active NaCl
   absorption
2. Low water
   permeability
3. Luminal dilution

FIGURE 6 Countercurrent multiplication in renal medulla. Loops of Henle are hairpin-shaped structures that have descending limbs and ascending limbs with different properties. A gradient of increasing osmolality results from counterflow between the interacting descending and ascending limbs. Energy for generation of axial osmolality gradient is derived from active (ATP-dependent) transport of NaCl from lumen of ascending limbs (*right*). The low water permeability of the ascending limb due to absence of aquaporins allows active NaCl absorption to dilute the luminal fluid relative to surrounding interstitial fluid. Dilution of lumen allows loops of Henle to remove water from the medulla out of proportion to removal of NaCl. The descending limb has a high osmotic water permeability due to the presence of large amounts of aquaporin-1 in its plasma membranes. This allows the descending limb fluid to equilibrate osmotically with the hypertonic fluid of the interstitium by rapid water efflux (*left*). The efflux of water short circuits water back toward the renal cortex (*upper end of loop*), allowing delivery of concentrated solutes to the deep part of the medulla.

filtered at the glomerulus. This water-conserving mechanism greatly reduces the requirement for water intake in mammals and birds, permitting these species to exist in relatively dry environments.

Figure 6 summarizes, in simple terms, the concept of countercurrent multiplication, the process responsible for generation of the corticomedullary osmolality gradient in the renal medulla. Countercurrent multiplication is powered by active NaCl absorption from the thick ascending limb of Henle's loop. Because this renal tubule segment has a very low osmotic water permeability (due in part to a lack of aquaporins), the active salt transport results in a fall in the luminal osmolality relative to that of the surrounding interstitium. The low luminal osmolality provides the so-called "single-effect" (German: *Einzeleffekt*) for countercurrent multiplication (Kuhn and Ramel, 1959). This small transepithelial osmolality difference, when "multiplied" by the counterflow between ascending and descending limbs of Henle's loops, generates a large increase in osmolality in the deepest parts of the renal medulla. In rabbits, which have a concentrating capacity similar to humans, the osmolality in the deepest part of the inner medulla reaches a value of around 1200 mOsm/kg-water or four times the osmolality of systemic plasma (Knepper, 1982). Rodent species such as the rat can generate much higher osmolalities (Hai and Thomas, 1969).

While the single-effect NaCl transport process in the ascending limb of Henle's loop furnishes the energy for the medullary countercurrent multiplier mechanism, equally important is osmotic equilibration in the thin descending limb. Mass balance calculations have demonstrated that failure of osmotic equilibration would have a severe damping effect on the overall concentrating process (Knepper *et al.*, 1993). Because the flow rate of tubule fluid into the thin descending limb is very rapid, approximately one-third of the filtered load of water, the requirement for osmotic equilibration demands extraordinarily rapid water efflux. Thus, a high water permeability in the descending limb is essential for generation of a hypertonic medullary interstitium and therefore is critical for the water conserving process of the kidney.

B. Evidence That Aquaporin-1 Is Responsible for the High Water Permeability of the Thin Descending Limb of Henle's Loop

Osmotic water permeability measurements in isolated perfused thin descending limbs have demonstrated the high levels of osmotic water permeability required for countercurrent multiplication (Chou and Knepper, 1992; Imai *et al.*, 1984; Kokko, 1970). In fact, osmotic water permeability values in this segment are the largest seen along the renal tubule with values in excess of 2000 μm/s, at least 2 orders of magnitude greater than values determined in the ascending limb of Henle's loop. Descending limb water permeability measurements have been accomplished almost exclusively in descending limbs from long-loop nephrons, that is, nephrons whose loops of Henle reach the inner medulla. The technical difficulty of dissecting the descending limbs of short-loop nephrons has precluded comprehensive study of their transport properties. These short loops of Henle bend in the outer medulla and have a much different epithelial structure than the long-loop descending limbs (Kriz, 1981).

Immunocytochemical and immunoelectron microscopic studies in the descending limbs of Henle's loops have revealed that aquaporin-1 is very abundant in both the apical and basolateral plasma membranes of the long-loop descending limbs, both in the outer medullary and inner medullary portions of these loops (Chou *et al.*, 1993; Nielsen *et al.*, 1993c, 1995c). Descending limbs from short-looped nephrons were found to have abundant aquaporin-1 in their early portions (first 750 μm after the transition from the proximal tubule). However, the late portion of the short-looped descending limbs was found to be devoid of aquaporin-1 labeling (Wade *et al.*, 2000). This part of the descending limb has been found to have very high levels of the urea transporter UTA-2, which has been proposed to mediate osmotic equilibration by rapid urea efflux into the specialized blood vessels of the renal medulla, called the *vasa recta* (Wade *et al.*, 2000).

ELISA measurements of aquaporin-1 abundance in microdissected renal tubules (Maeda *et al.*, 1995) revealed that aquaporin-1 abundance is even higher in descending limb segments from long-looped nephrons than in S-1, S-2, or S-3 portions of the proximal tubule. Calculations based on the measured abundance of aquaporin-1 and the unit conductance of aquaporin-1 indicate that the level of aquaporin-1 expression is sufficient to account for the enormously high osmotic water permeabilities measured in descending limb segments.

C. Osmotic Water Transport in Descending Limbs of Aquaporin-1 Knockout Mice

In isolated perfused tubule experiments using descending limbs of long-looped nephrons of aquaporin-1 mice, the osmotic water permeability was reduced by an average of 88% relative to wild-type controls (Chou *et al.*, 1999a). Such a profound reduction in water permeability would be expected to impair the countercurrent multiplier process by interfering with osmotic equilibration between the tubule fluid of the descending limb and the surrounding interstitium as described above. When concentrating capacity was measured in the aquaporin-1 knockout mice, an almost complete failure to concentrate the urine above plasma osmolality was noted (Ma *et al.*, 1998). These results are therefore consistent with a critical role for aquaporin-1 in the concentrating mechanism. Freeze-fracture studies revealed that one population of intramembrane particles present in the apical plasma membrane of wild-type descending limbs was absent in descending limbs from aquaporin-1 mice, supporting the view that aquaporin-1 is a major protein component of the plasma membrane in these cells (Chou *et al.*, 1999a).

VI. AQUAPORIN-1 ALLOWS OSMOTIC EQUILIBRATION IN THE DESCENDING VASA RECTA

Delivery of oxygen and nutrients to the renal medulla requires a substantial rate of blood flow. Rapid flow of blood to the medulla would tend to wash out the high osmolality generated by countercurrent multipication. However, specialized blood vessels called *vasa recta* carry blood into and out of the renal medulla and are arranged in tightly opposed fashion in vascular bundles in the outer medulla, permitting conservation of medullary osmolality by a process called *countercurrent exchange*. The endothelial cells of the ascending vasa recta are fenestrated, permitting rapid movement of water and solutes by bulk flow. However, the descending vasa recta are not fenestrated, and their permeability properties are determined by the transporters present in their plasma membranes. Water permeability measurements in isolated perfused descending vasa recta have revealed high values (Turner and Pallone, 1997). Immunocytochemistry has demonstrated abundant aquaporin-1 expression in endothelial cells of the vasa recta (Kim *et al.,* 1999; Nielsen *et al.,* 1995c). ELISA measurements using microdissected descending limbs demonstrated that there is a sufficient level of aquaporin-1 expression to account for the high water permeabilities measured in perfused vasa recta (Pallone *et al.,* 1997). Recent measurements of osmotic water permeability in descending vasa recta from aquaporin-1 knockout mice have revealed a marked decrease relative to descending vasa recta from wild-type control mice (Pallone *et al.,* 2000). Thus the existing evidence supports the view that aquaporin-1 present in the descending vasa recta plays an important role in the countercurrent exchange process of the renal medulla. Undoubtedly, the decrease in water permeability of the descending vasa recta contributes to the failure of aquaporin-1 knockout mice to concentrate their urine (Ma *et al.,* 1998).

VII. AQUAPORINS PROVIDE MOLECULAR TARGETS FOR REGULATION OF WATER TRANSPORT IN THE RENAL COLLECTING DUCT

A. Regulation of Collecting Duct Water Permeability by Vasopressin

As discussed earlier, aquaporin-1 plays important roles in the proximal tubule and descending limb where rapid water transport is necessary for normal epithelial NaCl and water transport. In addition, three other aquaporins play critical roles in a different renal tubule epithelium, the collecting duct. Here, aquaporins-2, -3, and -4 mediate transepithelial water movement (Knepper, 1994, 1997; Nielsen *et al.,* 1998). It has long been recognized that the collecting duct is the site of control of epithelial water permeabilty by the peptide hormone vasopressin. Vasopressin secretion by the posterior pituitary gland increases when plasma osmolality increases above a threshold value of about 292 mOsm/kg-water (Verbalis, 1993). An increase in circulating vasopressin signals the kidney to reduce water excretion, largely as a result of the ability of vasopressin to increase the epithelial water permeability of the collecting duct (Knepper *et al.,* 1994, 2000). The increase in collecting duct water permeability permits the tubule fluid to partially or completely equilibrate osmotically with the hypertonic medullary interstitium generated by countercurrent multiplication. As reviewed in the remainder of this section, vasopressin increases water permeability of the collecting duct by increasing the plasma membrane localization of cellular aquaporin-2 through regulated trafficking of aquaporin-2-containing intracellular vesicles, and by increasing the cellular abundance of both aquaporin-2 and aquaporin-3. Thus, aquaporins-2 and -3 function as molecular targets for regulation by vasopressin.

B. Cellular Localization of Aquaporin-2, -3, and -4

Cellular and subcellular localization of the collecting duct aquaporins has been carried out using peptide-directed polyclonal antibodies that specifically recognize each aquaporin protein.

Aquaporin-2 is located in the apical plasma membrane and small intracellular vesicles of principal cells in the renal collecting duct (Nielsen *et al.*, 1993a). ELISA measurements in microdissected rat collecting ducts showed that aquaporin-2 is an extraordinarily abundant protein in principal cells (more than 6 million copies of aquaporin-2 per cell) (Kishore *et al.*, 1996b). Aquaporin-2 is recognized to provide the major means of water entry into the principal cells. This view was firmly established by studies showing a direct correlation between aquaporin-2 abundance and water permeability in rat collecting ducts (DiGiovanni *et al.*, 1994) and by studies showing that human patients with deletions of the aquaporin-2 gene experience severe vasopressin-resistant nephrogenic diabetes insipidus (Deen *et al.*, 1994).

Studies using the single-tubule ELISA method and immunocytochemistry also showed that aquaporin-2 is expressed in connecting tubule arcades of rats, suggesting that regulated water transport may occur in the connecting tubule in addition to the collecting duct (Kishore *et al.*, 1996a). This localization could be physiologically important because the arcades are localized in the cortical labyrinth where blood flow is relatively rapid and reabsorbed water could be rapidly returned to the systemic circulation without interstitial dilution. However, recent studies in rabbit, a species with a lower urinary concentrating ability, show that rabbit connecting tubules lack aquaporin-2 (Loffing *et al.*, 2000).

Aquaporins-3 and -4 are located in the basolateral plasma membranes of collecting duct principal cells (Ecelbarger *et al.*, 1995; Frigeri *et al.*, 1995; Terris *et al.*, 1995). Aquaporin-3 is the chief basolateral aquaporin in the cortical and outer medullary collecting duct but aquaporin-4 predominates in the inner medullary portion of the collecting duct. Isolated perfused tubule studies of inner medullary collecting ducts from aquaporin-4 knockout mice revealed osmotic water permeabilities that were reduced by about 77% relative to inner medullary collecting ducts from wild-type mice (Chou *et al.*, 1998a). Thus aquaporin-4 provides the major route of water transport across the basolateral plasma membrane in the inner medullary collecting duct.

C. Vasopressin-Mediated Intracellular Signaling in the Collecting Duct

Most of the renal effects of vasopressin on the collecting duct, including regulation of aquaporins-2 and -3, are mediated by the V2 vasopressin receptor, a seven-membrane-spanning receptor that activates adenylyl cyclase via the heterotrimeric protein Gs. The type V and type VI isoforms of adenylyl cyclase predominate in collecting duct cells (Chabardes *et al.*, 1996). These two isoforms are distinguished by their direct inhibition by high intracellular calcium, absence of an effect of the $\beta\gamma$ G-protein subunit complex on activity, absence of stimulation by Ca^{2+}/calmodulin, and absence of inhibition by protein kinase C (Cooper *et al.*, 1995). Most intracellular effects of cyclic AMP are thought to be secondary to activation of protein kinase A (PKA) (Miyamoto *et al.*, 1969). Activated PKA phosphorylates various proteins on specific threonine or serine residues. Among the proteins that are phosphorylated by PKA in the collecting duct is aquaporin-2 (phosphorylated at serine-256) (Kuwahara *et al.*, 1995; Lande *et al.*, 1996). The likely significance of this post-translational modification in the regulation of osmotic water permeability of the collecting duct principal cells is discussed below. Selective effects of PKA for specific proteins are facilitated by anchor proteins, so-called "AKAPs," which hold the kinase in proximity with its specific substrate. One or more AKAPs have been demonstrated in aquaporin-2-containing vesicles (Klussmann *et al.*, 1999; Ward *et al.*, 1999) and may play a role in regulation of aquaporin-2 by phosphorylation.

Vasopressin also triggers an increase in intracellular calcium in the inner medullary collecting duct (Star *et al.*, 1988b). This calcium mobilization response is mediated by the V2 vasopressin receptor (Champigneulle *et al.*, 1993; Ecelbarger *et al.*, 1996a; Maeda *et al.*, 1993; Star *et al.*, 1988b). Vasopressin, at physiological concentrations, does not activate the phosphoinositide signaling pathway, indicating that the increase in intracellular calcium mobilization is not mediated by inositol 1,4,5-tris-phosphate receptors (Chou *et al.*, 1998b). Instead, a preliminary report has suggested that vasopressin-elicited calcium mobilization in the inner medullary collecting duct may be mediated by release of calcium from ryanodine-sensitive calcium stores, presumably involving the type 1 ryanodine receptor (Chou *et al.*, 2000a; Chou *et al.*, 2000b). Furthermore, studies in isolated perfused collecting ducts (Chou *et al.*, 1999b) have supported the view that the calcium mobilization is necessary for the water permeability response. Furthermore inhibitors of calmodulin block the water permeability response to vasopressin (Chou *et al.*, 1999b; Chou *et al.*, 2000b). Therefore, it appears possible that regulation of collecting duct aquaporins may be more complex than currently believed, including steps that could be regulated by intracellular calcium rather than solely via PKA activation.

D. Two Modes of Regulation of Collecting Duct Water Permeability by Vasopressin

Vasopressin regulates the water permeability of collecting duct principal cells in two distinct response modes, namely, a short-term and a long-term regulatory mode (Fig. 7). Short-term regulation is the widely recognized process by which vasopressin rapidly increases the osmotic water permeability of the collecting duct

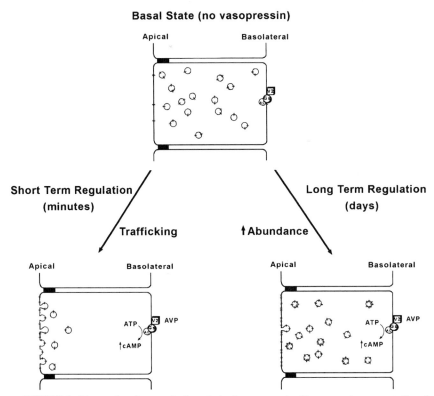

FIGURE 7   Two modes of aquaporin-2 regulation by vasopressin. Diagrammatic representation of collecting duct principal cells showing distribution and relative abundance of aquaporin-2 represented by small particles (*bars*) in intracellular vesicles and plasma membrane (*upper cell*). Short-term regulation by vasopressin results from exocytosis of aquaporin-2 vesicles to plasma membrane, increasing cell water permeability (*lower left*). Long-term regulation by vasopressin results from an increase in the number of aquaporin-2 protein copies per cell (*lower right*). Vasopressin also has similar effects to increase aquaporin-3 abundance in basolateral plasma membrane of principal cells (not shown).

after binding to V2 vasopressin receptors. At 37°C, the osmotic water permeability increase is seen over a 40-min period following addition of vasopressin (Kuwahara and Verkman, 1989; Nielsen and Knepper, 1993; Wall *et al.*, 1992). The initial increase can be observed about 40 s after vasopressin exposure (Wall *et al.*, 1992). The water permeability increase is mimicked by application of cyclic AMP analogs (Nielsen and Knepper, 1993; Star *et al.*, 1988a), suggesting that the vasopressin effect on water permeability is mediated by cyclic AMP. This short-term mode of water permeability regulation is a result of trafficking of aquaporin-2-containing intracellular vesicles to the apical plasma membrane, resulting in an increase in the number of aquaporin-2 water channels in the apical plasma membrane (see below).

Long-term regulation of osmotic water permeability in the collecting duct results from an adaptation in the collecting duct principal cells that occurs when circulating vasopressin levels are increased for 24 h or more. This is a consequence of an increase in the maximal osmotic water permeability of the collecting duct (Lankford *et al.*, 1991). This response to vasopressin occurs through an increase in the abundances of the aquaporin-2 (DiGiovanni *et al.*, 1994) and aquaporin-3 (Ecelbarger *et al.*, 1995) water channels in collecting duct cells and is thought to be a consequence of increased transcription of the aquaporin-2 and aquaporin-3 genes (see below).

1. Short-Term Regulation of Collecting Duct Water Permeability via Regulated Trafficking of Aquaporin-2

The first direct evidence that vasopressin increases the water permeability of the collecting duct epithelium was derived from the studies of Grantham and Burg (1966) and Morgan and Berliner (1968) in the 1960s. These studies showed that application of vasopressin to the basolateral aspect of collecting ducts *in vitro* increases epithelial water permeability within a few minutes of exposure to vasopressin. The cloning of the complementary DNA for aquaporin-2 (Fushimi *et al.*, 1993) allowed the preparation of polyclonal antibodies to aquaporin-2, the key reagents necessary to discovery of the mechanism of the short-term response to vasopressin. Initial immunolocalization studies at light microscopic and electron microscopic levels showed that aquaporin-2 is present in both the apical plasma membrane and in small intracellular vesicles of collecting duct principal cells (DiGiovanni *et al.*, 1994; Nielsen *et al.*, 1993a). This finding is consistent with the view first presented almost 20 years ago by James B. Wade and his colleagues (Wade *et al.*, 1981) that the vasopressin-induced increase in water permeability of the collecting duct cells could be the consequence of translocation of water-channel-containing vesicles to the plasma membrane, the so-called "shuttle hypothesis." The shuttle hypothesis was tested directly in isolated perfused tubules by measuring osmotic water permeability *in vitro* in the presence and absence of vasopressin and then fixing the tubules for immunoelectron microscopy immediately after the permeability measurements (Nielsen *et al.*, 1995a). As seen in

Immuno-Gold Localization of Aquaporin-2
in Inner Medullary Collecting Duct Cells

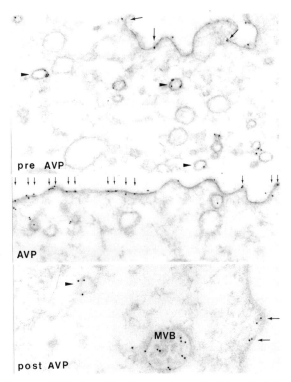

FIGURE 8 Immunogold labeling with anti-aquaporin-2 antibody in epithelial cells from isolated perfused inner medullary collecting ducts. The three panels show sections from tubules fixed before (A), during (B), and after (C) exposure to arginine vasopressin (AVP). Small arrows point to sites of apical plasma membrane labeling and arrowheads show sites of vesicular labeling. Note the shift of labeling from intracellular vesicles to plasma membrane in response to arginine vasopressin exposure of perfused inner medullary collecting ducts. Statistical analysis of labeling distribution in 30 perfused tubules showed a highly significant shift from vesicles to plasma membrane in response to arginine vasopressin. MVB, multivesicular body. (From Nielsen *et al.,* 1995a.)

Fig. 8, immunogold labeling of the perfusion-fixed collecting ducts revealed that vasopressin induces a strong redistribution of aquaporin-2 labeling from intracellular vesicles to the apical plasma membrane. When the vasopressin was withdrawn, labeling of the plasma membrane was again decreased in association with reappearance of aquaporin-2 labeling in intracellular vesicles. These changes in the cellular distribution of aquaporin-2 were paralleled by changes in osmotic water permeability of the collecting duct epithelium (Nielsen *et al.,* 1995a) and provided a direct demonstration of vasopressin-induced trafficking of aquaporin-2-bearing intracellular vesicles to the plasma membrane. In addition, several investigators showed that injection of vasopressin into live rats resulted in an *in vivo* redistribution of aquaporin-2 to the plasma membrane of collecting duct principal cells (Marples *et al.*, 1995; Sabolic *et al.*, 1995; Yamamoto, N., *et al.*, 1995). More recently, surface biotinylation studies in collecting duct suspensions from rat inner medulla have demonstrated increased surface labeling of aquaporin-2 in response to vasopressin exposure (Inoue *et al.*, 1999), confirming the conclusion that vasopressin stimulates aquaporin-2 trafficking to the plasma membrane.

Theoretically, vasopressin could increase the abundance of aquaporin-2 in the apical plasma membrane either by increasing the rate of aquaporin-2 transfer to the plasma membrane by exocytosis, decreasing the rate of retrieval of aquaporin-2 from the apical plasma membrane by endocytosis, or through both mechanisms. Measurements of the kinetics of the water permeability response in isolated perfused rat inner medullary collecting ducts following vasopressin addition and withdrawal demonstrated that vasopressin both increases the rate of water-channel exocytosis and decreases the rate of water-channel endocytosis (Knepper and Nielsen, 1993; Nielsen and Knepper, 1993). Also providing evidence for the view that vasopressin inhibits aquaporin-2 endocytosis, Strange *et al.* (1988) used isolated perfused rabbit cortical collecting ducts to show that the removal of vasopressin results in a marked increase in the rate of endocytic uptake of horseradish peroxidase (HRP) from the lumen, presumably associated with the accelerated retrieval of apical aquaporin-2. Similarly, luminal uptake of cationic ferritin and albumin gold into small apical vesicles and multivesicular bodies was seen when vasopressin was removed from the peritubular bath in experiments carried out in isolated perfused inner medullary collecting ducts from rats (Nielsen *et al.*, 1993b). This effect was quite rapid, corresponding to the rapid initial phase of water permeability decrease seen in the first 3–5 min after vasopressin washout (Nielsen and Knepper, 1993). The mechanism of the regulation of endocytosis by vasopressin has not been determined.

In apparent contrast, sustained exposure of collecting ducts to vasopressin also increases apical endocytosis. Following intravenous injection of HRP, principal cells took up the HRP from the collecting duct lumen into small vesicles and multivesicular bodies (Brown *et al.*, 1988). The rate of uptake in vasopressin-deficient Brattleboro rats was much less than in normal rats, and vasopressin administration to Brattleboro rats normalized the uptake. These and subsequent studies using fluorescein-labeled dextrans (Lencer *et al.*, 1990) lead the authors to conclude that vasopressin *enhances* the rate of endocytosis of the apical plasma membrane. The increase in the rate of endocytosis in response to vasopressin has been proposed to be due to a mass action effect due to rapid addition of membrane surface when water-channel vesicles fuse with the apical plasma membrane (Knepper and

Nielsen, 1993). This mass action effect may overcome the direct inhibitory action of vasopressin on apical endocytosis.

Regulation of aquaporin-2 vesicle exocytosis by vasopressin has received greater attention. Vasopressin-stimulated exocytosis of aquaporin-2 vesicles involves several steps including (1) translocation of vesicles from a diffuse distribution throughout the cell to the apical region of the cell, (2) translocation of aquaporin-2 across the apical cortex of the cell composed of a dense filamentous actin network, (3) priming of vesicles for docking, (4) docking of vesicles, and (5) fusion of vesicles with the apical plasma membrane. Theoretically each of these steps could be targets for regulation by vasopressin. The first step, translocation of vesicles from a diffuse distribution throughout the cell to the apical region of the cell, has been proposed to be mediated by the molecular motor protein dynein, which is believed to convey the aquaporin-2 vesicles along individual microtubules in the direction of the cell apex (Marples *et al.,* 1998). Thus far, there is no direct evidence that this step is regulated by vasopressin. The next step, translocation of aquaporin-2 across the apical cortex of the cell composed of a dense filamentous actin network, has been studied by Hays and colleagues (Ding *et al.,* 1991; Gao *et al.,* 1992; Hays and Lindberg, 1991; Holmgren *et al.,* 1992; Simon *et al.,* 1993). These investigators have presented evidence that vasopressin stimulates conversion of filamentous actin to globular actin (depolymerization of actin), hypothetically removing a barrier to diffusion of the aquaporin-2 vesicles to the plasma membrane. An alternative idea is that the aquaporin-2 vesicles may be conveyed through the cortical actin network via actin-based molecular motors, for example, nonmuscle myosins. Many myosins are targets for calmodulin-mediated regulation and therefore are potentially regulated via V2-receptor-mediated intracellular calcium mobilization (see above).

The final three steps in aquaporin-2 vesicle trafficking—vesicle priming, docking with the plasma membrane, and fusion with the plasma membrane—are believed to involve vesicle targeting receptors, the so-called "SNARE" proteins. These proteins have been identified chiefly in the context of regulated exocytosis of synaptic vesicles (Bajjalieh and Scheller, 1995; Söllner *et al.,* 1993). The SNARE proteins have been proposed to mediate specific interaction between a given vesicle and its target membrane with which it is destined to fuse. One class of targeting receptors in the vesicles (v-SNAREs) has been called "VAMPs" (or "synaptobrevins"). Another class of targeting receptor found predominantly in the target membrane (t-SNAREs) is the syntaxins. The third family of SNARE proteins is the SNAP-25-like proteins. These were initially thought to function as t-SNAREs, but are now recognized to reside in translocating vesicles as well as the target membrane. Members of all three of these families have been found in collecting duct cells with a subcellular localization consistent with a role in aquaporin-2 vesicle exocytosis. First, VAMP-2 has been shown to be present in aquaporin-2 vesicles and to be virtually absent from the apical plasma membrane of principal cells of the inner medullary collecting ducts (Nielsen *et al.,* 1995b). Similar evidence for an intracellular localization of VAMP-2 in collecting duct cells has been provided by Jo *et al.* (1995), Liebenhoff and Rosenthal (1995), and Mandon *et al.* (1996). Second, syntaxin-4 has been localized to the apical plasma membrane of collecting duct principal cells (Mandon *et al.,* 1996). Finally, SNAP-23 has been found to be present in both the apical plasma membrane of principal cells and in aquaporin-2 vesicles (Inoue *et al.,* 1998). The establishment of a role for these SNARE proteins in regulated trafficking of aquaporin-2 will most likely depend on the preparation of targeted knockouts for each of these genes in the collecting duct.

If the SNARE proteins are involved in the vasopressin-induced trafficking of aquaporin-2 to the apical plasma membrane, the regulatory mechanism might involve selective phosphorylation of one of the SNARE proteins or an ancillary protein that binds to them, possibly via protein kinase A or calmodulin-dependent kinase II. Although the SNARE proteins are recognized as potential targets for phosphorylation (Foster *et al.,* 1998; Hirling and Scheller, 1996; Risinger and Bennett, 1999; Shimazaki *et al.,* 1996), there is presently no evidence for their phosphorylation in the collecting duct. As noted above, however, aquaporin-2 itself appears to be a target for PKA-mediated phosphorylation at a serine in position 256 (Kuwahara *et al.,* 1995). The phosphorylation does not modify the single-channel water permeability of aquaporin-2 (Lande *et al.,* 1996), and instead the phosphorylation is believed to play a critical role in regulation of aquaporin-2 trafficking to and/or from the plasma membrane (Fushimi *et al.,* 1997; Katsura *et al.,* 1997). Cultured cell studies of aquaporin-2 trafficking are discussed more thoroughly in another chapter of this volume. Nishimoto and colleagues (1999) have developed a polyclonal antibody that specifically recognizes the phosphorylated form of aquaporin-2. Studies in isolated inner medullary tissue using this antibody demonstrated that vasopressin induces maximum phosphorylation of aquaporin-2 within 1 min of the initial exposure, that is, before the maximal increase in water permeability as would be required for a key role in vasopressin-induced exocytosis (Nishimoto *et al.,* 1999). Additional studies using the same preparation (Zelenina *et al.,* 2000) demonstrated that prostaglandin E_2 added after vasopressin did not decrease aquaporin-2 phosphorylation but reversed vasopressin-induced translocation of aquaporin-2 to the plasma membrane. As noted earlier, both the exocytosis and endocytosis of aquaporin-2 are believed to be regulated by vasopressin. Consequently, the dissociation between aquaporin-2 trafficking and aquaporin-2 phosphorylation in the study of Zelenina *et al.* (2000) could be indicative of a lack of a role for phosphorylation in aquaporin-2 trafficking in general or possibly in just one component of aquaporin-2 trafficking, that is, either exo- or endocytosis. Although the abundance of phosphorylated aquaporin-2 is markedly increased by vasopressin, there is thus far no evidence that the ratio of the phosphorylated form to the total aquaporin-2 is greater in the plasma membrane than in intracellular vesicles, as would be expected if phosphorylation triggered exocytosis.

2. Long-Term Regulation of Collecting Duct Water Permeability via Regulated Transcription of Aquaporin-2 Gene

The second mode of aquaporin-2 regulation by vasopressin is the recently described long-term regulatory mechanism involving transcriptionally mediated upregulation of aquaporin-2 abundance in collecting duct principal cells in response to sustained increases in circulating vasopressin levels. Interest in this problem dates from work done in the 1950s by Jones and De Wardener (1956) and Epstein *et al.* (1957) showing long-term enhancement of urinary concentrating ability as a result of prolonged antidiuresis. These studies showed that maximal urinary osmolality is greater after a long-term antidiuretic stimulus than after a short-term stimulus. The identification of the aquaporins has led to a much more complete understanding of this phenomenon.

Experiments by Lankford and colleagues (1991) in rats showed that water restriction of the rats (a maneuver that increases circulating vasopressin levels) causes a large stable increase in water permeability of inner medullary collecting ducts of the rats when these ducts are dissected from the kidneys and perfused *in vitro*. The osmotic water permeabilities of the inner medullary collecting duct segments from water-restricted rats was 5-fold greater than that found in hydrated rats. Subsequent studies have confirmed this long-term conditioning effect (Flamion *et al.*, 1995; Han *et al.*, 1994; Wade *et al.*, 1994). In contrast to the effect on water permeability, water restriction of rats did not increase the urea permeability of isolated perfused inner medullary collecting ducts, establishing that the effect is selective for the water transport in these tubules (Lankford *et al.*, 1991). Studies using polyclonal antibodies to each of the renal aquaporins have demonstrated that increases in osmotic water permeability in response to water restriction in rats are associated with increased abundance of the aquaporin-2 and aquaporin-3 water channels, with no effect on aquaporin-1 and aquaporin-4 (Ecelbarger *et al.*, 1995; Nielsen *et al.*, 1993a; Terris *et al.*, 1996). In studies of aquaporin-2, water restriction for 24–48 h resulted in an approximately 5-fold increase in the abundance of aquaporin-2 protein in the rat renal inner medulla (Nielsen *et al.*, 1993a; Terris *et al.*, 1996). This effect paralleled the increase in collecting duct water permeability seen in response to thirsting in the isolated perfused tubule studies of Lankford and colleagues (1991). A similar effect of water restriction to increase aquaporin-2 abundance in rat kidney was also demonstrated by immunocytochemistry (Hayashi *et al.*, 1994).

Additional studies of the long-term regulation of collecting duct water permeability have shown a direct role for vasopressin in the regulation of aquaporin-2 water channel abundance. For example, a continuous 5-day infusion of arginine vasopressin into Brattleboro rats produced both a 3-fold increase in aquaporin-2 abundance and a 3-fold increase in maximal water permeability of inner medullary collecting ducts from these rats (DiGiovanni *et al.*, 1994). Time-course studies using single-tubule ELISA techniques showed that the effect of vasopressin infusions to increase abundance of aquaporin-2 protein in the IMCD required about 5 days for full development (Kishore *et al.*, 1996b). Beyond this, in rats that manifest an absolute lack of circulating vasopressin (Brattleboro strain), restriction of water intake did not increase aquaporin-2 abundance, leading to the conclusion that vasopressin is necessary for the long-term response to water restriction (Chou *et al.*, 1995). The increase in aquaporin-2 abundance normally seen with restriction of water intake in normal rats was blocked by simultaneous administration of a V2-selective receptor antagonist (Hayashi *et al.*, 1994). Administration of the V2-selective agonist dDAVP to rats elicited a large increase in renal aquaporin-2 protein (Ecelbarger *et al.*, 1996b) and mRNA levels (Fujita *et al.*, 1995). Finally, 5-day infusion of vasopressin stimulated increases in aquaporin-2 abundance in the cortex of the rat kidney that were similar to those found in the medulla, showing that the vasopressin-induced increase in aquaporin-2 abundance does not depend on increased solute concentrations in the tissue surrounding the collecting ducts (Terris *et al.*, 1996). Thus, the evidence, viewed together, establishes that the increase in aquaporin-2 water channel abundance resulting from water restriction is mediated largely if not entirely by binding of vasopressin to the V2 receptor in collecting duct principal cells.

Aquaporin-3 is also a target for long-term regulation by vasopressin. A large increase in aquaporin-3 protein abundance was found in rats following restriction of water intake (Ecelbarger *et al.*, 1995) and after a 5-day infusion of arginine vasopressin (Terris *et al.*, 1996). Thus, long-term increases in circulating vasopressin levels result in an adaptive increase in maximal water permeability in collecting ducts, which is a consequence of increases in the abundance of both the aquaporin-2 and aquaporin-3 water channels in collecting duct principal cells.

A substantial quantity of evidence supports the conclusion that increases in aquaporin-2 and aquaporin-3 abundance in the kidney in response to long-term increases in plasma vasopressin levels are the consequences of increases in aquaporin-2 and aquaporin-3 mRNA levels in collecting duct cells. Northern blotting studies have shown that water restriction in rats is associated with increased renal medullary aquaporin-2 mRNA (Ma *et al.*, 1994b; Yamamoto, T., *et al.*, 1995), and similar increases in aquaporin-2 mRNA have been seen in response to dDAVP infusion (Ecelbarger *et al.*, 1997; Fujita *et al.*, 1995). Theoretically, the vasopressin-induced increase in aquaporin-2 mRNA expression could be a result of either vasopressin-induced increases in the rate of aquaporin-2 gene transcription or vasopressin-induced decreases in the rate of aquaporin-2 mRNA degradation. The 3'-untranslated region of the aquaporin-2 mRNA does not contain motifs that suggest regulation of mRNA stability, however. Therefore, it is very likely that vasopressin directly increases aquaporin-2 gene transcription. However, at this writing there is no direct evidence to support this conclusion. Transcriptional regulation is thought to be a result of vasopressin-induced increases in intracellular cyclic AMP with concomitant increases in activation of PKA. This view is supported by the results of molecular cloning and sequencing of the 5'-flanking region of the aquaporin-2 gene, which has demonstrated the presence of a putative cAMP-response element (CRE) motif (Uchida *et al.*, 1994). Studies involving expression of promoter/luciferase-reporter constructs in cultured renal epithelial cells indicate that activation of the CRE can increase aquaporin-2 gene transcription in the response to increased intracellular cyclic AMP levels (Hozawa *et al.*, 1996; Matsumura *et al.*, 1997; Yasui *et al.*, 1997).

Increased aquaporin-3 protein abundance in response to dDAVP infusion has been demonstrated to be associated with increased aquaporin-3 mRNA levels in the kidney (Ecelbarger *et al.*, 1996b). In contrast to the aquaporin-2 gene, however, no CRE has been identified in the 5' regulatory region of the aquaporin-3 gene (Inase *et al.*, 1995). Nevertheless, the 5'-flanking region of the aquaporin-3 gene contains potential AP2 and Sp1 regulatory elements, both of which have been previously demonstrated to be associated with cyclic AMP-mediated transcriptional regulation under some circumstances (Chang *et al.*, 1992; Imagawa *et al.*, 1987; Roesler *et al.*, 1988; Shiotani and Merchant, 1995).

E. Various Water Balance Abnormalities Are Associated with Altered Regulation of Both the Short- and Long-Term Modes of Aquaporin Regulation in the Collecting Duct

The availability of antibodies and cDNA probes for the aquaporins expressed in the kidney has permitted extensive study of the role of aquaporins in the pathophysiology of these disorders. In general, most of the disorders studied have been demonstrated to be due to altered regulation of aquaporin-2, either due to abnormalities in the short-term or long-term regulatory processes associated with vasopressin action. A treatment of these abnormalities is provided in another chapter in this volume by Nielsen and his colleagues (Chapter 4).

Acknowledgments

The work reported herein was supported by the intramural budget of the National Heart, Lung and Blood Institute (to M.A.K., Project Z01-HL-01282-KE). The authors thank Maurice B. Burg for his critical review of the manuscript.

References

Agre, P., Saboori, A. M., Asimos, A., and Smith, B. L. (1987). Purification and partial characterization of the M_r 30,000 integral membrane protein associated with the erythrocyte Rh(D) antigen. *J. Biol. Chem.* **262**, 17497–17503.

Andreoli, T. E., Schafer, J. A., and Troutman, S. L. (1978). Perfusion rate-dependence of transepithelial osmosis in isolated proximal convoluted tubules: Estimation of the hydraulic conductance. *Kidney Int.* **14,** 263–269.

Bajjalieh, S. M., and Scheller, R. H. (1995). The biochemistry of neurotransmitter secretion. *J. Biol. Chem.* **270,** 1971–1974.

Brown, D., Weyer, P., and Orci, L. (1988). Vasopressin stimulates endocytosis in kidney collecting duct principal cells. *Eur. J. Cell Biol.* **46,** 336–341.

Burg, M. B., and Green, N. (1973). Function of the thick ascending limb of Henle's loop. *Am. J. Physiol.* **224,** 659–668.

Chabardes, D., Firsov, D., Aarab, L., Clabecq, A., Bellanger, A. C., Siaume-Perez, S., and Elalouf, J. M. (1996). Localization of mRNAs encoding Ca^{2+}-inhibitable adenylyl cyclases along the renal tubule. Functional consequences for regulation of the cAMP content. *J. Biol. Chem.* **271,** 19264–19271.

Champigneulle, A., Siga, E., Vassent, G., and Imbert-Teboul, M. (1993). V_2-like vasopressin receptor mobilizes intracellular Ca^{2+} in medullary collecting tubules. *Am. J. Physiol.* **265,** F35–F45.

Chang, C. Y., Huang, C., Guo, I. C., Tsai, H. M., Wu, D. A., and Chung, B. C. (1992). Transcription of the human ferredoxin gene through a single promoter which contains the $3',5'$-cyclic adenosine monophosphate-responsive sequence and Sp 1-binding site. *Mol. Endocrinol.* **6,** 1362–1370.

Chou, C.-L., and Knepper, M. A. (1992). *In vitro* perfusion of chinchilla thin limb segments: Segmentation and osmotic water permeability. *Am. J. Physiol.* **263,** F417–F426.

Chou, C.-L., Nielsen, S., and Knepper, M. A. (1993). Structural–functional correlation in thin limbs of the chinchilla long-loop of Henle: Evidence for a novel papillary subsegment. *Am. J. Physiol.* **265,** F863–F874.

Chou, C.-L., DiGiovanni, S. R., Mejia, R., Nielsen, S., and Knepper, M. A. (1995). Oxytocin as an antidiuretic hormone: I. Concentration dependence of action. *Am. J. Physiol.* **268,** F70–F77.

Chou, C. L., Ma, T., Yang, B., Knepper, M. A., and Verkman, A. S. (1998a). Fourfold reduction of water permeability in inner medullary collecting duct of aquaporin-4 knockout mice. *Am. J. Physiol.* **274,** C549–C554.

Chou, C.-L., Rapko, S. I., and Knepper, M. A. (1998b). Phosphoinositide signaling in rat inner medullary collecting duct. *Am. J. Physiol.* **274,** F564–F572.

Chou, C. L., Knepper, M. A., van Hoek, A. N., Brown, D., Ma, T., and Verkman, A. S. (1999a). Reduced water permeability and altered ultrastructure in thin descending limb of Henle in aquaporin-1 null mice. *J. Clin. Invest.* **103,** 491–496.

Chou, C.-L., Yip, K.-P., and Knepper, M. A. (1999b). Role of Ca/calmodulin in vasopressin-stimulated aquaporin-2 trafficking in rat collecting duct. *J. Am. Soc. Nephrol.* **10,** 13A.

Chou, C.-L., Kador, K., Wade, J., and Knepper, M. A. (2000a). Colocalization of type 1 ryanodine receptor with aquaporin-2 in rat inner medullary collecting duct. *FASEB J.* **14,** A345.

Chou, C.-L., Yip, K.-P., Michea, L., Kador, K., Ferraris, J., Wade, J. B, and Knepper, M. A. (2000b). Regulation of aquaporin-2 trafficking by vasopressin in renal collecting duct: Roles of ryanodine-sensitive. Ca^{2+} stores and calmodulin. *J. Biol. Chem.* **275,** 36839–36846.

Cooper, D. M. F., Mons, N., and Karpen, J. W. (1995). Adenylyl cyclases and the interaction between calcium and cAMP signalling. *Nature* **374,** 421–424.

Deen, P. M., Verdijk, M. A., Knoers, N. V., Wieringa, B., Monnens, L. A., van Os, C. H., and van Oost, B. A. (1994). Requirement of human renal water channel aquaporin-2 for vasopressin-dependent concentration of the urine. *Science* **264,** 92–95.

Denker, B. M., Smith, B. L., Kuhajda, F. P., and Agre, P. (1988). Identification, purification, and partial characterization of a novel M_r 28,000 integral membrane protein from erythrocytes and renal tubules. *J. Biol. Chem.* **263,** 15634–15642.

DiGiovanni, S. R., Nielsen, S., Christensen, E. I., and Knepper, M. A. (1994). Regulation of collecting duct water channel expression by vasopressin in Brattleboro rat. *Proc. Natl. Acad. Sci. USA* **91**, 8984–8988.

Ding, G. H., Franki, N., Condeelis, J., and Hays, R. M. (1991). Vasopressin depolymerizes F-actin in toad bladder epithelial cells. *Am. J. Physiol.* **260**, C9–C16.

Ecelbarger, C. A., Terris, J., Frindt, G., Echevarria, M., Marples, D., Nielsen, S., and Knepper, M. A. (1995). Aquaporin-3 water channel localization and regulation in rat kidney. *Am. J. Physiol.* **269**, F663–F672.

Ecelbarger, C. A., Chou, C.-L., Lolait, S. J., Knepper, M. A., and DiGiovanni, S. R. (1996a). Evidence for dual signaling pathways for V_2 vasopressin receptor in rat inner medullary collecting duct. *Am. J. Physiol.* **270**, F623–F633.

Ecelbarger, C. A., Knepper, M. A., Nielsen, S., Olsen, B. R., Baker, E. A., and Verbalis, J. G. (1996b). The vasopressin-escape phenomenon is associated with marked downregulation of aquaporin-2 water channel expression (abstract). *J. Am. Soc. Nephrol.* **7**, 1266.

Ecelbarger, C. A., Nielsen, S., Olson, B. R., Murase, T., Baker, E. A., Knepper, M. A., and Verbalis, J. G. (1997). Role of renal aquaporins in escape from vasopressin-induced antidiuresis in rat. *J. Clin. Invest.* **99**, 1852–1863.

Epstein, F. H., Kleeman, C. R., and Hendrikx, A. (1957). The influence of bodily hydration on the renal concentrating process. *J. Clin. Invest.* **36**, 629–634.

Finkelstein, A. (1976). Water and nonelectrolyte permeability of lipid bilayer membranes. *J. Gen. Physiol.* **68**, 127–135.

Flamion, B., Spring, K. R., and Abramow, M. (1995). Adaptation of inner medullary collecting duct to dehydration involves a paracellular pathway. *Am. J. Physiol.* **268**, F53–F63.

Foster, L. J., Yeung, B., Mohtashami, M., Ross, K., Trimble, W. S., and Klip, A. (1998). Binary interactions of the SNARE proteins syntaxin-4, SNAP23, and VAMP-2 and their regulation by phosphorylation. *Biochemistry* **37**, 11089–11096.

Frigeri, A., Gropper, M. A., Turck, C. W., and Verkman, A. S. (1995). Immunolocalization of the mercurial-insensitive water channel and glycerol intrinsic protein in epithelial cell plasma membranes. *Proc. Natl. Acad. Sci. USA* **92**, 4328–4331.

Fujita, N., Ishikawa, S. E., Sasaki, S., Fujisawa, G., Fushimi, K., Marumo, F., and Saito, T. (1995). Role of water channel AQP-CD in water retention in SIADH and cirrhotic rats. *Am. J. Physiol.* **269**, F926–F931.

Fushimi, K., Uchida, S., Hara, Y., Hirata, Y., Marumo, F., and Sasaki, S. (1993). Cloning and expression of apical membrane water channel of rat kidney collecting tubule. *Nature* **361**, 549–552.

Fushimi, K., Sasaki, S., and Marumo, F. (1997). Phosphorylation of serine 256 is required for cAMP-dependent regulatory exocytosis of the aquaporin-2 water channel. *J. Biol. Chem.* **272**, 14800–14804.

Gao, Y., Franki, N., Macaluso, F., and Hays, R. M. (1992). Vasopressin decreases immunogold labeling of apical actin in the toad bladder granular cell. *Am. J. Physiol.* **263**, C908–C912.

Grantham, J. J., and Burg, M. B. (1966). Effect of vasopressin and cyclic AMP on permeability of isolated collecting tubules. *Am. J. Physiol.* **211**, 255–259.

Green, R., and Giebisch, G. (1984). Luminal hypotonicity: A driving force for fluid absorption from the proximal tubule. *Am. J. Physiol.* **246**, F167–F174.

Green, R., and Giebisch, G. (1989). Reflection coefficients and water permeability in rat proximal tubule. *Am. J. Physiol.* **257**, F658–F668.

Gross, J. B., Imai, M., and Kokko, J. P. (1975). A functional comparison of the cortical collecting tubule and the distal convoluted tubule. *J. Clin. Invest.* **55**, 1284–1294.

Hai, M. A., and Thomas, S. (1969). The time-course of changes in renal composition during lysine vasopressin infusion in the rat. *Pfluegers Arch.* **310**, 297–319.

Han, J. S., Maeda, Y., Ecelbarger, C., and Knepper, M. A. (1994). Vasopressin-independent regulation of collecting duct water permeability. *Am. J. Physiol.* **266**, F139–F146.

Hayashi, M., Sasaki, S., Tsuganezawa, H., Monkawa, T., Kitajima, W., Konishi, K., Fushimi, K., Marumo, F., and Saruta, T. (1994). Expression and distribution of aquaporin of collecting duct are regulated by V_2 receptor in rat kidney. *J. Clin. Invest.* **94,** 1778–1783.

Hays, R. M., and Lindberg, U. (1991). Actin depolymerization in the cyclic AMP-stimulated toad bladder epithelial cell, determined by the DNAse method. *FEBS Lett.* **280,** 397–399.

Hirling, H., and Scheller, R. H. (1996). Phosphorylation of synaptic vesicle proteins: Modulation of the alpha SNAP interaction with the core complex. *Proc. Natl. Acad. Sci. USA* **93,** 11945–11949.

Holmgren, K., Magnusson, K. E., Franki, N., and Hays, R. M. (1992). ADH-induced depolymerization of F-actin in the toad bladder granular cell: A confocal microscope study. *Am. J. Physiol.* **262,** C672–C677.

Hozawa, S., Holtzman, E. J., and Ausiello, D. A. (1996). cAMP motifs regulating transcription in the aquaporin-2 gene. *Am. J. Physiol.* **270,** C1695–C1702.

Imagawa, M., Chiu, R., and Karin, M. (1987). Transcription factor AP-2 mediates induction by two different signal-transduction pathways: Protein kinase C and cAMP. *Cell* **51,** 251–260.

Imai, M. (1979). The connecting tubule: A functional subdivision of the rabbit distal nephron segments. *Kidney Int.* **15,** 346–356.

Imai, M., and Kokko, J. P. (1976). Mechanism of sodium and chloride transport in the thin ascending limb of Henle. *J. Clin. Invest.* **58,** 1054–1060.

Imai, M., Hayashi, M., and Araki, M. (1984). Functional heterogeneity of the descending limbs of Henle's loops. I. Internephron heterogeneity in the hamster kidney. *Pflügers Arch.* **402,** 385–392.

Inase, N., Fushimi, K., Ishibashi, K., Uchida, S., Ichioka, M., Sasaki, S., and Marumo, F. (1995). Isolation of human aquaporin 3 gene. *J. Biol. Chem.* **270,** 17913–17916.

Inoue, T., Nielsen, S., Mandon, B., Terris, J., Kishore, B. K., and Knepper, M. A. (1998). SNAP-23 in rat kidney: Colocalization with aquaporin-2 in collecting duct vesicles. *Am. J. Physiol.* **275,** F752–F760.

Inoue, T., Terris, J., Ecelbarger, C. A., Chou, C.-L., Nielsen, S., and Knepper, M. A. (1999). Vasopressin regulates apical targeting of aquaporin-2 but not of UT1 urea transporter in renal collecting duct. *Am. J. Physiol.* **276,** F559–F566.

Ishibashi, K., Sasaki, S., Fushimi, K., Uchida, S., Kuwahara, M., Saito, H., Furukawa, T., Nakajima, K., Yamaguchi, Y., Gojobori, T., and Marumo, F. (1994). Molecular cloning and expression of a member of the aquaporin family with permeability to glycerol and urea in addition to water expressed at the basolateral membrane of kidney collecting duct cells. *Proc. Natl. Acad. Sci. USA* **91,** 6269–6273.

Jamison, R. L. (1970). Micropuncture study of superficial and juxtamedullary nephrons in the rat. *Am. J. Physiol.* **218,** 46–55.

Jo, I., Harris, H. W., Amendt-Raduege, A. M., Majewski, R. R., and Hammond, T. G. (1995). Rat kidney papilla contains abundant synaptobrevin protein that participates in the fusion of antidiuretic hormone-regulated water channel-containing endosomes in vitro. *Proc. Natl. Acad. Sci. USA* **92,** 1876–1880.

Jones, R. V. H., and De Wardener, H. E. (1956). Urine concentration after fluid deprivation or pitressin tannate in oil. *Br. Med. J.* **1,** 271–274.

Katsura, T., Gustafson, C. E., Ausiello, D. A., and Brown, D. (1997). Protein kinase A phosphorylation is involved in regulated exocytosis of aquaporin-2 in transfected LLC-PK1 cells. *Am. J. Physiol.* **272,** F817–F822.

Kim, J., Kim, W.-Y., Han, K.-H., Knepper, M. A., Nielsen, S., and Madsen, K. M. (1999). Developmental expression of aquaporin-1 in the rat renal vasculature. *Am. J. Physiol.* **276,** F498–F509.

Kishore, B. K., Mandon, B., Oza, N. B., DiGiovanni, S. R., Coleman, R. A., Ostrowski, N. L., Wade, J. B., and Knepper, M. A. (1996a). Rat renal arcade segment expresses vasopressin-regulated water channel and vasopressin V_2 receptor. *J. Clin. Invest.* **97,** 2763–2771.

Kishore, B. K., Terris, J. M., and Knepper, M. A. (1996b). Quantitation of aquaporin-2 abundance in microdissected collecting ducts: Axial distribution and control by AVP. *Am. J. Physiol.* **271,** F62–F70.

Klussmann, E., Maric, K., Wiesner, B., Beyermann, M., and Rosenthal, W. (1999). Protein kinase A anchoring proteins are required for vasopressin-mediated translocation of aquaporin-2 into cell membranes of renal principal cells. *J. Biol. Chem.* **274,** 4934–4938.

Knepper, M. A. (1982). Measurement of osmolality in kidney slices using vapor pressure osmometry. *Kidney Int.* **21,** 653–655.

Knepper, M. A. (1994). The aquaporin family of molecular water channels. *Proc. Natl. Acad. Sci. USA* **91,** 6255–6258.

Knepper, M. A. (1997). Molecular physiology of urinary concentrating mechanism: Regulation of aquaporin water channels by vasopressin. *Am. J. Physiol.* **272,** F3–F12.

Knepper, M., and Burg, M. (1983). Organization of nephron function. *Am. J. Physiol.* **244,** F579–F589.

Knepper, M. A., and Nielsen, S. (1993). Kinetic model of water and urea permeability regulation by vasopressin in collecting duct. *Am. J. Physiol.* **265,** F214–F224.

Knepper, M. A., Chou, C.-L., and Layton, H. E. (1993). How is urine concentrated by the renal inner medulla? *Contr. Nephrol.* **102,** 144–160.

Knepper, M. A., Nielsen, S., Chou, C.-L., and DiGiovanni, S. R. (1994). Mechanism of vasopressin action in the renal collecting duct. *Semi. Nephrol.* **14,** 302–321.

Knepper, M. A., Valtin, H., and Sands, J. M. (2000). Renal actions of vasopressin. *In* "Handbook of Physiology: The Endocrine System (Section 7)" (J. C. S. Fray, and H. M. Goodman, eds.), pp. 496–529. Oxford, New York.

Kokko, J. P. (1970). Sodium chloride and water transport in the descending limb of Henle. *J. Clin. Invest.* **49,** 1838–1846.

Kriz, W. (1981). Structural organization of the renal medulla: Comparative and functional aspects. *Am. J. Physiol.* **241,** R3–R16.

Kuhn, W., and Ramel, A. (1959). Activer Salztransport als möglicher (und wahrscheinlicher) Einzel-effekt bei der Harnkonzentrierung in der Niere. *Helv. Chim. Acta* **42,** 628–660.

Kuwahara, M., and Verkman, A. S. (1989). Pre-steady-state analysis of the turn-on and turn-off of water permeability in the kidney collecting tubule. *J. Membr. Biol.* **110,** 57–65.

Kuwahara, M., Fushimi, K., Terada, Y., Bai, L., Marumo, F., and Sasaki, S. (1995). cAMP-dependent phosphorylation stimulates water permeability of aquaporin-collecting duct water channel protein expressed in *Xenopus* oocytes. *J. Biol. Chem.* **270,** 10384–10387.

Lande, M. B., Jo, I., Zeidel, M. L., Somers, M., and Harris Jr., H. W. (1996). Phosphorylation of aquaporin-2 does not alter the membrane water permeability of rat papillary water channel-containing vesicles. *J. Biol. Chem.* **271,** 5552–5557.

Lankford, S. P., Chou, C.-L., Terada, Y., Wall, S. M., Wade, J. B., and Knepper, M. A. (1991). Regulation of collecting duct water permeability independent of cAMP-mediated AVP response. *Am. J. Physiol.* **261,** F554–F566.

Lassiter, W. E., Gottschalk, C. W., and Mylle, M. (1961). Micropuncture study of net transtubular movement of water and urea in nondiuretic mammalian kidney. *Am. J. Physiol.* **200,** 1139–1146.

Lencer, W. I., Brown, D., Ausiello, D. A., and Verkman, A. S. (1990). Endocytosis of water channels in rat kidney: Cell specificity and correlation with *in vivo* antidiuresis. *Am. J. Physiol.* **259,** C920–C932.

Liebenhoff, U., and Rosenthal, W. (1995). Identification of Rab3-, Rab5a and synaptobrevin II-like proteins in a preparation of rat kidney vesicles containing the vasopressin-regulated water channel. *FEBS Lett.* **365,** 209–213.

Loffing, J., Loffing-Cueni, D., Macher, A., Hebert, S. C., Olson, B., Knepper, M. A., Rossier, B. C., and Kaissling, B. (2000). Localization of epithelial sodium channel and aquaporin-2 in rabbit kidney cortex. *Am. J. Physiol.* **278,** F530–F539.

Ma, T., Frigeri, A., Hasegawa, H., and Verkman, A. S. (1994a). Cloning of a water channel homolog expressed in brain meningeal cells and kidney collecting duct that functions as a stilbene-sensitive glycerol transporter. *J. Biol. Chem.* **269,** 21845–21849.

Ma, T., Hasegawa, H., Skach, W. R., Frigeri, A., and Verkman, A. S. (1994b). Expression, functional analysis, and *in situ* hybridization of a cloned rat kidney collecting duct water channel. *Am. J. Physiol.* **266,** C189–C197.

Ma, T., Yang, B., Gillespie, A., Carlson, E. J., Epstein, C. J., and Verkman, A. S. (1998). Severely impaired urinary concentrating ability in transgenic mice lacking aquaporin-1 water channels. *J. Biol. Chem.* **273,** 4296–4299.

Maeda, Y., Han, J. S., Gibson, C. C., and Knepper, M. A. (1993). Vasopressin and oxytocin receptors coupled to Ca^{2+} mobilization in rat inner medullary collecting duct. *Am. J. Physiol.* **265,** F15–F25.

Maeda, Y., Smith, B. L., Agre, P., and Knepper, M. A. (1995). Quantification of Aquaporin-CHIP water channel protein in microdissected renal tubules by fluorescence-based ELISA. *J. Clin. Invest.* **95,** 422–428.

Mandon, B., Chou, C.-L., Nielsen, S., and Knepper, M. A. (1996). Syntaxin-4 is localized to the apical plasma membrane of rat renal collecting duct cells: Possible role in aquaporin-2 trafficking. *J. Clin. Invest.* **98,** 906–913.

Marples, D., Knepper, M. A., Christensen, E. I., and Nielsen, S. (1995). Redistribution of aquaporin-2 water channels induced by vasopressin in rat kidney inner medullary collecting duct. *Am. J. Physiol.* **269,** C655–C664.

Marples, D., Schroer, T. A., Ahrens, N., Taylor, A., Knepper, M. A., and Nielsen, S. (1998). Dynein and dynactin colocalize with AQP2 water channels in intracellular vesicles from kidney collecting duct. *Am. J. Physiol.* **274,** F384–F394.

Matsumura, Y., Uchida, S., Rai, T., Sasaki, S., and Marumo, F. (1997). Transcriptional regulation of aquaporin-2 water channel gene by cAMP. *J. Am. Soc. Nephrol.* **8,** 861–867.

Miyamoto, E., Kuo, J. F., and Greengard, P. (1969). Cyclic nucleotide-dependent protein kinases. III. Purification and properties of adenosine $3',5'$-monophosphate-dependent protein kinase from bovine brain. *J. Biol. Chem.* **244,** 6395–6402.

Morgan, T., and Berliner, R. W. (1968). Permeability of the loop of Henle, vasa recta, and collecting duct to water, urea, and sodium. *Am. J. Physiol.* **215,** 108–115.

Morgan, T., Sakai, F., and Berliner, R. W. (1968). *In vitro* permeability of medullary collecting ducts to water and urea. *Am. J. Physiol.* **214,** 574–581.

Nielsen, S., and Knepper, M. A. (1993). Vasopressin activates collecting duct urea transporters and water channels by distinct physical processes. *Am. J. Physiol.* **265,** F204–F213.

Nielsen, S., DiGiovanni, S. R., Christensen, E. I., Knepper, M. A., and Harris, H. W. (1993a). Cellular and subcellular immunolocalization of vasopressin-regulated water channel in rat kidney. *Proc. Natl. Acad. Sci. USA* **90,** 11663–11667.

Nielsen, S., Muller, J., and Knepper, M. A. (1993b). Vasopressin- and cAMP-induced changes in ultrastructure of isolated perfused inner medullary collecting ducts. *Am. J. Physiol.* **265,** F225–F238.

Nielsen, S., Smith, B. L., Christensen, E. I., Knepper, M. A., and Agre, P. (1993c). CHIP28 water channels are localized in constitutively water-permeable segments of the nephron. *J. Cell. Biol.* **120,** 371–383.

Nielsen, S., Chou, C.-L., Marples, D., Christensen, E. I., Kishore, B. K., and Knepper, M. A. (1995a). Vasopressin increases water permeability of kidney collecting duct by inducing translocation of aquaporin-CD water channels to plasma membrane. *Proc. Natl. Acad. Sci. USA* **92,** 1013–1017.

Nielsen, S., Marples, D., Mohtashami, M., Dalby, N. O., Trimble, W., and Knepper, M. (1995b). Expression of VAMP2-like protein in kidney collecting duct intracellular vesicles: Co-localization with aquaporin-2 water channels. *J. Clin. Invest.* **96,** 1834–1844.

Nielsen, S., Pallone, T., Smith, B. L., Christensen, E. I., Agre, P., and Maunsbach, A. B. (1995c). Aquaporin-1 water channels in short and long loop descending thin limbs and in descending vasa recta in rat kidney. *Am. J. Physiol.* **268**, F1023–F1037.

Nielsen, S., Frokiaer, J., and Knepper, M. A. (1998). Renal aquaporins: Key roles in water balance and water balance disorders. *Curr. Opin. Nephrol. Hypertension* **7**, 509–516.

Nishimoto, G., Zelenina, M., Li, D., Yasui, M., Aperia, A., Nielsen, S., and Nairn, A. C. (1999). Arginine vasopressin stimulates phosphorylation of aquaporin-2 in rat renal tissue. *Am. J. Physiol.* **276**, F254–F259.

Nonoguchi, H., Sands, J. M., and Knepper, M. A. (1988). Atrial natriuretic factor inhibits vasopressin-stimulated osmotic water permeability in rat inner medullary collecting duct. *J. Clin. Invest.* **82**, 1383–1390.

Pallone, T. L., Kishore, B. K., Nielsen, S., Agre, P., and Knepper, M. A. (1997). Evidence that aquaporin-1 mediates NaCl-induced water flux across descending vasa recta. *Am. J. Physiol.* **272**, F587–F596.

Pallone, T. L., Edwards, A., Ma, T., Silldorff, E. P., and Verkman, A. S. (2000). Requirement of aquaporin-1 for NaCl-driven water transport across descending vasa recta. *J. Clin. Invest.* **105**, 215–222.

Pennell, J. P., Lacy, F. B., and Jamison, R. L. (1974). An *in vivo* study of the concentrating process in the descending limb of Henle's loop. *Kidney Int.* **5**, 337–347.

Preisig, P. A., and Barry, C. A. (1985). Evidence for transcellular osmotic water flow in rat proximal tubules. *Am. J. Physiol.* **249**, F124–F131.

Preston, G. M., and Agre, P. (1991). Isolation of the cDNA for erythrocyte integral membrane protein of 28 kilodaltons: Member of an ancient channel family. *Proc. Natl. Acad. Sci. USA* **88**, 11110–11114.

Reif, M. C., Troutman, S. L., and Schafer, J. A. (1984). Sustained response to vasopressin in isolated rat cortical collecting tubule. *Kidney Int.* **26**, 725–732.

Risinger, C., and Bennett, M. K. (1999). Differential phosphorylation of syntaxin and synaptosome-associated protein of 25 kDa (SNAP-25) isoforms. *J. Neurochem.* **72**, 614–624.

Rocha, A. S., and Kokko, J. P. (1973). Sodium chloride and water transport in the medullary thick ascending limb of Henle. *J. Clin. Invest.* **52**, 612–624.

Rocha, A. S., and Kokko, J. P. (1974a). Permeability of medullary nephron segments to urea and water: Effect of vasopressin. *Kidney Int.* **6**, 379–387.

Rocha, A. S., and Kokko, J. P. (1974b). Permeability of medullary nephron segments to urea and water—Effect of vasopressin. *Kidney Int.* **6**, 397–387.

Roesler, W. J., Vandenbark, G. R., and Hanson, R. W. (1988). Cyclic AMP and the induction of eukaryotic gene transcription. *J. Biol. Chem.* **263**, 9063–9066.

Sabolic, I., Valenti, G., Verbabatz, J.-M., Van Hoek, A. N., Verkman, A. S., Ausiello, D. A., and Brown, D. (1992). Localization of the CHIP28 water channel in rat kidney. *Am. J. Physiol.* **263**, C1225–C1233.

Sabolic, I., Katsura, T., Verbabatz, J. M., and Brown, D. (1995). The AQP2 water channel: effect of vasopressin treatment, microtubule disruption, and distribution in neonatal rats. *J. Membr. Biol.* **143**, 165–177.

Sands, J. M., Nonoguchi, H., and Knepper, M. A. (1987). Vasopressin effects on urea and H_2O transport in inner medullary collecting duct subsegments. *Am. J. Physiol.* **253**, F823–F832.

Schafer, J. A., and Andreoli, T. E. (1972). Cellular constraints to diffusion. The effect of antidiuretic hormone on water flows in isolated mammalian collecting tubules. *J. Clin. Invest.* **51**, 1264–1278.

Schafer, J. A., Patlak, C. S., Troutman, S. L., and Andreoli, T. E. (1978). Volume absorption in the pars recta. II. Hydraulic conductivity coefficient. *Am. J. Physiol.* **234**, F340–F348.

Schnermann, J., Chou, C.-L., Ma, T., Traynor, T., Knepper, M. A, and Verkman, A. S. (1998). Defective proximal tubular fluid reabsorption in transgenic aquaporin-1 null mice. *Proc. Natl. Acad. Sci. USA* **95**, 9660–9664.

Shimazaki, Y., Nishiki, T., Omori, A., Sekiguchi, M., Kamata, Y., Kozaki, S., and Takahashi, M. (1996). Phosphorylation of 25-kDa synaptosome-associated protein. Possible involvement in protein kinase C-mediated regulation of neurotransmitter release. *J. Biol. Chem.* **271**, 4548–4553.

Shiotani, A., and Merchant, J. L. (1995). cAMP regulates gastrin gene expression. *Am. J. Physiol.* **269**, G458–G464.

Simon, H., Gao, Y., Franki, N., and Hays, R. M. (1993). Vasopressin depolymerizes apical F-actin in rat inner medullary collecting duct. *Am. J. Physiol.* **265**, C757–C762.

Söllner, T., Whiteheart, S. W., Brunner, M., Erdjument-Bromage, H., Geanos, S., Tempst, P., and Rothman, J. E. (1993). SNAP receptors implicated in vesicle targeting and fusion. *Nature* **362**, 318–324.

Star, R. A., Nonoguchi, H., Balaban, R., and Knepper, M. A. (1988a). Calcium and cyclic adenosine monophosphate as second messengers for vasopressin in the rat inner medullary collecting duct. *J. Clin. Invest.* **81**, 1879–1888.

Star, R. A., Nonoguchi, H., Balaban, R., and Knepper, M. A. (1988b). Calcium and cyclic adenosine monophosphate as second messengers for vasopressin in the rat inner medullary collecting duct. *J. Clin. Invest.* **81**, 1879–1888.

Strange, K., Willingham, M. C., Handler, J. S., and Harris, H. W. Jr. (1988). Apical membrane endocytosis via coated pits is stimulated by removal of antidiuretic hormone from isolated, perfused rabbit cortical collecting tubule. *J. Membr. Biol.* **103**, 17–28.

Terris, J., Ecelbarger, C. A., Marples, D., Knepper, M. A., and Nielsen, S. (1995). Distribution of aquaporin-4 water channel expression within rat kidney. *Am. J. Physiol.* **269**, F775–F785.

Terris, J., Ecelbarger, C. A., Nielsen, S., and Knepper, M. A. (1996). Long-term regulation of four renal aquaporins in rat. *Am. J. Physiol.* **271**, F414–F422.

Turner, M. R., and Pallone, T. L. (1997). Hydraulic and diffusional permeabilities of isolated outer medullary descending vasa recta from the rat. *Am. J. Physiol.* **272**, H392–H400.

Uchida, S., Sasaki, S., Fushimi, K., and Marumo, F. (1994). Isolation of human aquaporin-CD gene. *J. Biol. Chem.* **269**, 23451–23455.

Vallon, V., Verkman, A. S., and Schnermann, J. (2000). Luminal hypotonicity in proximal tubule and flow dependence of loop of Henle fluid absorption in aquaporin-1 knockout mice. *FASEB J.* **14**, A345.

van Hoek, A. N., and Verkman, A. S. (1992). Functional reconstitution studies of isolated erythrocyte water channel CHIP28. *J. Biol. Chem.* **267**, 18267–18269.

van Hoek, A. N., Ma, T., Yang, B., Verkman, A. S., and Brown, D. (2000). Aquaporin-4 is expressed in basolateral membranes of proximal tubule S3 segments in mouse kidney. *Am. J. Physiol.* **278**, F310–F316.

Verbalis, J. G. (1993). Osmotic inhibition of neurohypophysial secretion. *Ann. N.Y. Acad. Sci.* **689**, 146–160.

Wade, J. B., Stetson, D. L., and Lewis, S. A. (1981). ADH action: Evidence for a membrane shuttle mechanism. *Ann. N.Y. Acad. Sci.* **372**, 106–117.

Wade, J. B., Nielsen, S., Coleman, R. A., and Knepper, M. A. (1994). Long-term regulation of collecting duct water permeability: Freeze-fracture analysis of isolated tabales. *Am. J. Physiol.* **266**, F223–F230.

Wade, J. B., Lee, A. J., Liu, J., Ecelbarger, C. A., Mitchell, C., Bradford, A. D., Terris, J., Kim, G.-H., and Knepper, M. A. (2000). UTA-2: A 55 kDa urea transporter in thin descending limb whose abundance is regulated by vasopressin. *Am. J. Physiol.* **278**, F52–F62.

Wall, S. M., Han, J. S., Chou, C.-L., and Knepper, M. A. (1992). Kinetics of urea and water permeability activation by vasopressin in rat terminal IMCD. *Am. J. Physiol.* **262**, F989–F998.

Ward, D. T., Hammond, T. G., and Harris, H. W. (1999). Modulation of vasopressin-elicited water transport by trafficking of aquaporin2-containing vesicles. *Ann. Rev. Physiol.* **61**, 683–697.

Yamamoto, N., Sasaki, S., Fushimi, K., Ishibashi, K., Yaiota, E., Kawasaki, K., Maurmo, F., and Kihara, I. (1995). Vasopressin increases AQP-CD water channel in the apical membrane of collecting duct cells without affecting AQP3 distribution in Brattleboro rat. *Am. J. Physiol.* **268,** C1546–C1551.

Yamamoto, T., Sasaki, S., Fushimi, K., Kawasaki, K., Yaoita, E., Oota, K., Hirata, K., Marumo, F., and Kihara, I. (1995). Localization and expression of a collecting duct water channel, aquaporin, in hydrated and dehydrated rats. *Exp. Nephrol.* **3,** 193–201.

Yasui, M., Zelenin, S. M., Celsi, G., and Aperia, A. (1997). Adenylate cyclase-coupled vasopressin activates AQP2 promoter via a dual effect on CRE and AP1 elements. *Am. J. Physiol.* **272,** F443–F450.

Zeidel, M. L., Ambudkar, S. V., Smith, B. L., and Agre, P. (1992). Reconstitution of functional water channels in liposomes containing purified red cell CHIP28 protein. *Biochemistry* **31,** 7436–7440.

Zelenina, M., Christensen, B. M., Palmer, J., Nairn, A. C., Nielsen, S., and Aperia, A. (2000). Prostaglandin E_2 interaction with AVP: Effects on AQP2 phosphorylation and distribution. *Am. J. Physiol. Renal Physiol.* **278,** F388–F394.

CHAPTER 4

Pathophysiology of Renal Aquaporins

Søren Nielsen,[*] Tae-Hwan Kwon,[†] Henrik Hager,[*] Mark A. Knepper,[+] David Marples,[#] and Jørgen Frøkiaer[*]

[*]Department of Cell Biology, University of Aarhus, DK-8000 Aarhus, Denmark; [†]Dongguk University, Kyungju, South Korea; [+]National Heart, Lung and Blood Institute, National Institutes of Health, Bethesda, Maryland, 20892; [#]University of Leeds, Leeds LS2 9JT, United Kingdom

I. INTRODUCTION

The seminal discovery of aquaporin water channels by Agre and colleagues answered a long-standing question about how water specifically crosses biological membranes (Preston and Agre, 1991; Preston *et al.*, 1992). This provided insight, at the molecular and cellular level, into the fundamental physiology of water balance regulation and into the pathophysiology of congenital and acquired water balance disorders. Out of at least 10 aquaporin isoforms at least 6 are known to be present in the kidney at distinct sites along the nephron and collecting duct. AQP1 is extremely abundant in the proximal tubule and descending thin limb where it appears to be the main site for proximal nephron water reabsorption (Nielsen *et al.*, 1993a). It is also present in the descending vasa recta (Nielsen *et al.*, 1995a). AQP2 (Fushimi *et al.*, 1993) is abundant in the collecting duct principal cells and is the chief target for the regulation of collecting duct water reabsorption by vasopressin (Nielsen *et al.*, 1993b, 1995b). Acute regulation involves vasopressin-induced trafficking of AQP2 between an intracellular reservoir in vesicles and the apical plasma membrane (Fig. 1). In addition AQP2 is involved in chronic/adaptational control of body water balance, which is achieved through regulation of AQP2 expression (Marples *et al.*, 1999). Importantly, multiple studies have now underscored a critical role of AQP2 in several inherited and acquired water balance disorders (Table I). This includes inherited forms of nephrogenic diabetes insipidus (NDI), acquired states of NDI, and other diseases associated with urinary concentrating defects where AQP2 expression and targeting is affected (Fig. 2). Conversely, AQP2 expression and targeting appear to be increased in some conditions with water retention such as pregnancy and congestive heart failure (Fig. 2). AQP3 (Echevarria *et al.*, 1994; Ishibashi *et al.*, 1994) and AQP4 (Hasegawa *et al.*, 1994; Jung *et al.*, 1994) are basolateral water channels located in the kidney collecting duct and representing exit pathways for water reabsorbed via AQP2 (Ecelbarger

---

FIGURE 1 Regulation of AQP2 trafficking and expression in collecting duct principal cells. Vasopressin (AVP) acts on V_2-receptors (V_2R) in the basolateral plasma membrane (Knepper *et al.*, 1994; Marples *et al.*, 1999; Nielsen *et al.*, 1999). Through the GTP-binding protein G_s adenylyl cyclase (A.C.) is activated, which accelerates the production of cyclic AMP (cAMP) from ATP. Then cAMP binds to the regulatory subunit of protein kinase A (PKA), which activates a catalytic subunit of PKA. PKA phosphorylates AQP2 in intracellular vesicles and possible other cytosolic or membrane proteins. Specifically cAMP participates in the long-term regulation of AQP2 by increasing the levels of the catalytic subunit of PKA in the nuclei, which is thought to phosphorylate transcription factors such as CREB-P (cyclic AMP responsive element binding protein) and C-Jun/c-Fos. Binding of these factors is thought to increase gene transcription of AQP2 resulting in synthesis of AQP2 protein, which in turn enters the regulated trafficking system. In parallel, AQP3 synthesis and trafficking to the basolateral plasma membrane takes place (not indicated). AQP2 is also excreted into urine or recycled from the apical plasma membrane.

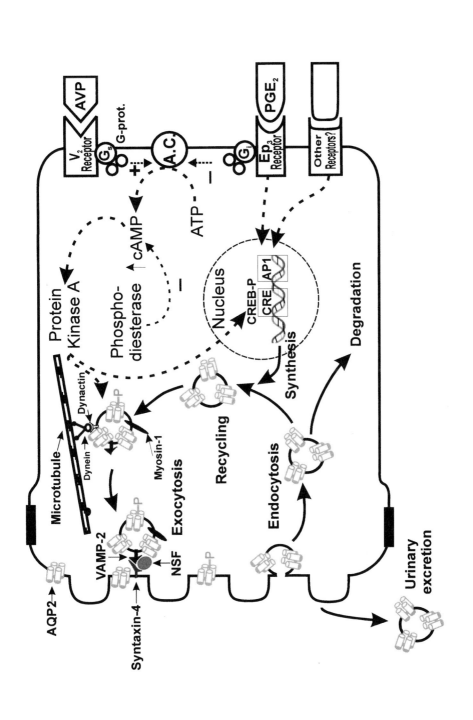

TABLE I

Water Balance Disorders or Conditions with Altered Water Balance Which Are Associated with Dysregulation (Altered Abundance and/or Targeting) of AQP2

Reduced abundance of AQP2	Increased abundance of AQP2
Conditions demonstrating reduced AQP2 abundance and polyuria	**Conditions demonstrating increased AQP2 abundance with expansion of extracellular fluid volume**
Genetic defects	Vasopressin infusion (SIADH)
Brattleboro rats (central DI)	Congestive heart failure
DI +/+ Severe mice (low cAMP)	Hepatic cirrhosis (CCL$_4$-induced,
AQP2 mutants (human)	noncompensated) ?
V$_2$-receptor variants (human)[a]	Pregnancy
Acquired NDI (rat models)	**Conditions demonstrating maintained or increased AQP2 abundance with polyuria**
Lithium	Diabetes mellitus (osmotic diuresis)
Hypokalemia	Furosemide treatment (1–5 days)
Hypercalcemia	
Postobstructive NDI	
Bilateral	
Unilateral	
Water loading (compulsive water drinking)	
Chronic renal failure (5/6 nephrectomy model)	
Ischemia or cisplatin-induced acute renal failure (polyuric phase in rat model)	
Calcium channel blocker (nifedipine) treatment (rat model)	
Age-induced NDI[b]	
Conditions demonstrating reduced AQP2 abundance and altered urinary concentration without polyuria	
Nephrotic syndrome models (rat models)	
PAN-induced	
Adriamycin-induced	
Hepatic cirrhosis (CBDL, compensated)	
Ischemia-induced acute renal failure (oliguric phase in rat model)	
Low protein diet (urinary concentrating defect without polyuria)	

Key: CBDL, common bile duct ligation; CCl$_4$, carbon tetrachloride; DI, diabetes insipidus; DM, diabetes mellitus; NDI, nephrogenic diabetes insipidus; PAN, puromycin aminonucleoside; SIADH, syndrome of inappropriate secretion of antidiuretic hormone.

[a]Reduced V$_2$-receptor density has profound effect on the AQP2 targeting and expression.

[b]Mild increase in urine production rates.

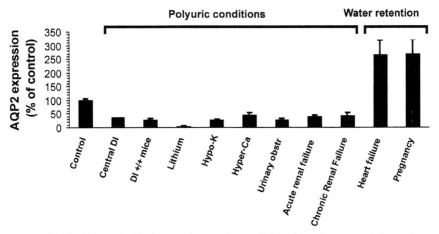

FIGURE 2 Changes in AQP2 expression seen in association with different water balance disorders. Levels are expressed as a percentage of control levels (*leftmost bar*). AQP2 expression is reduced in a wide range of hereditary and acquired forms of diabetes insipidus characterized by different degrees of polyuria. Conversely, congestive heart failure and pregnancy are conditions associated with increased expression of AQP2 levels and excessive water retention.

et al., 1995; Terris *et al.,* 1995). Several additional aquaporins have been identified in rat kidney, including AQP6, which is expressed at lower abundance at several sites including the collecting duct intercalated cells and in the proximal tubule, in an entirely intracellular distribution (Yasui *et al.,* 1999a, 1999b). Four additional aquaporins [AQP7 (Ishibashi *et al.,* 1997a), AQP8 (Ishibashi *et al.,* 1997b; Koyama *et al.,* 1997), AQP9 (Ishibashi *et al.,* 1998; Tsukaguchi *et al.,* 1998), and AQP10 (Ishibashi *et al.,* 1999)] are also expressed in kidney, but less is known about expression sites and their pathophysiological function. In this review we focus mainly on the role of collecting duct aquaporins in the pathophysiology of water balance disorders. This review represents an update of several previous reviews focusing on these issues (Agre, 2000; Agre *et al.,* 1998, 1999; Borgnia *et al.,* 1999; Deen and Knoers, 1998; Frokiaer *et al.,* 1998; King *et al.,* 2000; Knepper *et al.,* 1997; Marples *et al.,* 1999; Nielsen *et al.,* 1999).

II. PATHOPHYSIOLOGY OF RENAL AQUAPORINS

There are a variety of disorders in which water handling is abnormal. Some of these are primary renal disorders, whereas others reflect changes in other organs or systems, but the renal handling of water may be considered abnormal in all of them, since there is a breakdown of the normal homeostatic mechanisms. The role of changes in the expression and/or function of aquaporins has been investigated in a range of these conditions, including genetic defects, acquired defects of renal responsiveness (acquired NDI), and conditions in which there is an inappropriate retention of water.

III. INHERITED NDI AND CDI

Two forms of inherited diabetes insipidus (DI) can be encountered: central and nephrogenic. In central or neurogenic DI there is a complete or partial defect in vasopressin production or release. Central diabetes insipidus (CDI) is rarely hereditary in man, usually occurring as a consequence of head trauma or disease in the hypothalamus or pituitary. A rat strain was discovered more than 30 years ago (the Brattleboro rat) by Valtin and colleagues, and this rat model provides an important model of this condition. These animals have a total or near-total lack of vasopressin production that is aggravated with age, although even at a very young age these rats display a dramatic polyuria. In a study by DiGiovanni *et al.* (1994) it was demonstrated that AQP2 levels in Brattleboro rats were substantially lower than those in the parent strain (Long Evans), and that this deficit was reversed by chronic vasopressin infusion. Thus in parallel with a restoration of urinary concentration, AQP2 levels were increased, AQP2 targeting was increased, and the osmotic water permeability was increased. Thus this strongly supports the view that patients lacking vasopressin are likely to have decreased AQP2 expression and especially reduced targeting. Later immunoelectron microscopic studies revealed that the major defect in these rats is the absence of vasopressin-mediated AQP2 targeting to the apical plasma membrane. Less than a few percent of total AQP2 levels is present in the apical plasma membrane, and this is many-fold lower than the fraction present in the apical plasma membrane in vasopressin-treated Brattleboro rats or in normal rat strains (Gronbeck *et al.,* 1998; Sabolic *et al.,* 1995; Yamamoto *et al.,* 1995).

It has also become clear that AQP3 is regulated by vasopressin (Ecelbarger *et al.,* 1995) and recent studies in transgenic mice lacking AQP3 revealed that these mice exhibit a major urinary concentrating defect (Ma *et al.,* 2000). Thus it is entirely possible that there will also be a reduced expression of AQP3 in vasopressin-deficient states such as CDI. However, since the rate-limiting barrier is the apical plasma membrane, the major defect is likely to be caused by the reduced trafficking of AQP2. Obviously treatment with exogenous vasopressin is a very effective cure of the polyuria.

The second form of DI is called nephrogenic DI and is caused by the inability of the kidney to respond to vasopressin stimulation. Both acquired and inherited forms exist with the acquired forms being much more common than the relatively rare inherited forms. The most common hereditary cause of NDI is due to mutations in the vasopressin V_2 receptor, rendering the collecting duct cells insensitive to the vasopressin. This gene is found on the X chromosome in man and shows the classical pattern expected, with males affected much more often and with females being asymptomatic carriers because they have inherited a normal gene from their father. Although there is no direct evidence, it is likely that this form of NDI will be associated with decreased expression of AQP2 expression and targeting because the cells are unable to respond to circulating vasopressin. Consistent with this, urinary AQP2 levels are very low in patients with X-linked NDI (Deen *et al.,* 1995; Kanno *et al.,* 1995). However, because the amount of AQP2 in the urine appears to be determined largely by the response of the collecting duct cells to vasopressin (Wen *et al.,* 1999) rather than their content of AQP2, these data must be interpreted with caution with respect to predicting AQP2 expression levels. Rather this reflects absence of trafficking of AQP2 and hence less excretion. It may be possible to increase AQP2 expression by stimulating non-vasopressin-mediated pathways, but the delivery of AQP2 is likely to remain low, so this may be of limited value. Furthermore, such pathways may well be maximally activated in these patients anyway because they will tend to be rather dehydrated much of the time.

About 10% of hereditary NDI is caused by mutations in the aquaporin-2 gene. In most cases, this results in an autosomal recessive condition (Deen *et al.,* 1994; van Lieburg *et al.,* 1994), although a few cases have been reported that show an autosomal dominant inheritance (Mulders *et al.,* 1998). It is thought that the recessive forms on NDI are due to mutations in which either the mutant protein is unable to form tetramers with the normal form, or in which the normal protein functions normally. In contrast, in the dominant cases, it has been shown that heterotetramers between the mutant and normal forms can be created, but that they are unable to be targeted to the plasma membrane (Kamsteeg *et al.,* 1999). Thus the mutant protein can prevent the function of the normal form. It has raised a potential possibility that mutant AQP2 being impaired in trafficking could be rescued by chemical treatment with chaperone acting substances (Tamarappoo *et al.,* 1999). More work is necessary on this.

IV. ACQUIRED NDI

Under a variety of circumstances the responsiveness of the kidney to vasopressin can become impaired. Often this is a side effect of drug treatment, with lithium being the commonest example, but it can also be caused by electrolyte disturbances or dietary effects, and it can also arise in association with other urinary tract disorders. The defining characteristic of such acquired NDI is an increased urine output and reduced urine osmolality, despite normal or elevated levels of circulating vasopressin. Although these forms of NDI are rarely as extreme as the hereditary forms, they can be very disruptive of patients' lifestyles and can be severe enough to necessitate a change in treatment. It is also worth remembering that, unlike hereditary NDI, which is rare, this is a fairly widespread problem in clinical practice. An interesting aspect of many of these conditions is that the concentrating defect may persist for weeks or even months after correction of the underlying defect, and the changes seen in animal models go some way to explain this.

To investigate the role of aquaporins in such conditions, various rat models have been used. A general finding has been that AQP2 levels are decreased, sometimes dramatically so. However, the degree to which the acute shuttling response is disrupted is quite variable, and the degree of polyuria produced by different treatments broadly reflects the combination of the changes in total AQP2 and the efficiency with which it can be inserted into the apical plasma membrane. These results are interesting both for the implication that different therapeutic approaches may be needed for different conditions, and because they indicate that the signals involved in the down-regulation and trafficking of AQP2 are different and may indeed be different in the different conditions. Some of these lessons may also be valuable in interpreting other conditions in which urine output is normal, but maximal concentrating capacity is impaired (see Table I); these will be discussed in a later section.

A. Lithium-Induced NDI

The use of lithium is very common in the treatment of manic-depressive disorder, and about 1 in 1000 of the population will receive lithium at some time during their lives. Of these, some 20–30% develop clinically significant polyuria (Boton *et al.*, 1987; Christensen *et al.*, 1985), which has been shown to be vasopressin resistant. Rats provide a good, if extreme, model for this condition: After a month on lithium, and with serum lithium levels in the therapeutic range, rats produced a daily urine output that matched their own weight (Marples *et al.*, 1995). In such animals the levels of both AQP2 and AQP3 were progressively reduced to approximately 5% of the levels in control rats after 25 days of lithium treatment (Fig. 3) (Kwon *et al.*, 2000; Marples *et al.*, 1995). In this model the delivery of AQP2 to the apical plasma membrane was also impaired: Quantitative immunoelectron microscopy showed that a smaller fraction of the little AQP2 remaining than in the controls was in the plasma membrane, as is required if it is to mediate water reabsorption (Marples *et al.*, 1995). Immunocytochemistry also confirmed the reduction in AQP3, which provides the major exit pathway for water across the basolateral membrane in cortical collecting ducts (Kwon *et al.*, 2000). Thus, in lithium-induced polyuria the permeability of both apical and basolateral membranes is greatly reduced. The severity of the polyuria is a consequence of the combination of decreased aquaporin expression, and less efficient utilization of the remaining channels.

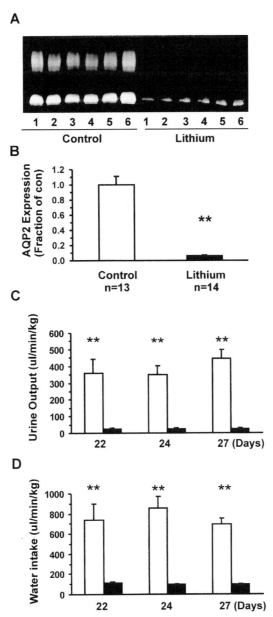

FIGURE 3 Effect of chronic lithium treatment on whole-kidney AQP2 expression levels, urine output, and water intake in rats. (A) The immunoblots were reacted with anti-AQP2 and revealed 29-kDa and 35- to 50-kDa AQP2 bands, representing nonglycosylated and glycosylated forms of AQP2, respectively. (B) Densitometric analyses revealed a dramatic decrease in whole-kidney AQP2 expression in lithium-treated rats. (C and D) The urine output and water intake in lithium-treated rats was significantly increased at day 22, 24, and 27 after initiation of lithium treatment. **, $P < 0.01$.

The reduction in AQP2 and AQP3 expression during lithium treatment is a relatively specific effect: Neither AQP1 nor AQP4 levels change, and there are only modest changes in sodium transporters (Kwon *et al.,* 2000). Lithium has been shown to impair the production of cAMP in collecting duct principal cells (Christensen, 1988; Christensen *et al.,* 1985), and this explains the inhibition of targeting to the plasma membrane in response to lithium treatment. It may also be responsible, at least in part, for the reduction in AQP2 expression, because there is a cAMP-responsive element in the 5′ untranslated region of the AQP2 gene (Hozawa *et al.,* 1996; Matsumura *et al.,* 1997). Consistent with this, mice with inherently low cAMP levels (the DI+/+ severe strain) have low expression of AQP2 (Frokiaer *et al.,* 1999). This is also consistent with the observation that chronic vasopressin infusion, or dehydration to cause an increase in endogenous vasopressin, increases AQP2 expression (DiGiovanni *et al.,* 1994; Ecelbarger *et al.,* 1995; Terris *et al.,* 1996). In an attempt to overcome this, lithium-diuretic rats were treated with high doses of the specific V_2 receptor agonist dDAVP. This was able to induce efficient delivery of AQP2, as judged by the large fraction of the total AQP2 that was present in the apical plasma membrane, but caused only a modest increase in AQP2 expression relative to animals treated with lithium alone, and levels remained far below that seen in the controls. In contrast, 2 days of water deprivation resulted in a much larger increase in AQP2 protein levels, but targeting to the apical plasma membrane remained very poor, suggesting that the effect was not mediated by increased circulating vasopressin and cytosolic cAMP. These results indicate that dehydration was a more potent stimulus for increased AQP2 expression than dDAVP administration under these conditions and provided evidence for the presence of a vasopressin-independent regulation of AQP2 expression levels. The existence of such a signal transduction pathway has recently gained support from an SIADH escape model, as discussed below (Ecelbarger *et al.,* 1997).

An interesting observation in this study was that after a week of recovery from the lithium diet, rats still had less than half the AQP2 seen in controls. This is consistent with the slow recovery of urinary concentrating capacity seen in patients who have been on lithium treatment (Marples *et al.,* 1995).

B. Electrolyte Disturbances

Both hypokalemia and hypercalcemia are associated with a vasopressin-resistant polyuria and urinary concentrating defect. Recent studies using rat models have shown that AQP2 expression is decreased in both cases, but the pattern of AQP2 distribution is strikingly different in the two cases (Earm *et al.,* 1998; Marples *et al.,* 1996; Sands *et al.,* 1998), indicating that the cellular mechanism underlying them must be different.

C. Hypokalemia-Induced NDI

Rats given a potassium-deficient diet for 11 days became hypokalemic, and urine production increased moderately from 11 ± 1 to 30 ± 4 ml/day. Furthermore, in response to 12 hr of water deprivation, urine osmolality was significantly lower in hypokalemic rats compared with controls. This modest polyuria, together with a defect in urinary concentrating capacity, was associated with a decrease in AQP2 expression in both inner medulla and cortex (to $27 \pm 3\%$ and $34 \pm 15\%$ of control levels, respectively) (Marples *et al.*, 1996). Immunocytochemistry revealed that in this model delivery of AQP2 to the apical plasma membrane appeared to occur normally. This probably explains why there is only a modest polyuria in this condition, despite a striking drop in total AQP2 levels. In contrast to the results in the lithium study, after 7 days back on a normal diet rats showed normal urinary concentrating capacity and normalization of AQP2 levels.

D. Hypercalcemia-Induced NDI

Hypercalcemia, induced by oral administration of dihydrotachysterol for 7 days caused a 3-fold increase in urine output. In parallel with this, urine osmolality decreased from 2007 to 925 mOsm/kg-water (Earm *et al.*, 1998; Sands *et al.*, 1998). These results are very similar to those obtained in the hypokalemia study (Marples *et al.*, 1996). In these animals AQP2 expression in kidney inner medulla was only reduced to about half that found in the controls. However, immunocytochemistry showed that the delivery process was also markedly impaired. Thus, as with lithium, it is a combination of decreased expression and decreased shuttling to the membrane that underlies the polyuria.

Recently another aspect of the vitamin-D-induced polyuria was demonstrated. Using the same model Wang *et al.* (unpublished) showed that there was also a significant down-regulation of the bumetanide-sensitive Na-K-2Cl cotransporter BSC-1 in membranes from inner stripe of the outer medulla. This is likely to result in decreased salt accumulation in the inner medulla and, hence, a decreased osmotic gradient to drive water reabsorption from the collecting duct.

E. NDI Caused by Urinary Tract Obstruction

Urinary tract obstruction is a frequent condition in both children and adults with an estimated prevalence of 3–4%. Urinary tract obstruction may be unilateral or bilateral depending on the site of obstruction. Urinary tract obstruction is associated with multiple changes in renal function including alterations in both glomerular and tubular function, ultimately resulting in impaired renal function,

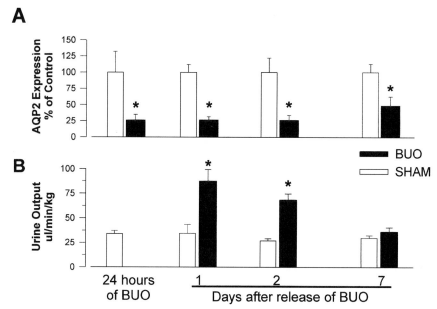

FIGURE 4   Changes in AQP2 expression levels and urine output in rats with bilateral ureteral obstruction (BUO) and SHAM operated control animals. (A) Densitometric analysis showed that 24 h of BUO was associated with a dramatic reduction of AQP2 expression to 25% of the levels in SHAM operated controls. One and 2 days after release of BUO, AQP2 expression remained down-regulated at this level, and 7 days after release of BUO AQP2 expression levels were only partially normalized (50% of SHAM levels). (B) During 24 h of BUO, there was no urine production. Release of BUO was characterized by a dramatic polyuria during day 1 and 2 after release. Seven days after release of BUO, urine production was back to normal.

and bilateral urinary tract obstruction may be associated with long-term impairment in the ability to concentrate urine (Yarger, 1991). The pathophysiological importance of the vasopressin-regulated water channel AQP2 located to the principal cells of the collecting duct was examined in a rat model in which both ureters were reversibly obstructed (Frokiaer *et al.*, 1996). To examine whether AQP2 expression levels were altered, urinary output was determined after 24 h of bilateral ureteral obstruction and 24 h, 48 h, and 7 days after release of bilateral ureteral obstruction. Twenty-four hours after bilateral obstruction, AQP2 expression levels were markedly reduced, even before obstruction was released (Fig. 4). During this period of obstruction, urine production was zero and supported the view that diuresis per se does not decrease AQP2 levels consistent with the lack of down-regulation of AQP2 in response to short-term (24 h) or long-term (5 days) polyuria induced by furosemide treatment (Marples *et al.*, 1996, 1998).

Release of the bilateral obstruction is associated with a dramatic polyuria, which initially is primarily osmotic, due to washout of the accumulated waste products (creatinine and urea). However, the polyuria persists for several days after the plasma biochemistry has normalized, and there is an increased solute-free water clearance. Despite the fact that urine output in this particular study normalized 7 days after release of bilateral obstruction, the animals had an impaired urinary concentrating capacity when subjected to a 24-h period of thirsting. These results were entirely consistent with the measurements of AQP2, which were reduced to about 20% of control levels 2 days after release of obstruction before increasing to about 50% 7 days after release of obstruction. Thus, the persisting urinary concentrating defect is likely to be related to the continued reduction in AQP2 levels. Therefore, it seems very likely that the reduction in total AQP2 available combined with a reduced osmotic gradient driving the water reabsorption is responsible for the concentrating defect seen in these animals. In addition to the reduced AQP2 expression, AQP3 and AQP1 expression levels have also been shown to be down-regulated in response to BUO. Expression of AQP2 and AQP3 tends to normalize within 30 days after release of BUO, whereas the urinary concentrating capacity remains impaired. Interestingly, AQP1 remains down-regulated coinciding with a significant impairment in urinary concentrating capacity (Frøkiær and associates, unpublished) supporting the view that AQP1 may play a major role in maintaining a normal urinary concentrating capacity (Chou *et al.,* 1999; Schnermann *et al.,* 1998).

In contrast to cases with bilateral ureteral obstruction, conditions with unilateral ureteral obstruction are not associated with changes in the absolute excretion of sodium and water since the nonobstructed kidney compensates for the reduced ability of the obstructed kidney to excrete solutes. Thus, in order to examine whether the down-regulation previously shown in rats with bilateral ureteral obstruction was caused by local or systemic factors, we examined AQP2 expression levels in a rat model with unilateral ureteral obstruction where the effect of local factors (increased tissue pressure, changes in renal hemodynamics or parenchymal biochemistry) could be determined in response to unilateral ureteral obstruction for 24 h (Frokiaer *et al.,* 1997). Similar to the results observed in bilateral ureteral obstruction, unilateral ureteral obstruction was associated with profound down-regulation of AQP2 levels in the obstructed kidney (23%) and only a moderate AQP2 reduction (75%) in the nonobstructed kidney, suggesting that local factors play a major role, whereas systemic factors may induce a change in the nonobstructed kidney. Consistent with this, urine production increased by 150% from the nonobstructed kidney, suggesting that relative down-regulation of AQP2 levels in the nonobstructed kidney may be of a compensatory nature. Importantly, further experiments using the same model revealed that changes in AQP2 expression are reciprocal to the changes in free water clearance, demonstrating a functional association between these two parameters. These results support the view that local factors play an important role in the down-regulation of AQP2 expression during obstruction, but the signals leading to this decrease remain to be determined. However, the reduction in AQP2 expression in the contralateral, nonobstructed kidney may suggest a systemic effect, which may potentially involve decreased circulating vasopressin, washout of metabolites from the obstructed kidney, or may be a consequence of reno-renal nerve activity, known to play a role in the compensation for unilateral obstruction.

V. URINARY CONCENTRATING DEFECTS

A. *Experimental Acute Renal Failure Induced by Renal Ischemia and Reperfusion Injury in Rat*

Renal failure, both acute and chronic, is associated with polyuria and a urinary concentrating defect. In both cases, a wide range of glomerular and tubular abnormalities result in the overall renal dysfunction. Bilateral renal ischemia-induced experimental acute renal failure (ARF) in rats is a model that has widely been used. In this model there are structural alterations in renal tubule, in association with an impairment in urinary concentration (Brezis *et al.,* 1984; Venkatachalam *et al.,* 1978).

Urinary concentration and dilution depend on the presence of a discrete segmental distribution of transport properties along the renal tubule and urinary concentration also depends on (1) the hypertonic medullary interstitium, which is generated by active NaCl reabsorption as a consequence of countercurrent multiplication in water-impermeable nephron segments, and (2) the high water permeability (constitutive or vasopressin regulated) in other renal tubular segments for osmotic equilibration, which chiefly depends on aquaporins. Thus, defects in any of these mechanisms would be predicted to be associated with urinary concentrating defects.

In postischemic kidneys, several studies have shown defects both in collecting duct water reabsorption and proximal tubule water reabsorption as well as defects in solute handling (Hanley, 1980; Johnston *et al.,* 1984; Tanner *et al.,* 1973; Venkatachalam *et al.,* 1978). In an isolated tubule microperfusion study of the rabbit, it was observed that water reabsorption in the proximal tubule and cortical collecting duct was significantly reduced following ischemia (Hanley, 1980). Moreover, it has been demonstrated that there are no differences in either basal or vasopressin-induced cAMP levels in outer or inner medulla in rats with ARF compared to SHAM operated rats (Anderson *et al.,* 1982), further supporting the view that there are defects in collecting duct water reabsorption in postischemic kidneys.

Moreover, it is also well known that kidneys subjected to injury by ischemia are unable to establish or maintain a high medullary solute content (Beck *et al.,*

1995). A decreased ability of the thick ascending limb to lower perfusate chloride ion concentration was observed as well (Hanley, 1980). Therefore, these studies suggest that there are defects in both countercurrent multiplication and collecting duct water permeability in response to ischemic injury, hence resulting in the reduced urinary concentration.

Consistent with these findings, it was recently demonstrated that the levels of AQP2 and AQP3 in the collecting duct as well as AQP1 in proximal tubule are significantly reduced in response to renal ischemia (Fernandez-Llama *et al.,* 1999a; Kwon *et al.,* 1999). The decreased levels of aquaporins were associated with impaired urinary concentration in rats with both oliguric or nonoliguric ARF. To further examine the role of aquaporin down-regulation in the development of a urinary concentrating defect, rats with ischemia-induced ARF were treated with alpha-melanocyte-stimulating hormone (α-MSH) to prevent the establishment of a urinary concentrating defect (Chiao *et al.,* 1997). Both the reduced expression of AQP1, -2, and -3 and the reduced urinary concentration capacity was significantly prevented by cotreatment with α-MSH (Fig. 5) (Kwon *et al.,* 1999). The results suggest that decreased levels of aquaporins in both the proximal tubule and collecting duct in postischemic kidneys may play a significant role in the impairment of urinary concentration encountered in the oliguric, maintenance, and polyuric phases of experimental ischemia-induced ARF.

B. Experimental Chronic Renal Failure Induced by 5/6 Nephrectomy in Rat

Patients with advanced chronic renal failure (CRF) have urine that remains hypotonic to plasma despite the administration of supramaximal doses of vasopressin (Tannen *et al.,* 1969). This vasopressin-resistant hyposthenuria specifically implies abnormalities in collecting duct water reabsorption in CRF patients. Consistently, Fine *et al.* (1978) observed that isolated and perfused cortical collecting ducts dissected from remnant kidneys of severely uremic rabbits exhibited a significantly decreased water flux and adenylate cyclase activity in the response to vasopressin. Importantly, they demonstrated that the 8-bromo-cAMP failed to induce a normal hydro-osmotic response in cortical collecting duct from remnant kidneys. Several micropuncture and microcatheterization studies have also indicated that the impaired urinary concentrating ability may be caused, at least partly, by impairment of vasopressin-stimulated water reabsorption in the collecting duct in CRF (Buerkert *et al.,* 1979; Wilson and Sonnenberg, 1979). As an extension of these observations, Teitelbaum and McGuinness demonstrated that RT–PCR of total RNA from the inner medulla of CRF rat kidneys revealed virtual absence of V_2 receptor mRNA (Teitelbaum and McGuinness, 1995). Thus, these studies provide firm evidences

A

AQP2 (2 days after 40min ischemia)

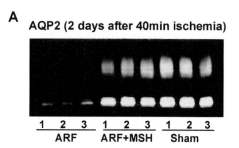

$$\underset{\text{ARF}}{\underline{1 \quad 2 \quad 3}} \quad \underset{\text{ARF+MSH}}{\underline{1 \quad 2 \quad 3}} \quad \underset{\text{Sham}}{\underline{1 \quad 2 \quad 3}}$$

B

AQP2 Expression (Fraction of Sham)

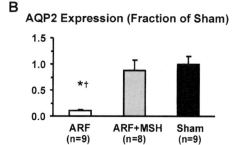

C

Urine output (ul/min/kg)

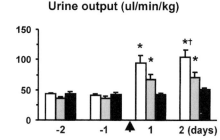

FIGURE 5 Effect of α-MSH treatment of AQP2 levels and changes in urine output and urine osmolality in rats with ARF (2 days after 40-min bilateral ischemia). (A) Immunoblot is reacted with anti-AQP2. (B) Densitometric analysis of all samples from ARF (ARF 40/2), either in the absence of α-MSH treatment (ARF) or with α-MSH treatment (ARF+MSH), and SHAM-operated rats. In the absence of α-MSH treatment, rats with ARF have markedly decreased AQP2 expression levels (13″3% of sham levels, $P < 0.05$). AQP2 expression is 7-fold higher in response to α-MSH of ARF rats compared with untreated rats. (C) Time courses of the changes in urine output. Urine output is significantly increased after 40-min bilateral renal ischemia in ARF rats (both α-MSH treated and nontreated). ARF rats treated with α-MSH showed a reduced urine output compared with untreated ARF rats. *, $P < 0.05$, ARF compared with SHAM-operated rats. [†]$P < 0.05$, nontreated ARF rats compared with α-MSH-treated ARF rats.

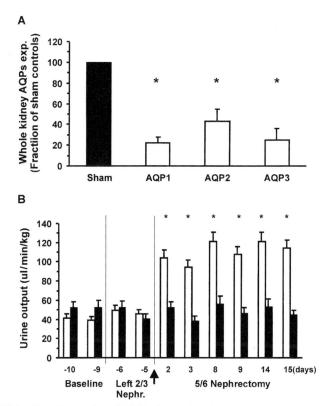

FIGURE 6 Altered expression of AQP1, -2, and -3 and urine output in rats with chronic renal failure induced by 5/6 nephrectomy. (A) Semiquantitative immunoblotting of membrane fractions of whole kidneys revealed a marked decrease in AQP1, -2, and -3 abundance in rats with CRF, compared to SHAM operated control rats. *, $P < 0.05$. (B) Time course of the changes in urine output in CRF rats (*open bars*) and SHAM operated rats (*solid bars*). Urine output significantly increased after induction of 5/6 nephrectomy in rats with CRF, whereas there were no changes in urine output after sham operation in control rats. *, $P < 0.05$. Arrow indicates time point of induction of 5/6 nephrectomy.

for significant defects in the collecting duct water permeability. Consistent with these observations, recent studies have shown both decreased collecting duct water channel AQP2 and AQP3 expression and a vasopressin-resistant down-regulation of AQP2 in a 5/6 nephrecomy-induced CRF rat model (Fig. 6) (Kwon *et al.*, 1998). Immunocytochemistry and immunoelectron microscopy confirmed a marked reduction in AQP2 and AQP3 expression in the inner medullary principal cells. This suggests that reduced AQP2 and AQP3 expression levels may be important factors involved in the impaired collecting duct water permeability and in reduced vasopressin responsiveness in CRF.

C. Other Models

Three other conditions are associated with impairment of maximal urinary concentrating capacity. Although none of them are, strictly speaking, pathological, it seems likely that the reduction in concentrating capacity may be mechanistically related to changes seen in the other pathological states discussed here.

1. Aging

In man there is a decline in maximal urinary concentrating capacity with age. Rodents show a similar decline and also become polyuric. It has been shown (Preisser *et al.,* 2000; Teillet *et al.,* 1999) that this is associated with a striking decline in AQP2 expression in the inner medulla, but not in the cortex. AQP3 levels are also somewhat decreased in the inner medulla, while AQP1, AQP4, and vasopressin receptor levels are unchanged. These results may partially explain the impaired concentrating capacity in old age. They are also interesting because they show that the regulation of AQP2 expression in the cortex and medulla can be separated, further indicating that there must be non-vasopressin-mediated signals involved, and supporting the evidence from the UUO studies described earlier that these may be local factors.

2. Low Protein Diet

A low protein diet results in reduced production of urea by the body. Because urea is an important medullary osmolyte, a reduction in urea levels will reduce the maximum osmolarity of the urine that can be excreted, and indeed a low protein diet has been shown to result in such a deficit. However, in a rat model (Sands *et al.,* 1996) in which animals were given a diet containing 8% protein instead of 18% protein, AQP2 expression in the terminal inner medulla was decreased significantly, and this resulted in decreased vasopressin-induced water permeability of tubules isolated from these rats. This study showed even more specific local regulation of AQP2 expression, since AQP2 levels were unchanged in the first part of the inner medulla. There were also no changes in AQP3 expression. Thus the urinary concentrating defect observed during a low protein diet is likely to be the consequence of a combination of decreased medullary interstitial tonicity and decreased collecting duct water permeability caused by low AQP2 levels.

3. Polydipsia

It has been known for many years that chronic high water intake causes an impairment of maximal urinary concentrating capacity (De Wardener and Herxheimer, 1957). In part, this may be a consequence of washout of the medullary osmotic gradient, but this should be rapidly reversed during dehydration, unlike the concentrating defect. Recent studies in rats have shown that water loading decreases AQP2 expression at both the protein and mRNA levels (Nielsen *et al.*, 1993b; Saito *et al.*, 1997). Because water loading will suppress endogenous vasopressin release, this effect is probably partly mediated by a decrease in circulating vasopressin activity. However, as discussed below, AQP2 levels also fall following overhydration in models in which vasopressin levels are kept high by a continuous infusion. Thus hydration status seems to affect AQP2 expression by a vasopressin-independent pathway.

VI. STATES OF WATER RETENTION

A. SIADH and Vasopressin Escape

In the syndrome of inappropriate ADH secretion (SIADH), circulating levels of vasopressin (ADH) are higher than expected for a given plasma osmolality. This can arise because of ADH-secreting tumors, brain injury, or a variety of other causes. Under these circumstances the kidney will retain water inappropriately, and the subject becomes hyponatremic. However, after a few days, a phenomenon called *escape* occurs, in which the ability to excrete water returns partially. To examine the mechanisms underlying vasopressin escape, rats were infused with dDAVP in osmotic minipumps and then given either a water load (incorporated with their diet) or allowed free access to water (Ecelbarger *et al.*, 1997). Although all rats had the same levels of circulating dDAVP, the expression of AQP2 in the water-loaded rats decreased dramatically, in parallel with an increase in water excretion compared to the rats with free access to water. In contrast, levels of AQP1, AQP3, and AQP4 remained constant (Verbalis *et al.*, 1998). Thus the rats were able to decrease AQP2 levels despite a continued high level of circulating dDAVP. This suggests that changes in hydration status can influence AQP2 expression independent of vasopressin, consistent with the results of the study on lithium treatment described earlier. This down-regulation of AQP2 represents a physiologically appropriate way to reduce the reabsorption of water (Ecelbarger *et al.*, 1997).

B. Congestive Heart Failure

Severe chronic heart failure is characterized by defects in renal handling of water and sodium resulting in extracellular fluid expansion and hyponatremia, which is thought to be critically dependent on an increased baroreceptor-mediated vasopressin release. Recently, two studies have examined the renal aquaporin expression in rats with congestive heart failure (CHF) induced by ligation of the left coronary artery (Nielsen *et al.*, 1997; Xu *et al.*, 1997) in order to test whether up-regulation of AQP2 expression and targeting may play a role in the development of CHF. These studies showed uniformly that renal water retention in severe congestive heart failure in rats is associated with marked dysregulation of AQP2 in the renal collecting duct involving both an increase in the abundance of AQP2 water channel protein in the collecting duct principal cells and a marked change in the distribution of AQP2 water channels in the collecting duct principal cells with most channels located to the apical plasma membrane (Nielsen *et al.*, 1997; Xu *et al.*, 1997). Immunoblotting revealed a 3-fold increase in AQP2 expression compared to SHAM operated animals. Importantly, these changes were markedly related to elevated left ventricular end-diastolic pressure (LVEDP) or hyponatremia, since animals without changes in LVEDP and plasma sodium did not have increased AQP2 levels compared to SHAM operated controls (Nielsen *et al.*, 1997). Rats with severe heart failure had significantly elevated LVEDP compared with SHAM operated animals (26.9″ 3.2 vs. 4.1″ 0.3 mmHg) and had reduced plasma sodium concentrations (Nielsen *et al.*, 1997). In addition this study also demonstrated a marked increase in plasma membrane targeting of AQP2 providing an explanation for the increased permeability of the collecting duct causing an increase in water reabsorption. This may provide an explanation for excess free water retention in severe CHF and for the development of hyponatremia. In parallel, up-regulation of both AQP2 protein and AQP2 mRNA levels in kidney inner medulla and cortex was demonstrated in rats with chronic heart failure (Xu *et al.*, 1997). These rats had significantly decreased cardiac output and, importantly, increased plasma vasopressin levels. Importantly, in this study administration of the V2 antagonist OPC 31260 was associated with a significant increase in diuresis, a decrease in urine osmolality, a rise in plasma osmolality, and a significant reduction in AQP2 protein and AQP2 mRNA levels compared with untreated rats with CHF. Consequently, there is now clear evidence for a major role for vasopressin in the up-regulation of AQP2 in experimental CHF in rats.

C. Hepatic Cirrhosis

Hepatic cirrhosis is another serious chronic condition associated with water retention. An important pathophysiological mechanism has been suggested to be impaired ability to excrete water due to increased levels of plasma vasopressin. However, unlike congestive heart failure the changes in expression of AQP2 protein levels vary considerably between fundamentally different models of hepatic cirrhosis. Several studies have looked at changes in renal aquaporin expression in rats with cirrhosis induced by common bile duct ligation (CBDL) (Fernandez-Llama *et al.*, 1999b, 2000; Jonassen *et al.*, 1998). Hepatic cirrhosis was associated with significant sodium retention (Jonassen *et al.*, 1998) consistent with significant hypertrophy of the thick ascending limb (Jonassen *et al.*, 1997). The rats displayed impaired vasopressin-regulated water reabsorption despite normal plasma vasopressin levels evidenced as an impaired effect of the vasopressin-V2-receptor antagonist OPC 31260. Consistent with this, immunoblotting and semiquantitative densitometry showed a significant decrease in AQP2 abundance in rats with CBDL-induced hepatic cirrhosis (Fernandez-Llama et al., 1999b; Jonassen et al., 1998). Furthermore, the two aquaporins located to the basolateral plasma membrane of the collecting duct principal cells, AQP3 and AQP4, were also down-regulated in CBDL rats, which may predict a reduced water permeability of the collecting duct independent of the effects due to AQP2 dysregulation in this model (Fernandez-Llama et al., 1999b). Thus, these results support the view that AQP2, AQP3, and AQP4 play important roles for abnormal collecting duct water transport as evidenced from studies using AQP3 and AQP4 knockouts (Chou et al., 1998; Ma et al., 2000). In contrast, the expression levels of aquaporin-1 were not decreased (Fernandez-Llama et al., 1999b). Thus, dysregulation of multiple water channels may play a role in water balance abnormalities associated with CBDL-induced cirrhosis in rat. The results with CBDL-induced cirrhosis are similar to what has been demonstrated in rats with nephrotic syndrome, supporting the view that increased expression of AQP2 is not a uniform finding in rat models of volume-expanded states. In a study by Fernandez-Llama et al. (1999b), CBDL was also associated with hyponatremia in response to water loading, demonstrating that in case of extracellular fluid volume expansion, excessive water retention can occur in the absence of increased AQP2 levels.

In contrast to the studies with CBDL-induced hepatic cirrhosis, Fujita and coworkers (1995) showed that hepatic cirrhosis induced by intraperitoneal administration of carbon tetrachloride and olive oil twice a week for 12 weeks was associated with a significant increase in both AQP2 protein levels and AQP2 mRNA expression. Interestingly, AQP2 mRNA levels correlated with the amount of ascites, suggesting that AQP2 may play a role in the abnormal water retention followed by the development of ascites in hepatic cirrhosis (Asahina et al., 1995). Although the explanation for the differences between cirrhosis induced by common bile duct ligation and carbon tetrachloride remains to be determined, it is well known that the dysregulation of body water balance depends on the severity of cirrhosis (Gines et al., 1998; Wood et al., 1988). CBDL results in a compensated cirrhosis characterized by peripheral vasodilation and increased cardiac output, whereas cirrhosis induced by 12 weeks of carbon tetrachloride administration may be associated with the late decompensated state of liver cirrhosis characterized by sodium retention, edema, and ascites (Gines et al., 1998; Levy and Wexler, 1987). Thus, down-regulation of AQP2 in compensated cirrhosis may represent a physiological down-regulation to prevent development of water retention, whereas increased vasopressin levels in cirrhosis with severe acites may be responsible for up-regulation of AQP2 and hence increased water reabsorption.

D. Pregnancy

Pregnancy is characterized by a 30–50% increase in extracellular fluid, plasma, and blood volume in different mammalian species, including humans and rats (Schrier and Briner, 1991). These changes are associated with arterial vasodilatation together with sodium and water retention and a decrease in plasma osmolality both in rats and humans. In parallel with the arterial vasodilation the renin-angiotensin-aldosterone system is activated and it has been suggested that the decrease in plasma osmolality is caused by resetting of the threshold for vasopressin secretion during pregnancy (Durr *et al.*, 1981). Therefore the hypothesis that AQP2 expression could be increased during pregnancy was tested in rats on days 7, 14, and 20 of pregnancy (Ohara *et al.*, 1998). AQP2mRNA and AQP2 levels increased significantly during pregnancy. Administration of the V_2-receptor agonist suppressed the increase in both AQP2mRNA and AQP2 despite normal levels of plasma vasopressin (Ohara *et al.*, 1998). Thus, this study suggests that the up-regulation of AQP2 contributes to water retention during pregnancy in part through a V_2-receptor-mediated effect, but AVP independent factors (such as oxytocin) may also be important for this up-regulation (Schrier *et al.*, 1998).

Circulating levels of sex hormones are also altered in response to ovariectomy, which has been shown to reduce plasma vasopressin concentrations in rats and suppress diurnal changes in hormone concentrations (Peysner and Forsling, 1990). Sardeli and coworkers (2000) therefore hypothesized that AQP2 dysregulation could be associated with changes in renal water handling. One and 2 weeks after OVX a significant increase in AQP2 and phosphorylated-AQP2 levels in kidneys of pregnant rats was demonstrated, which was associated with a reduction in urine output and a significant increase in body weight (Sardeli *et al.*, 2000). Thus, these results suggest that AQP2 dysregulation may play an important role in renal water handling after OVX.

E. Experimental Nephrotic Syndrome

Disturbed renal water handling is one of the main characteristics of nephrotic syndrome. This is caused by defects in both the urinary diluting mechanism and urinary concentrating capacity. The reasons for these disturbances are incompletely understood. The decrease in urinary diluting ability is thought to be the result of a nonosmotic elevation in plasma vasopressin levels, which then may result in an enhanced free water reabsorption. Using a rat model with puromycin aminonucleoside-induced nephrotic (PAN) syndrome, Apostel *et al.* (1997) demonstrated an 87% reduction in AQP2 and a 70% reduction in AQP3 expression, suggesting that the impaired urinary concentrating capacity in nephrotic syndrome could be the result of extensive down-regulation of both AQP2 and AQP3 in the collecting duct. This response seems to be physiologically appropriate in order to reduce a further extracellular fluid volume expansion.

Thus, common for both hepatic cirrhosis induced by ligation of the common bile duct and puromycin aminonucleoside-induced nephrotic syndrome is down-regulation of AQP2 levels. Therefore, it could be speculated that AQP2 down-regulation occurs as a physiological compensatory response.

Nephrotic syndrome (NS) is characterized by extracellular volume expansion with excessive renal water and sodium reabsorption. The mechanisms of water and sodium retention are poorly understood; however, it was expected to be associated with dysregulation of aquaporins and sodium transporters (Apostol et al., 1997; Fernandez-Llama et al., 1998a). In contrast to congestive heart failure and liver cirrhosis, where extracellular volume expansion and hyponatremia have been reported to be associated with upregulation of AQP2 (Asahina et al., 1995; Fujita et al., 1995; Nielsen et al., 1997), rats with NS do not develop hyponatremia despite extensive extracellular fluid volume expansion. This absence of hyponatremia may reflect an absence of up-regulation of AQP2 expression in the collecting duct. Indeed, a marked down-regulation of AQP2 and AQP3 expression was demonstrated in the renal inner medulla in the rats with PAN-mediated NS and adriamycin-induced NS (Apostol et al., 1997; Fernandez-Llama et al., 1998b). This reduced expression of collecting water channels could represent a physiologically appropriate response to extracellular volume expansion. Circulating vasopressin levels are high in rats with PAN-induced NS. Nevertheless, AQP2 levels are dramatically reduced in the experimental NS, thus suggesting that this represents an "escape" phenomenon. Certainly, it indicates that there is a signal other than vasopressin than can cause changes of AQP2 expression.

VII. CONCLUSIONS

The seminal discovery of aquaporins by Agre and colleagues has led to dramatic progress in the understanding of how water is transported across epithelial cells, and how renal water handling occurs at the physiological level as well as in multiple diseases associated with severe derangement of body water balance. Currently more than 10 mammalian aquaporins have been identified and at least 5 renal aquaporins have been well characterized. Future studies will concentrate on defining the physiological and pathophysiological roles at the integrated level of each aquaporin, including identification of novel aquaporins and establishment of their regulation, and physiological/pathophysiological roles at the cellular and integrated level.

References

Agre, P. (2000). Homer W. Smith award lecture. Aquaporin water channels in kidney. *J. Am. Soc. Nephrol.* **11**, 764–777.

Agre, P., Bonhivers, M., and Borgnia, M. J. (1998). The aquaporins, blueprints for cellular plumbing systems. *J. Biol. Chem.* **273**, 14659–14662.

Agre, P., Mathai, J. C., Smith, B. L., and Preston, G. M. (1999). Functional analyses of aquaporin water channel proteins. *Methods Enzymol.* **294**, 550–572.

Anderson, R. J., Gordon, J. A., Kim, J., Peterson, L. M., and Gross, P. A. (1982). Renal concentration defect following nonoliguric acute renal failure in the rat. *Kidney Int.* **21**, 583–591.

Apostol, E., Ecelbarger, C. A., Terris, J., Bradford, A. D., Andrews, P., and Knepper, M. A. (1997). Reduced renal medullary water channel expression in puromycin aminonucleoside-induced nephrotic syndrome. *J. Am. Soc. Nephrol.* **8**, 15–24.

Asahina, Y., Izumi, N., Enomoto, N., Sasaki, S., Fushimi, K., Marumo, F., and Sato, C. (1995). Increased gene expression of water channel in cirrhotic rat kidneys. *Hepatology* **21**, 169–173.

Beck, F. X., Ohno, A., Dorge, A., and Thurau, K. (1995). Ischemia-induced changes in cell element composition and osmolyte contents of outer medulla. *Kidney Int.* **48**, 449–457.

Borgnia, M., Nielsen, S., Engel, A., and Agre, P. (1999). Cellular and molecular biology of the aquaporin water channels. *Annu. Rev. Biochem.* **68**, 425–458.

Boton, R., Gaviria, M., and Batlle, D. C. (1987). Prevalence, pathogenesis, and treatment of renal dysfunction associated with chronic lithium therapy. *Am. J. Kidney Dis.* **10**, 329–345.

Brezis, M., Rosen, S., Silva, P., and Epstein, F. H. (1984). Selective vulnerability of the medullary thick ascending limb to anoxia in the isolated perfused rat kidney. *J. Clin. Invest.* **73**, 182–190.

Buerkert, J., Martin, D., Prasad, J., Chambless, S., and Klahr, S. (1979). Response of deep nephrons and the terminal collecting duct to a reduction in renal mass. *Am. J. Physiol.* **236**, F454–F464.

Chiao, H., Kohda, Y., McLeroy, P., Craig, L., Housini, I., and Star, R. A. (1997). Alpha-melanocyte-stimulating hormone protects against renal injury after ischemia in mice and rats. *J. Clin. Invest.* **99**, 1165–1172.

Chou, C. L., Ma, T., Yang, B., Knepper, M. A., and Verkman, A. S. (1998). Fourfold reduction of water permeability in inner medullary collecting duct of aquaporin-4 knockout mice. *Am. J. Physiol.* **274**, C549–C554.

Chou, C. L., Knepper, M. A., Hoek, A. N., Brown, D., Yang, B., Ma, T., and Verkman, A. S. (1999). Reduced water permeability and altered ultrastructure in thin descending limb of Henle in aquaporin-1 null mice. *J. Clin. Invest.* **103**, 491–496.

Christensen, S. (1988). Vasopressin and renal concentrating ability. In "Lithium Therapy Monographs," Vol. 2 (F. N. Johnson, ed.), pp. 20–34. S. Karger, Basel.

Christensen, S., Kusano, E., Yusufi, A. N., Murayama, N., and Dousa, T. P. (1985). Pathogenesis of nephrogenic diabetes insipidus due to chronic administration of lithium in rats. *J. Clin. Invest.* **75**, 1869–1879.

De Wardener, H. E., and Herxheimer, A. (1957). The effect of a high water intake on the kidney's ability to concentrate the urine in man. *J. Physiol.* **139**, 42–52.

Deen, P. M., and Knoers, N. V. (1998). Physiology and pathophysiology of the aquaporin-2 water channel. *Curr. Opin. Nephrol. Hypertens.* **7**, 37–42.

Deen, P. M., Verdijk, M. A., Knoers, N. V., Wieringa, B., Monnens, L. A., van-Os, C. H., and van-Oost, B. A. (1994). Requirement of human renal water channel aquaporin-2 for vasopressin-dependent concentration of urine. *Science* **264**, 92–95.

Deen, P. M., Aubel, R., Lieburg, A. F., and Os, C. H. V. (1995). Urinary content of aquaporin 1 and 2 in nephrogenic diabetes insipidus. *J. Am. Soc. Nephrol.* **7**, 836–841.

DiGiovanni, S. R., Nielsen, S., Christensen, E. I., and Knepper, M. A. (1994). Regulation of collecting duct water channel expression by vasopressin in Brattleboro rat. *Proc. Natl. Acad. Sci. USA* **91**, 8984–8988.

Durr, J. A., Stamoutsos, B., and Lindheimer, M. D. (1981). Osmoregulation during pregnancy in the rat. Evidence for resetting of the threshold for vasopressin secretion during gestation. *J. Clin. Invest.* **68**, 337–346.

Earm, J. H., Christensen, B. M., Frokiaer, J., Marples, D., Han, J. S., Knepper, M. A., and Nielsen, S. (1998). Decreased aquaporin-2 expression and apical plasma membrane delivery in kidney collecting ducts of polyuric hypercalcemic rats. *J. Am. Soc. Nephrol.* **9**, 2181–2193.

Ecelbarger, C. A., Terris, J., Frindt, G., Echevarria, M., Marples, D., Nielsen, S., and Knepper, M. A. (1995). Aquaporin-3 water channel localization and regulation in rat kidney. *Am. J. Physiol.* **269**, F663–F672.

Ecelbarger, C. A., Nielsen, S., Olson, B. R., Murase, T., Baker, E. A., Knepper, M. A., and Verbalis, J. G. (1997). Role of renal aquaporins in escape from vasopressin-induced antidiuresis in rat. *J. Clin. Invest.* **99**, 1852–1863.

Echevarria, M., Windhager, E. E., Tate, S. S., and Frindt, G. (1994). Cloning and expression of AQP3, a water channel from the medullary collecting duct of rat kidney. *Proc. Natl. Acad. Sci. USA* **91**, 10997–11001.

Fernandez-Llama, P., Andrews, P., Nielsen, S., Ecelbarger, C. A., and Knepper, M. A. (1998a.). Impaired aquaporin and urea transporter expression in rats with adriamycin-induced nephrotic syndrome. *Kidney Int.* **53**, 1244–1253.

Fernandez-Llama, P., Andrews, P., Ecelbarger, C. A., Nielsen, S., and Knepper, M. (1998b). Concentrating defect in experimental nephrotic syndrome: Altered expression of aquaporins and thick ascending limb Na$^+$ transporters. *Kidney Int.* **54**, 170–179.

Fernandez-Llama, P., Andrews, P., Turner, R., Saggi, S., Dimari, J., Kwon, T. H., Nielsen, S., Safirstein, R., and Knepper, M. A. (1999a). Decreased abundance of collecting duct aquaporins in postischemic renal failure in rats. *J. Am. Soc. Nephrol.* **10**, 1658–1668.

Fernandez-Llama, P., Turner, R., Dibona, G., and Knepper, M. A. (1999b). Renal expression of aquaporins in liver cirrhosis induced by chronic common bile duct ligation in rats. *J. Am. Soc. Nephrol.* **10**, 1950–1957.

Fernandez-Llama, P., Jimenez, W., Bosch-Marce, M., Arroyo, V., Nielsen, S., and Knepper, M. A. (2000). Dysregulation of renal aquaporins and Na-Cl cotransporter in CCl4-induced cirrhosis. *Kidney Int.* **58**, 216–228.

Fine, L. G., Schlondorff, D., Trizna, W., Gilbert, R. M., and Bricker, N. S. (1978). Functional profile of the isolated uremic nephron. Impaired water permeability and adenylate cyclase responsiveness of the cortical collecting tubule to vasopressin. *J. Clin. Invest.* **61**, 1519–1527.

Frokiaer, J., Marples, D., Knepper, M. A., and Nielsen, S. (1996). Bilateral ureteral obstruction downregulates expression of vasopressin-sensitive AQP-2 water channel in rat kidney. *Am. J. Physiol.* **270**, F657–F668.

Frokiaer, J., Christensen, B. M., Marples, D., Djurhuus, J. C., Jensen, U. B., Knepper, M. A., and Nielsen, S. (1997). Downregulation of aquaporin-2 parallels changes in renal water excretion in unilateral ureteral obstruction. *Am. J. Physiol.* **273**, F213–F223.

Frokiaer, J., Marples, D., Knepper, M. A., and Nielsen, S. (1998). Pathophysiology of aquaporin-2 in water balance disorders. *Am. J. Med. Sci.* **316**, 291–299.

Frokiaer, J., Marples, D., Valtin, H., Morris, J. F., Knepper, M. A., and Nielsen, S. (1999). Low aquaporin-2 levels in polyuric DI +/+ severe mice with constitutively high cAMP-phosphodiesterase activity. *Am. J. Physiol.* **276**, F179–F190.

Fujita, N., Ishikawa, S. E., Sasaki, S., Fujisawa, G., Fushimi, K., Marumo, F., and Saito, T. (1995). Role of water channel AQP-CD in water retention in SIADH and cirrhotic rats. *Am. J. Physiol.* **269**, F926–F931.

Fushimi, K., Uchida, S., Hara, Y., Hirata, Y., Marumo, F., and Sasaki, S. (1993). Cloning and expression of apical membrane water channel of rat kidney collecting tubule. *Nature* **361**, 549–552.

Gines, P., Berl, T., Bernardi, M., Bichet, D. G., Hamon, G., Jimenez, W., Liard, J. F., Martin, P. Y., and

Schrier, R. W. (1998). Hyponatremia in cirrhosis: From pathogenesis to treatment. *Hepatology* **28**, 851–864.

Gronbeck, L., Marples, D., Nielsen, S., and Christensen, S. (1998). Mechanism of antidiuresis caused by bendroflumethiazide in conscious rats with diabetes insipidus. *Br. J. Pharmacol.* **123**, 737–745.

Hanley, M. J. (1980). Isolated nephron segments in a rabbit model of ischemic acute renal failure. *Am. J. Physiol.* **239**, F17–F23.

Hasegawa, H., Ma, T., Skach, W., Matthay, M. A., and Verkman, A. S. (1994). Molecular cloning of a mercurial-insensitive water channel expressed in selected water-transporting tissues. *J. Biol. Chem.* **269**, 5497–5500.

Hozawa, S., Holtzman, E. J., and Ausiello, D. A. (1996). cAMP motifs regulating transcription in the aquaporin 2 gene. *Am. J. Physiol.* **270**, C1695–C1702.

Ishibashi, K., Sasaki, S., Fushimi, K., Uchida, S., Kuwahara, M., Saito, H., Furukawa, T., Nakajima, K., Yamaguchi, Y., Gojobori, T., and Marumo, F. (1994). Molecular cloning and expression of a member of the aquaporin family with permeability to glycerol and urea in addition to water expressed at the basolateral membrane of kidney collecting duct cells. *Proc. Natl. Acad. Sci. USA* **91**, 6269–6273.

Ishibashi, K., Kuwahara, M., Gu, Y., Kageyama, Y., Tohsaka, A., Suzuki, F., Marumo, F., and Sasaki, S. (1997a). Cloning and functional expression of a new water channel abundantly expressed in the testis permeable to water, glycerol, and urea. *J. Biol. Chem.* **272**, 20782–20786.

Ishibashi, K., Kuwahara, M., Kageyama, Y., Tohsaka, A., Marumo, F., and Sasaki, S. (1997b). Cloning and functional expression of a second new aquaporin abundantly expressed in testis. *Biochem. Biophys. Res. Commun.* **237**, 714–718.

Ishibashi, K., Kuwahara, M., Gu, Y., Tanaka, Y., Marumo, F., and Sasaki, S. (1998). Cloning and functional expression of a new aquaporin (AQP9) abundantly expressed in the peripheral leukocytes permeable to water and urea, but not to glycerol. *Biochem. Biophys. Res. Commun.* **244**, 268–274.

Ishibashi, K., Suzuki, M., and Imai, M. (1999). Molecular cloning of a first member of new aquaporin (AQP) superfamily from rat testis. *J. Am. Soc. Nephrol.* **10**, 16A.

Johnston, P. A., Rennke, H., and Levinsky, N. G. (1984). Recovery of proximal tubular function from ischemic injury. *Am. J. Physiol.* **246**, F159–F166.

Jonassen, T. E., Marcussen, N., Haugan, K., Skyum, H., Christensen, S., Andreasen, F., and Petersen, J. S. (1997). Functional and structural changes in the thick ascending limb of Henle's loop in rats with liver cirrhosis. *Am. J. Physiol.* **273**, R568–R577.

Jonassen, T. E., Nielsen, S., Christensen, S., and Petersen, J. S. (1998). Decreased vasopressin-mediated renal water reabsorption in rats with compensated liver cirrhosis. *Am. J. Physiol.* **275**, F216–F225.

Jung, J. S., Bhat, R. V., Preston, G. M., Guggino, W. B., Baraban, J. M., and Agre, P. (1994). Molecular characterization of an aquaporin cDNA from brain: Candidate osmoreceptor and regulator of water balance. *Proc. Natl. Acad. Sci. USA* **91**, 13052–13056.

Kamsteeg, E. J., Wormhoudt, T. A., Rijss, J. P., van Os, C. H., and Deen, P. M. (1999). An impaired routing of wild-type aquaporin-2 after tetramerization with an aquaporin-2 mutant explains dominant nephrogenic diabetes insipidus. *EMBO J.* **18**, 2394–2400.

Kanno, K., Sasaki, S., Hirata, Y., Ishikawa, S.-E., Fushimi, K., Nakanishi, S., Bichet, D. G., and Marumo, F. (1995). Urinary excreation of aquaporin-2 in patients with diabetes insipidus. *N. Engl. J. Med.* **332**, 1540–1545.

King, L. S., Yasui, M., and Agre, P. (2000). Aquaporins in health and disease. *Mol. Med. Today* **6**, 60–65.

Knepper, M. A., Nielsen, S., Chou, C. L., and DiGiovanni, S. R. (1994). Mechanism of vasopressin action in the renal collecting duct. *Semin. Nephrol.* **14**, 302–321.

Knepper, M. A., Verbalis, J. G., and Nielsen, S. (1997). Role of aquaporins in water balance disorders. *Curr. Opin. Nephrol. Hypertens.* **6**, 367–371.

Koyama, Y., Yamamoto, T., Kondo, D., Funaki, H., Yaoita, E., Kawasaki, K., Sato, N., Hatakeyama, K., and Kihara, I. (1997). Molecular cloning of a new aquaporin from rat pancreas and liver. *J. Biol. Chem.* **272,** 30329–30333.

Kwon, T. H., Frokiaer, J., Knepper, M. A., and Nielsen, S. (1998). Reduced AQP1, -2, and -3 levels in kidneys of rats with CRF induced by surgical reduction in renal mass. *Am. J. Physiol.* **275,** F724–F741.

Kwon, T. H., Frokiaer, J., Fernandez-Llama, P., Knepper, M. A., and Nielsen, S. (1999). Reduced abundance of aquaporins in rats with bilateral ischemia-induced acute renal failure: Prevention by alpha-MSH. *Am. J. Physiol.* **277,** F413–F427.

Kwon, T. H., Laursen, U. H., Marples, D., Maunsbach, A. B., Knepper, M. A., Frokiaer, J., and Nielsen, S. (2000). Altered expression of renal aquaporins and Na^+ transporters in rats with lithium-induced nephrogenic diabetes insipidus. *Am. J. Physiol.* (in press)

Levy, M., and Wexler, M. J. (1987). Hepatic denervation alters first-phase urinary sodium excretion in dogs with cirrhosis. *Am. J. Physiol.* **253,** F664–F671.

Ma, T., Song, Y., Yang, B., Gillespie, A., Carlson, E. J., Epstein, C. J., and Verkman, A. S. (2000). Nephrogenic diabetes insipidus in mice lacking aquaporin-3 water channels. *Proc. Natl. Acad. Sci. USA* **97,** 4386–4391.

Marples, D., Christensen, S., Christensen, E. I., Ottosen, P. D., and Nielsen, S. (1995). Lithium-induced downregulation of aquaporin-2 water channel expression in rat kidney medulla. *J. Clin. Invest.* **95,** 1838–1845.

Marples, D., Frokiaer, J., Dorup, J., Knepper, M. A., and Nielsen, S. (1996). Hypokalemia-induced downregulation of aquaporin-2 water channel expression in rat kidney medulla and cortex. *J. Clin. Invest.* **97,** 1960–1968.

Marples, D., Christensen, B. M., Frokiaer, J., Knepper, M. A., and Nielsen, S. (1998). Dehydration reverses vasopressin antagonist-induced diuresis and aquaporin-2 downregulation in rats. *Am. J. Physiol.* **275,** F400–F409.

Marples, D., Frokiaer, J., and Nielsen, S. (1999). Long-term regulation of aquaporins in the kidney. *Am. J. Physiol.* **276,** F331–F339.

Matsumura, Y., Uchida, S., Rai, T., Sasaki, S., and Marumo, F. (1997). Transcriptional regulation of aquaporin-2 water channel gene by cAMP. *J. Am. Soc. Nephrol.* **8,** 861–867.

Mulders, S. M., Bichet, D. G., Rijss, J. P., Kamsteeg, E. J., Arthus, M. F., Lonergan, M., Fujiwara, M., Morgan, K., Leijendekker, R., van der, S. P., van Os, C. H., and Deen, P. M. (1998). An aquaporin-2 water channel mutant which causes autosomal dominant nephrogenic diabetes insipidus is retained in the Golgi complex. *J. Clin. Invest.* **102,** 57–66.

Nielsen, S., Smith, B., Christensen, E. I., Knepper, M. A., and Agre, P. (1993a). CHIP28 water channels are localized in constitutively water-permeable segments of the nephron. *J. Cell Biol.* **120,** 371–383.

Nielsen, S., DiGiovanni, S. R., Christensen, E. I., Knepper, M. A., and Harris, H. W. (1993b). Cellular and subcellular immunolocalization of vasopressin-regulated water channel in rat kidney. *Proc. Natl. Acad. Sci. USA* **90,** 11663–11667.

Nielsen, S., Pallone, T. L., Smith, B. L., Christensen, E. I., Agre, P., and Maunsbach, A. B. (1995a). Aquaporin-1 water channels in short and long loop descending thin limbs and in descending vasa recta in rat kidney. *Am. J. Physiol.* **268,** F1023–F1037.

Nielsen, S., Chou, C. L., Marples, D., Christensen, E. I., Kishore, B. K., and Knepper, M. A. (1995b). Vasopressin increases water permeability of kidney collecting duct by inducing translocation of aquaporin-CD water channels to plasma membrane. *Proc. Natl. Acad. Sci. USA* **92,** 1013–1017.

Nielsen, S., Terris, J., Andersen, D., Ecelbarger, C., Frokiaer, J., Jonassen, T., Marples, D., Knepper, M. A., and Petersen, J. S. (1997). Congestive heart failure in rats is associated with increased expression and targeting of aquaporin-2 water channel in collecting duct. *Proc. Natl. Acad. Sci. USA* **94,** 5450–5455.

Nielsen, S., Kwon, T. H., Christensen, B. M., Promeneur, D., Frokiaer, J., and Marples, D. (1999). Physiology and pathophysiology of renal aquaporins. *J. Am. Soc. Nephrol.* **10,** 647–663.

Ohara, M., Martin, P. Y., Xu, D. L., St John, J., Pattison, T. A., Kim, J. K., and Schrier, R. W. (1998). Upregulation of aquaporin 2 water channel expression in pregnant rats. *J. Clin. Invest.* **101,** 1076–1083.

Peysner, K., and Forsling, M. L. (1990). Effect of ovariectomy and treatment with ovarian steroids on vasopressin release and fluid balance in the rat. *J. Endocrinol.* **124,** 277–284.

Preisser, L., Teillet, L., Aliotti, S., Gobin, R., Berthonaud, V., Chevalier, J., Corman, B., and Verbavatz, J. (2000). Downregulation of aquaporin-2 and -3 in aging kidney is independent of V(2) vasopressin receptor. *Am. J. Physiol. Renal Physiol.* **279,** F144–F152.

Preston, G. M., and Agre, P. (1991). Isolation of the cDNA for erythrocyte integral membrane protein of 28 kilodaltons: Member of an ancient channel family. *Proc. Natl. Acad. Sci. USA* **88,** 11110–11114.

Preston, G. M., Carroll, T. P., Guggino, W. B., and Agre, P. (1992). Appearance of water channels in *Xenopus* oocytes expressing red cell CHIP28 protein. *Science* **256,** 385–387.

Sabolic, I., Katsura, T., Verbavatz, J. M., and Brown, D. (1995). The AQP2 water channel: Effect of vasopressin treatment, microtubule disruption, and distribution in neonatal rats. *J. Membr. Biol.* **143,** 165–175.

Saito, T., Ishikawa, S. E., Sasaki, S., Fujita, N., Fushimi, K., Okada, K., Takeuchi, K., Sakamoto, A., Ookawara, S., Kaneko, T., Marumo, F., and Saito, T. (1997). Alteration in water channel AQP-2 by removal of AVP stimulation in collecting duct cells of dehydrated rats. *Am. J. Physiol.* **272,** F183–F191.

Sands, J. M., Naruse, M., Jacobs, J. D., Wilcox, J. N., and Klein, J. D. (1996). Changes in aquaporin-2 protein contribute to the urine concentrating defect in rats fed a low-protein diet. *J. Clin. Invest.* **97,** 2807–2814.

Sands, J. M., Flores, F. X., Kato, A., Baum, M. A., Brown, E. M., Ward, D. T., Hebert, S. C., and Harris, H. W. (1998). Vasopressin-elicited water and urea permeabilities are altered in IMCD in hypercalcemic rats. *Am. J. Physiol.* **274,** F978–F985.

Sardeli, C., Li, C., Kwon, T. H., Knepper, M. A., Nielsen, S., and Frokiaer, J. (2000). Aquaporin-2 expression is increased in rat kidneys after ovariectomy. *FASEB J.* **14,** A345.

Schnermann, J., Chou, C. L., Ma, T., Traynor, T., Knepper, M. A., and Verkman, A. S. (1998). Defective proximal tubular fluid reabsorption in transgenic aquaporin-1 null mice. *Proc. Natl. Acad. Sci. USA* **95,** 9660–9664.

Schrier, R. W., and Briner, V. A. (1991). Peripheral arterial vasodilation hypothesis of sodium and water retention in pregnancy: Implications for pathogenesis of preeclampsia-eclampsia. *Obstet. Gynecol.* **77,** 632–639.

Schrier, R. W., Ohara, M., Rogachev, B., Xu, L., and Knotek, M. (1998). Aquaporin-2 water channels and vasopressin antagonists in edematous disorders. *Mol. Genet. Metab.* **65,** 255–263.

Tamarappoo, B. K., Yang, B., and Verkman, A. S. (1999). Misfolding of mutant aquaporin-2 water channels in nephrogenic diabetes insipidus. *J. Biol. Chem.* **274,** 34825–34831.

Tannen, R. L., Regal, E. M., Dunn, M. J., and Schrier, R. W. (1969). Vasopressin-resistant hyposthenuria in advanced chronic renal disease. *N. Engl. J. Med.* **280,** 1135–1141.

Tanner, G. A., Sloan, K. L., and Sophasan, S. (1973). Effects of renal artery occlusion on kidney function in the rat. *Kidney Int.* **4,** 377–389.

Teillet, L., Preisser, L., Verbavatz, J. M., and Corman, B. (1999). [Kidney aging: cellular mechanisms of problems of hydration equilibrium]. *Therapie* **54,** 147–154.

Teitelbaum, I., and McGuinness, S. (1995). Vasopressin resistance in chronic renal failure. Evidence for the role of decreased V2 receptor mRNA. *J. Clin. Invest.* **96,** 378–385.

Terris, J., Ecelbarger, C. A., Marples, D., Knepper, M. A., and Nielsen, S. (1995). Distribution of aquaporin-4 water channel expression within rat kidney. *Am. J. Physiol.* **269,** F775–F785.

Terris, J., Ecelbarger, C. A., Nielsen, S., and Knepper, M. A. (1996). Long-term regulation of four renal aquaporins in rats. *Am. J. Physiol.* **271,** F414–F422.

Tsukaguchi, H., Shayakul, C., Berger, U. V., Mackenzie, B., Devidas, S., Guggino, W. B., Van Hoek, A. N., and Hediger, M. A. (1998). Molecular characterization of a broad selectivity neutral solute channel. *J. Biol. Chem.* **273,** 24737–24743.

van Lieburg, A. F., Verdijk, M. A., Knoers, V. V., van Essen, A. J., Proesmans, W., Mallmann, R., Monnens, L. A., Van Oost, B. A., van Os, C. H., and Deen, P. M. (1994). Patients with autosomal nephrogenic diabetes insipidus homozygous for mutations in the aquaporin 2 water-channel gene. *Am. J. Hum. Genet.* **55,** 648–652.

Venkatachalam, M. A., Bernard, D. B., Donohoe, J. F., and Levinsky, N. G. (1978). Ischemic damage and repair in the rat proximal tubule: Differences among the S1, S2, and S3 segments. *Kidney Int.* **14,** 31–49.

Verbalis, J. G., Murase, T., Ecelbarger, C. A., Nielsen, S., and Knepper, M. A. (1998). Studies of renal aquaporin-2 expression during renal escape from vasopressin-induced antidiuresis. *Adv. Exp. Med. Biol.* **449,** 395–406.

Wen, H., Frokiaer, J., Kwon, T. H., and Nielsen, S. (1999). Urinary excretion of aquaporin-2 in rat is mediated by a vasopressin-dependent apical pathway. *J. Am. Soc. Nephrol.* **10,** 1416–1429.

Wilson, D. R., and Sonnenberg, H. (1979). Medullary collecting duct function in the remnant kidney before and after volume expansion. *Kidney Int.* **15,** 487–501.

Wood, L. J., Massie, D., McLean, A. J., and Dudley, F. J. (1988). Renal sodium retention in cirrhosis: Tubular site and relation to hepatic dysfunction. *Hepatology* **8,** 831–836.

Xu, D. L., Martin, P. Y., Ohara, M., St. John, J., Pattison, T., Meng, X., Morris, K., Kim, J. K., and Schrier, R. W. (1997). Upregulation of aquaporin-2 water channel expression in chronic heart failure rat. *J. Clin. Invest.* **99,** 1500–1505.

Yamamoto, T., Sasaki, S., Fushimi, K., Kawasaki, K., Yaoita, E., Oota, K., Hirata, Y., Marumo, F., and Kihara, I. (1995). Localization and expression of a collecting duct water channel, aquaporin, in hydrated and dehydrated rats. *J. Exp. Nephrol.* **3,** 193–201.

Yarger, W. E. (1991). Urinary tract obstruction. In "The Kidney" (B. M. Brenner and F. C. Rector, eds.), pp. 1768–1808. Saunders, Philadelphia.

Yasui, M., Kwon, T. H., Knepper, M. A., Nielsen, S., and Agre, P. (1999a). Aquaporin-6: An intracellular vesicle water channel protein in renal epithelia. *Proc. Natl. Acad. Sci. USA* **96,** 5808–5813.

Yasui, M., Hazama, A., Kwon, T. H., Nielsen, S., Guggino, W. B., and Agre, P. (1999b). Rapid gating and anion permeability of an intracellular aquaporin. *Nature* **402,** 184–187.

CHAPTER 5

Genetic and Biophysical Approaches to Study Water Channel Biology

A. S. Verkman,[*] Baoxue Yang,[*] William R. Skach,[†] Alok Mitra,[+]
Yuanlin Song,[*] Geoffrey T. Manley,[*] and Tonghui Ma[*]

[*]Departments of Medicine and Physiology, Cardiovascular Research Institute, University of California, San Francisco, California 94143-0521; [†]Division of Molecular Medicine, Oregon Health Sciences University, Portland, Oregon 97201; [+]Department of Cell Biology, Scripps Research Institute, La Jolla, California 92037

I. INTRODUCTION

There are at least 10 aquaporin water channels in mammals and many more in plants and lower organisms. There is an extensive body of information about aquaporin molecular genetics, tissue localization, developmental expression, molecular structure, and function. It has been assumed that aquaporins are centrally important in mammalian physiology based on their expression patterns and the high water permeabilities at sites of aquaporin expression. However, there has been little direct investigation about the physiological significance of aquaporins, in part because there are no aquaporin inhibitors suitable for *in vivo* use. In humans, mutations in the vasopressin-regulated water channel AQP2 in kidney collecting duct cause hereditary autosomal nephrogenic diabetes insipidus (NDI). Our laboratory has addressed the role of the other aquaporins in mammalian physiology by analyzing the phenotype of aquaporin knockout mice generated by targeted gene disruption. The first part of this review is focused on the general paradigms that have emerged from *in vivo* phenotype analysis of aquaporin knockout mice. General issues in the development and analysis of transgenic mouse models are discussed, recent phenotype data on aquaporin knockout mice are reviewed, and a working hypothesis is presented about the role of aquaporins in mammalian physiology.

The latter part of this review concerns the biophysical analysis of aquaporin structure and function. Our group has developed a number of quantitative biophysical methods to measure water permeability across cellular and organ barriers, which have been particularly useful in measurements of tissues from the aquaporin knockout mice. The biophysics of water transport is briefly reviewed, and new experimental strategies for water permeability measurements are described. Last, recent information from our group about aquaporin structure and function is summarized.

II. LESSONS FROM AQUAPORIN KNOCKOUT MICE

A. Generation and Phenotype Analysis of Transgenic Mice

Methodology has rapidly advanced for the generation of various transgenic mice involving gene transfer and conditional or organ-specific gene deletion/replacement. The generation of knockout mice by targeted gene disruption has become a routine procedure. After analysis of the exon–intron structure of the selected mouse gene, a targeting vector is constructed in which homologous recombination in embryonic stem (ES) cells produces a defective gene with partial deletion of the coding sequence. Practical features in vector design include sufficiently long arms of genomic sequence matching the ES cell genome (generally total arm length > 4 kb), the insertion of selection markers (generally neomycin for positive selection and thymidine kinase for negative selection) for ES cell clone screening, and the use of one relatively short arm for PCR screening. After ES cell transfection, selection of correctly integrated clones, and screening for genetic abnormalities, the ES cells are transferred to pseudopregnant female mice to produce chimeric offspring. Breeding of chimeric and wild-type mice produces heterozygous mice containing the modified gene in the germ line, which are then bred to produce homozygous mice with the null genotype. Because several breeding steps are required, the minimum time between the start of a transgenic mouse project and the birth of homozygous transgenic mice is at least 8 months.

After obtaining litters from intercross of germ line heterozygotes, the mice are genotyped and the null genotype is confirmed by the absence of full-length transcript and detectable protein. The generation of transgenic mice represents only the first step in what is often a challenging project in deducing the role of a protein in mammalian physiology. Initial phenotype analysis includes determination of genotype distribution of offspring from heterozygote matings and observations of gross mouse appearance, activity, survival, growth, and organ morphology. A myriad of specific physiological tests have been developed for analysis of organ-specific mouse phenotypes (for review, see Rao and Verkman, 2000).

There are many considerations in concluding that a gene is physiologically significant in mice and relevant to human biology. Defects in organ function can result directly from protein deletion or indirectly from secondary factors such as altered organ development/structure, hemodynamics, or electrolyte composition. For example, changes in a gross phenotype such as treadmill performance can result from a myriad of unrelated abnormalities in muscle protein function, mouse behavior, hemodynamics, lung gas exchange, serum electrolyte composition, and so on. There may be compensatory changes in the expression patterns of other proteins. Phenotype results can be influenced by genetic background, environmental factors (temperature, light cycle), dietary factors, and cage crowding. There may be diurnal variations and phenotypes may be influenced by mouse age and sex. Because infections and cage-specific factors can influence phenotype, phenotype comparisons are often done on litter-matched mice maintained in the same environment. Because of differences in mouse and human physiology, the extrapolation of data in mice to humans must be made with caution. For example, the maximum osmolality of mouse urine (>3000 mOsm) is much greater than that of human urine (~1000 mOsm). Protein expression patterns and thus the interpretation of phenotype studies may also be species dependent. For example, water channel AQP4 is expressed in both proximal tubule and collecting duct in mouse, but only in collecting duct in rats and humans (van Hoek *et al.*, 2000). Thus, the challenges in mouse phenotype analysis include not only technical factors in transgenic mouse generation and physiological recordings, but also numerous genetic, environmental, and species/strain-specific factors.

B. Neonatal Survival, Growth, and Appearance of Aquaporin Knockout Mice

Aquaporin null mice are produced by breeding of heterozygous mice in which one allele contains a disrupted gene. Homozygous knockout mice lacking aquaporins 1, 3, 4, and 5 are viable at birth, indicating that these aquaporins are not essential for fetal development. In addition, we have generated viable double knockout mice lacking pairs of aquaporins, AQP1/AQP3, AQP1/AQP4, AQP1/AQP5, and AQP3/AQP4. Genotype analysis is generally performed at age 5 days by PCR amplification using DNA extracted from tail fragments. A 1:2:1 Mendelian distribution of genotypes (wild-type:heterozygous:knockout) is predicted for autosomal genes like the aquaporins. An approximately 1:2:1 distribution was found for offspring from breeding of AQP3 and AQP4 heterozygous mice (Ma *et al.*, 1997, 2000b). However, there were significantly fewer knockout mice than predicted from breeding of AQP1 (18%; Ma *et al.*, 1998) and AQP5 (14%; Ma *et al.*, 1999) heterozygous mice, indicating prenatal or perinatal mortality. Growth of AQP3 and AQP4 knockout mice was similar to that of litter-matched wild-type mice, whereas AQP1 and AQP5 knockout mice were generally 10–15% smaller (by weight) than litter mates. These empirical observations provide no mechanistic information since growth retardation has numerous etiologies such as behavioral, endocrine, gastrointestinal, and renal abnormalities.

C. Renal Function in Aquaporin Null Mice

The kidney is an important site of aquaporin expression, with AQP1 in proximal tubule, thin descending limb of Henle, and descending vasa recta; AQP2, AQP3, and AQP4 in collecting duct principal cells; AQP6 in collecting duct intercalated cells; and AQP7 in the S3 segments of proximal tubule (Fig. 1A) (for review, see Nielsen *et al.*, 1998; Verkman, 2000b, Yamamoto and Sasaki, 1998). Integrated renal function was assessed in AQP1, AQP3, and AQP4 null mice by measurement of fluid input–output and urinary concentrating ability. AQP1 and AQP3 null mice were remarkably polyuric, consuming 2–3 (AQP1) to 8–10 (AQP3) times more fluid than wild-type mice, whereas the AQP4 null mice were not polyuric. Given free access to food and water, urine osmolalities of the AQP1 and AQP3 null mice were much lower than those of wild-type and AQP4 null mice (Fig. 1B). Urinary concentrating ability was determined by measuring mouse weight loss and urine osmolality in response to a 36-h water deprivation. A remarkable observation was that water-deprived AQP1 null mice continue to produce large volumes of urine and become severely dehydrated and lethargic (Ma *et al.*, 1998). Total body weight decreased by 35%, serum osmolality in many mice increased to >500 mOsm, and urinary osmolality did not increase

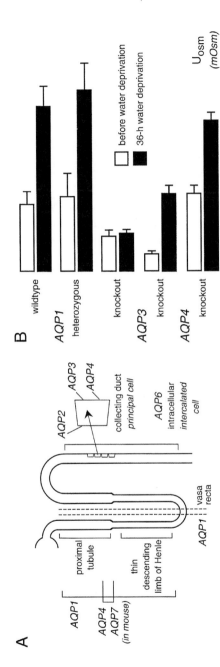

FIGURE 1 Urinary concentrating ability in aquaporin null mice. (A) Location of aquaporin water channels in kidney tubules. (B) Urine osmolality before and after a 36-h water deprivation in mice of indicated genotype. (Data summarized from Ma *et al.*, 1997, 1998, 2000b.)

above that before water deprivation. In contrast, wild-type mice remained active after water deprivation, body weight decreased by 20–22%, serum osmolality remained normal (310–330 mOsm), and urine osmolality rose to >2500 mOsm. Urine [Na^+] in water-deprived AQP1 knockout mice was <10 mM and urine osmolality was not increased by the V_2 agonist dDAVP. These findings suggested that the AQP1 knockout mice are unable to create a hypertonic medullary interstitium by countercurrent multiplication. The AQP4 knockout mice had a small but significant decrease in maximal urine osmolality after a 36-h water deprivation and did not manifest clinical signs of severe dehydration (Ma *et al.*, 1997). Interestingly, despite the marked polyuria in AQP3 null mice when given free access to water, these mice were able to partially concentrate their urine after water deprivation and did not manifest signs of severe dehydration (Ma *et al.*, 2000b). These results established an important role for aquaporins in the urinary concentrating mechanism.

1. Analysis of Kidney Tubule Function in AQP1 Null Mice

AQP1 is expressed at the apical and basolateral membranes of proximal tubule, where most of the fluid filtered by the glomerulus is reabsorbed by an active, near-isosmolar transport mechanism. Transepithelial osmotic water permeability (P_f) was measured in isolated microperfused S2 segments of proximal tubule using a raffinose gradient to drive water out of the tubule lumen (Schnermann *et al.*, 1998). P_f was decreased by nearly 5-fold in AQP1 knockout mice (0.15 ± 0.03 cm/s wild-type vs. 0.033 ± 0.005 cm/s knockout). The major pathway for osmotically driven water transport in the perfused S2 segment of proximal tubule is thus transcellular and mediated by AQP1 water channels. The low P_f after AQP1 deletion is consistent with water movement through the lipid bilayer, suggesting that other aquaporin-type water channels and nonaquaporin transporters make little or no contribution to proximal tubule water permeability, and that paracellular water movement is minimal.

The role of AQP1 in near-isosmolar fluid reabsorption in proximal tubule *in vivo* was studied by free-flow micropuncture (Schnermann *et al.*, 1998). Fluid samples in anesthetized mice were obtained from the end proximal tubule at surface-accessible sites for determination of percentage fluid absorption using iothalamate as an impermeant marker. Glomerular filtration rate was 317 ± 34 μl/min in wild-type and 230 ± 33 μl/min in AQP1 knockout mice. The fractional absorption of fluid along the proximal tubule, determined from end-proximal tubule vs. serum iothalamate concentrations, was remarkably reduced in AQP1 knockout mice ($48 \pm 3\%$ wild-type vs. $26 \pm 3\%$ knockout). These data indicate that the decreased proximal tubule water permeability in AQP1 knockout mice is associated with impaired isomolar absorption, which is consistent with three-compartment models in which mild luminal hypotonicity drives osmotic water movement through highly water-permeable cell membranes (Spring, 1998). Recent data indicate a marked decrease in luminal fluid osmolality in end-proximal tubule fluid from AQP1 knockout mice, consistent with active pumping of salt out of the tubule lumen without adequate water movement to dissipate the osmotic gradient.

Fluid collections made in the superficial distal tubule gave fractional fluid absorption values of $76 \pm 3\%$ in wild-type mice and $62 \pm 4\%$ in AQP1 knockout mice. Interestingly, flow rates in the distal nephron did not differ between wild-type and knockout mice despite reduced absorption because of a reduction in single nephron glomerular filtration rate (SNGFR) in the AQP1 knockout mice (11 ± 2 nl/min wild-type vs. 5.1 ± 0.4 nl/min knockout). It was proposed that activation of the tubuloglomerular feedback mechanism was responsible for the diminished SNGFR, an hypothesis that will require direct experiment verification. Whatever the mechanism, the decreased GFR in AQP1 knockout mice is an appropriate compensatory response to the threat of NaCl depletion caused by defective proximal tubule reabsorption. Further studies are needed to determine whether other compensatory responses occur in kidneys of AQP1 null mice, such as up-regulation of tubule ion transporters or changes in interstitial barrier properties.

The increased urinary flow rates despite normal distal fluid delivery suggest that the diuresis seen in AQP1 knockout mice results primarily from reduced fluid absorption in the collecting duct. Given the normally high expression of AQP1 in descending limb of Henle and vasa recta endothelia, it is likely that AQP1 deletion results in defective countercurrent multiplication, preventing the formation of a hypertonic medullary interstitium. This conclusion is supported by the finding that in water-deprived AQP1 knockout mice, dDAVP stimulation of collecting duct water permeability (that should nearly equalize urine and medullary interstitial osmolalities) did not increase urine osmolality (Ma *et al.*, 1998). It is noted that unlike nephrogenic diabetes insipidus due to AQP2 mutations, where urine osmolality is quite low, the urine can be mildly concentrated in AQP1 deficiency because salt transporters are functional and collecting duct water permeability can increase strongly. It may be for this reason that no overt abnormalities were found in AQP1-deficient humans not subject to a water deprivation stress (Preston *et al.*, 1994).

The above considerations implicate a defective countercurrent multiplication mechanism in the AQP1 null mice. The thin descending limb of Henle (TDLH) and outer medullary descending vasa recta (OMDVR) have critical roles in generating a hypertonic medullary interstitium. Measurements from several laboratories have indicated that transepithelial P_f in TDLH is exceptionally high. Measurement of AQP1 protein in microdissected tubule segments using a fluorescence-based ELISA method showed that TDLH of long-looped nephrons have the highest AQP1 content among nephron segments (Maeda *et al.*, 1995). To determine the contribution of AQP1 to TDLH water permeability, transepithelial P_f was measured in isolated perfused segments of TDLH at $37°C$ using a 100 mM bath-to-lumen osmotic gradient of raffinose (Chou *et al.*, 1999). P_f was remarkably decreased in the AQP1 knockout mice (0.26 ± 0.02 cm/s wild-type vs. 0.031 ± 0.007

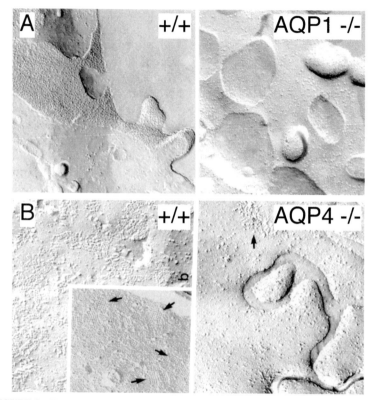

FIGURE 2 Ultrastructure analysis of kidney cell plasma membranes by freeze-fracture electron microscopy. (A) P-face plasma membranes of TDLH epithelial cells from wild-type mice (*left*) and AQP1 knockout mice (*right*). (B) P-face basolateral plasma membrane of collecting duct from wild-type mice (*left*) and AQP4 knockout mice (*right*). Complementary orthogonal array imprints were observed on the Ef-face (inset) in wild-type mice. See text for further explanation. Scale bars: 75 nm. (Adapted from Verbavatz *et al.*, 1997, and Chou *et al.*, 1999.)

knockout). These results indicate that AQP1 is the principal water channel in TDLH and support the view that osmotic equilibration along TDLH by water transport plays a key role in the renal countercurrent concentrating mechanism. The AQP1 null mice were also used to confirm that AQP1 is responsible for the high density of intramembrane particles (IMPs) observed in TDLH membranes by freeze-fracture electron microscopy (FFEM). Previously, comparisons of IMP ultrastructure in TDLH and AQP1 reconstituted proteoliposomes suggested that the majority of the IMPs in plasma membranes of rat TDLH consist of AQP1 tetramers (Verbavatz *et al.*, 1993). FFEM of TDLH showed an ~6-fold decreased IMP density and different IMP appearance in AQP1 knockout vs. wild-type mice (Fig. 2A)

(Chou *et al.*, 1999). Size distribution analysis indicated that the diameter of the majority of IMPs in TDLH of wild-type mice was 8.4 nm, in agreement with the size of tetrameric AQP1 IMPs reported by Verbavatz *et al.* Together these results indicate that AQP1 is responsible for the exceptionally high water permeability and membrane protein density in TDLH.

It is believed that the microcirculation of the renal medulla traps NaCl and urea by countercurrent exchange in order to preserve the medullary hypertonicity generated by the loops of Henle and collecting duct. The role of AQP1 in OMDVR water transport was determined by measurement of P_f in isolated microperfused OMDVR (Pallone *et al.*, 2000). In response to a 200 mM NaCl gradient (bath > lumen) P_f was dramatically reduced in OMDVR from AQP1 null mice (0.10 ± 0.02 cm/s wild-type vs. 0.009 ± 0.004 cm/s knockout). Interestingly, OMDVR diameters from AQP1 null mice were nearly 2-fold greater than those of wild-type mice, suggesting a chronic or developmental adaptive mechanism. When raffinose rather than NaCl was used to drive osmotic water movement, P_f values were 0.25 ± 0.04 cm/s (wild-type) vs. 0.10 ± 0.02 cm/s (knockout). Together with mercurial inhibition studies, it was concluded that solutes larger than NaCl are able to drive water movement both through AQP1 and by an AQP1-independent, mercurial-insensitive pathway that may involve paracellular transport. The results were incorporated into a mathematical model of the medullary microcirculation, which predicted that the decreased medullary hypertonicity in AQP1 null mice results from enhanced papillary blood flow.

2. Role of AQP3 and AQP4 in Collecting Duct

AQP3 and AQP4 are expressed at the basolateral membrane of collecting duct epithelium, with relatively greater expression of AQP3 in cortical and outer medullary collecting duct and AQP4 in inner medullary collecting duct (Ecelbarger *et al.*, 1995; Frigeri *et al.*, 1995a; Terris *et al.*, 1995). The data in Fig. 1B suggest a more important role for AQP3 than AQP4 with respect to the urinary concentrating mechanism. Isolated tubule perfusion measurements were done to determine the contribution of AQP4 to water permeability in the inner medullary collecting duct (IMCD) (Chou *et al.*, 1998). Transepithelial P_f was measured in microdissected IMCDs after water deprivation and in the presence of vasopressin to make basolateral membrane P_f rate limiting. P_f values at 37°C were 0.056 ± 0.008 cm/s in wild-type mice and 0.013 ± 0.003 cm/s in AQP4 knockout mice. Northern blot and immunoblot analysis of kidney showed that expression of other renal aquaporins was not affected by AQP4 deletion. These results indicated that AQP4 is responsible for the majority of basolateral membrane water movement in IMCD. The very mild defect in urinary concentrating ability in AQP4 null mice, despite their greatly reduced IMCD water permeability, supports the conclusion that the amount of water reabsorbed in the cortical portion of the collecting duct far exceeds that absorbed in the medullary collecting duct.

As discussed earlier, AQP3 null mice are remarkably polyuric and polydipsic when given free access to water (Ma *et al.,* 2000b). The normal functioning of the diluting segment, the high luminal flow, and the low collecting duct water permeability in the diuretic kidney permit the excretion of a hypo-osmolar urine. However, the AQP3 null mice are able to generate a partially concentrated urine in response to dDAVP administration or water deprivation. In contrast to AQP1 null mice, where the countercurrent exchange mechanism is defective, counter-current exchange in AQP3 null mice is basically intact, though medullary inter-stitial osmolalities are probably lower than in wild-type mice because of diuresis washout. The generation of a partially concentrated urine with a relatively water-impermeable collecting duct is a well-documented phenomenon that is related to decreased glomerular filtration rate and low luminal flow in collecting duct. The low flow and increased fluid contact time facilitate osmotic extraction of water from the collecting duct lumen. However, low luminal flow cannot fully explain the partial increase in urine osmolality in well-hydrated AQP3 null mice after dDAVP administration. It follows that water permeability of the collecting duct basolateral membrane is not an absolute rate-limiting barrier for transepithelial water permeability in the antidiuretic kidney. To test the hypothesis that the resid-ual concentrating ability of AQP3 null mice was due to the IMCD water channel AQP4, AQP3/AQP4 double knockout mice were generated. The double knockout mice had greater impairment of urinary concentrating ability than the AQP3 single knockout mice. Using a spatial filtering optical method, osmotic water permeabil-ity in cortical collecting duct was ~3-fold reduced by AQP3 deletion. These data establish a novel form of nephrogenic diabetes insipidus produced by impaired water permeability in collecting duct basolateral membrane. Basolateral mem-brane aquaporins may thus provide blood-accessible targets for drug discovery of aquaretic inhibitors.

The AQP4 knockout mouse was used to demonstrate that AQP4 is the "orthog-onal array protein" (OAP). OAPs are characteristic square arrays of IMPs that have been observed by FFEM in the basolateral membrane of kidney collecting duct epithelia and several other cell types. We initially proposed that AQP4 is the OAP protein based on the finding that AQP4 is present in cells in which OAPs had been found (Frigeri *et al.,* 1995b). Initial support for this hypothesis came from FFEM studies on AQP4 transfected CHO cells, where OAPs with large patch sizes were found with essentially identical ultrastructure to those seen in mammalian cells (Yang *et al.,* 1996). Definitive evidence that AQP4 is the OAP protein came from freeze-fracture studies on kidney, skeletal muscle, and brain from AQP4 knockout mice (Verbavatz *et al.,* 1997) (Fig. 2B). OAPs were identified in every sample from wild-type and heterozygous mice, and in no sample from AQP4 knockout mice. Label-fracture analysis showed immunogold labeling of OAPs in AQP4-expressing CHO cells. These results provided direct evidence that AQP4 is the OAP protein, thus establishing the identity and function of OAPs.

3. Adenoviral Gene Therapy in AQP1 Knockout Mice

We have exploited the unambiguous renal phenotype of AQP1 knockout mice—the development of lethargy after water deprivation and inability to concentrate their urine—to study the consequences of AQP1 delivery by an adenoviral vector (Yang *et al.*, 2000b). An Ad5 adenovirus containing the AQP1 coding sequence (Ad5-AQP1) was generated and shown to confer high-level expression of functional AQP1 water channels in cell culture. Ad5-AQP1 was delivered to AQP1 knockout mice by a single intravenous (tail-vein) infusion. Infection of CHO cells gave strong uniform AQP1 expression with plasma membrane localization and 8-fold increased water permeability over noninfected cells. AQP1-Ad5 was delivered to AQP1 null mice by tail-vein infusion. At 3–7 days, AQP1 protein expression was strongest in liver ($\sim$20 μg AQP1 protein) and next strongest in kidney, with expression in proximal tubule apical and basolateral membranes, and renal microvessels. Functional analysis showed increased water permeability in apical membrane vesicles from proximal tubule. AQP1 expression was not detected in glomerulus, limb of Henle, or collecting duct. In water-deprived null mice receiving 5×10^9 pfu AQP1-Ad5, U_{osm} increased by up to 510 mOsm (mean increase $\sim$225 mOsm in 33 mice). Whereas the control null mice became very lethargic, the virus-treated mice remained relatively active and were less dehydrated. Viral DNA and AQP1 transcript were detected in kidney and liver of null mice up to 17 weeks after virus infusion; partial correction of the urinary concentrating defect persisted for 3–5 weeks. These results demonstrated partial functional correction of a urinary concentrating defect by adenoviral delivery of the AQP1 gene. A single administration of adenovirus produced strong aquaporin expression in kidney and transient improvement in urinary concentrating ability. The kidney thus appears to be a suitable target organ for short-term expression of a target protein. We are also using aquaporin null mice to investigate the functional consequences of aquaporin gene delivery by viral and nonviral vectors to other organs including the salivary gland and lung.

4. Mouse Model of Non-X-Linked Nephrogenic Diabetes Insipidus

Vasopressin strongly increases transepithelial water permeability in kidney collecting duct. Mutations in AQP2 cause the rare non-X-linked form of hereditary NDI (Deen *et al.*, 1994), in which defective urinary concentrating ability causes hypernatremia and dehydration if adequate hydration is not provided. Heterologous expression studies in *Xenopus* oocytes suggested that impaired routing of some AQP2 mutants might contribute to the cellular pathogenesis of NDI (Deen *et al.*, 1997). We have used a mammalian cell transfection model to investigate the mechanisms responsible for defective cell water permeability in NDI caused by AQP2 mutations (Tamarappoo and Verkman, 1998). In a *Xenopus* oocyte expression system, it was found that the intrinsic water permeabilities were similar for wild-type AQP2 vs. the mutants L22V, T126M, and A147T, whereas the R187C and C181W AQP2 mutants were nonfunctional. In transiently transfected CHO cells that were pulse labeled with [^{35}S]-methionine, half-times for AQP2 degradation were $\sim$4 h for wild-type AQP2 and L22V, and mildly decreased by up to 50% for the NDI-causing mutants. Subcellular localization as determined by immunofluorescence and membrane fractionation indicated that a major reason for defective cell function was mistrafficking of AQP2 mutants with retention at the endoplasmic reticulum. Analysis of mutant AQP2 folding at the endoplasmic reticulum indicated remarkable resistance to membrane extraction by nonionic detergents compared to wild-type AQP2 (Tamarappoo *et al.*, 1999b). Interestingly, the endoplasmic reticulum-retained AQP2 mutants were functional as water channels, suggesting that maneuvers that correct AQP2 trafficking might restore normal phenotype in collecting duct cells.

We investigated one strategy to correct the trafficking defect in NDI involving the use of "chemical chaperones"—small molecules (such as glycerol) that facilitate protein folding *in vitro*. Glycerol has been shown in cell culture models to correct defective processing of ΔF508 CFTR in cystic fibrosis, the prion protein PrP, the viral oncogene protein pp60^{src}, and mutants of the tumor suppressor protein p53. It was found that growth of cells in media containing 1 *M* glycerol for 48 h resulted in correction of mutant AQP2 mistrafficking as assessed by localization studies and measurement of cell membrane water permeability (Tamarappoo and Verkman, 1998). The trafficking defect was also corrected by other chemical chaperones at lower concentrations including DMSO and trimethylamine oxide (TMAO, 50–100 m*M*). Proof-of-principle studies in mice showed that "therapeutic" TMAO concentrations could be obtained by intraperitoneal injections and that partial correction of the ΔF508 mistrafficking defect could be achieved in cystic fibrosis mice (Bai *et al.*, 1998). *In vivo* experiments are planned using a mouse model of human NDI. The mouse AQP2 cDNA and gene were analyzed, and a mouse ortholog of human NDI was characterized (Yang *et al.*, 1999b). Recently, we have generated homozygous AQP2-T126M knock-in mice (Yang *et al.*, 2000d). These mice have marked polyurla resulting in neonatal mortality.

D. Role of Aquaporins in Active Fluid Secretion and Absorption

The renal micropuncture data mentioned above indicate that the high water permeability in proximal tubule conferred by AQP1 is required for efficient fluid absorption (Schnermann *et al.*, 1998). The general paradigm supported by that study is that high water permeability is required for active, near-isosmolar fluid transport, where osmosis is driven by small osmotic gradients created by ion pumping. Similar fluid-transporting mechanisms are thought to occur in many absorptive and secretory epithelia such as exocrine glands, lung alveolus, choroid plexus, and ciliary body. To study near-isosmolar transport in a secretory epithelia, pilocarpine-stimulated saliva secretion was measured in mice lacking AQP5, the apical membrane water channel of epithelial cells in salivary gland acini (Ma *et al.,* 1999). As shown in Fig. 3A, wild-type and AQP5 heterozygous mice produce large amounts of clear, nonviscous saliva in the first 5 min after pilocarpine injection. In contrast, the AQP5 null mice produce little highly viscous fluid that remains stuck to the glass collection tube. We obtained similar results recently by direct cannulation of the parotid duct in wild-type and AQP5 null mice. Under normal conditions, primary saliva is produced by active salt pumping into the acinus of salivary gland, followed by secondary osmotic water transport. The near-isosmolar primary saliva then flows through a relatively water-impermeable salivary duct where ion pumping and exchange occur to produce a hypotonic saliva. In AQP5 null mice, fluid composition analysis indicated a hyperosmolar hypernatremic saliva compared to saliva from wild-type mice (Fig. 3B). These findings suggests that AQP5 null mice are able to pump salt actively into the acinar lumen, but that water permeability is too low to permit osmotic equilibration. These observations support the paradigm that high epithelial water permeability facilitates active fluid transport.

E. Aquaporins and Brain Edema

AQP4 protein is strongly expressed in astroglial cells at the blood–brain barrier and brain–cerebrospinal fluid (CSF) barrier. AQP4 protein is found in glial cells lining ependyma and pial surfaces in contact with the CSF. Initial evaluation of AQP4 null mice showed no overt neurologic abnormalities or defects in osmoregulation (Ma *et al.,* 1997). Recently, the hypothesis was tested that AQP4 plays a role in the accumulation of brain water in response to two established neurologic insults: acute water intoxication, producing hyponatremia and cellular brain edema, and ischemic stroke, producing brain swelling by a combination of cellular and vasogenic edema (Manley *et al.,* 2000). AQP4 deletion conferred remarkable protection from brain edema in these models. Figure 4A shows that survival of AQP4 null mice after acute water intoxication is greatly improved, which corresponded to significantly reduced brain swelling as quantified by transmission electron microscopy and parenchymal specific gravity measurements. At 24 h after ischemic stroke produced by permanent middle cerebral artery occlusion, there was improved clinical outcome and much less brain swelling (Fig. 4B). The data implicate a key role for AQP4 in modulating brain edema. AQP4 may be also be important in the pathophysiology of brain edema in other disease states such as tumor, infection, head trauma, and spinal cord injury. For the past 70 years, the only effective approaches to treat brain edema have been craniotomy (partial removal of the bony skull) to permit unimpeded brain swelling, and intravenous mannitol

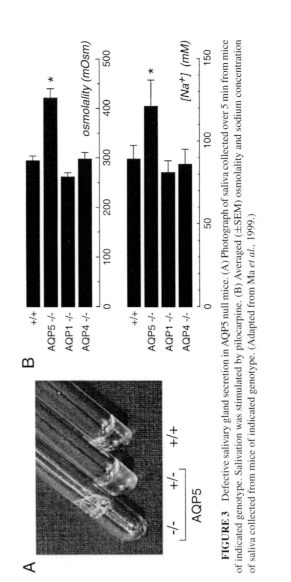

FIGURE 3 Defective salivary gland secretion in AQP5 null mice. (A) Photograph of saliva collected over 5 min from mice of indicated genotype. Salivation was stimulated by pilocarpine. (B) Averaged ($\pm$SEM) osmolality and sodium concentration of saliva collected from mice of indicated genotype. (Adapted from Ma *et al.*, 1999.)

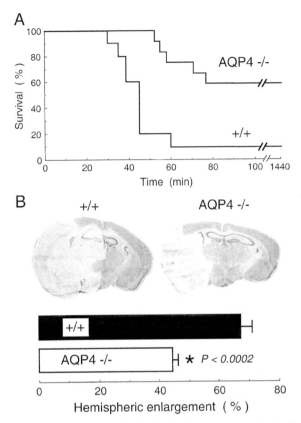

FIGURE 4 Reduced brain edema in AQP4 null mice after acute water intoxication and ischemic stroke. (A) Survival of wild-type vs. AQP4 knockout mice after acute water intoxication produced by intraperitoneal water infusion. (B, *top*) Brain sections of mice 24 h after ischemic stroke produced by permanent middle cerebral artery occlusion. Note midline shift and marked edema in brain from wild-type mice. (B, *bottom*) Averaged hemispheric enlargement expressed as a percentage determined by image analysis of brain sections. (Adapted from Manley *et al.*, 2000.)

administration to osmotically drive water out of the brain. Pharmacological inhibition of AQP4 might provide a novel therapy to minimize brain edema in stroke and other central nervous system disorders.

F. Aquaporins and the Lung

Fluid movement between the airspace and vascular compartments in lung plays an important physiological role in many processes such as regulation of airway

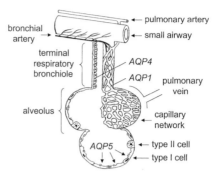

FIGURE 5 Barriers to water movement in lung. Anatomy of the distal lung showing blood supply, airspace epithelia, and locations of aquaporin water channel proteins. Transport of water between the airspace and capillary compartments involves movement across alveolar or airway epithelia, interstitium, and endothelium. Transcellular water transport across epithelial cells involves serial passage across apical and basolateral plasma membranes.

hydration, reabsorption of alveolar fluid in the neonatal period in preparation for alveolar respiration and the resolution of pulmonary edema. Fluid movement across epithelial and endothelial barriers occurs in interstitial and alveolar pulmonary edema resulting from numerous etiologies such as congestive heart failure, the acute respiratory distress syndrome, infection, acute lung injury following acid aspiration, and subacute hyperoxic lung injury. Figure 5 shows the barriers to the transport of water between airspace, interstitial, and vascular compartments in lung. The trachea and large airways contain an epithelial cell layer, but represent a small fraction of the total surface area available for movement of fluid into and out of the airspaces. The more numerous smaller airways also contain an epithelial cell layer, but still represent a relatively small surface area. The alveolar epithelium provides the major surface lining the airspace. The alveolar epithelium contains type I cells, which are flat cells comprising the majority of the alveolar epithelial surface, and type II cells, which transport salt actively and produce surfactant. Movement of water between the airspace and capillary compartments through the alveolar epithelium also encounters potential permeability barriers in the interstitium and capillary endothelia. As discussed further in Section III.D, osmotic water permeability of the airspace–capillary barrier is very high (Carter *et al.*, 1996; Folkesson *et al.*, 1994) as a result of high water permeabilities of type I alveolar epithelial cells (Dobbs *et al.*, 1998) and microvascular endothelial cells (Carter *et al.*, 1998). Water permeability across small airway epithelial cells is moderately high (Folkesson *et al.*, 1996).

Four members of the aquaporin protein family are expressed in the airways and lung: AQP1 on the plasma membrane of microvascular endothelial cells and to a lesser extent on some pneumocytes, AQP3 on basal epithelial cells in large airways and throughout the nasopharnyx, AQP4 at the basolateral plasma membrane surface in epithelial cells throughout the trachea and the small and large airways, and AQP5 at the apical membrane of type I alveolar epithelial cells (for review, see Verkman *et al.*, 2000). The specific expression pattern of aquaporins in lung corresponds to sites of high water permeability, suggesting that aquaporins play a role in water movement between the airspace, interstitial, and capillary compartments. Additional lines of indirect evidence supporting a role of aquaporins in lung physiology include increased aquaporin expression (Umenishi *et al.*, 1996; Yasui *et al.*, 1997) and airspace–capillary water permeability (Carter *et al.*, 1997) near the time of birth, and regulation of lung AQP1 by steroids (King *et al.*, 1996).

To determine whether aquaporins are required for the high water permeabilities in lung, we measured water transport in mice lacking AQP1, AQP4, or AQP5, and double knockout mice deficient in AQP1/AQP4 and AQP1/AQP5 (Bai *et al.*, 1999; Ma *et al.*, 2000a). As shown below, AQP1 or AQP5 alone decreased airspace–capillary water permeability by approximately 10-fold. Therefore AQP1 provides the major route for osmotically driven water movement across lung microvascular endothelia and AQP5 provides the major route for water movement across the alveolar epithelium. Because aquaporins are expressed in cell plasma membranes, osmotically driven water transport occurs mainly by a transcellular rather than paracellular pathway. These findings support the notion that the primary route for water movement between the capillary and airspace compartments is by serial passage across alveolar epithelial and microvascular endothelial barriers. The minimal residual water permeability across the alveolar apical surface after AQP5 deletion may involve type II alveolar epithelial cells, the apical membrane lipid of type I cells, and possibly as yet unidentified water transporters. The residual osmotic water permeability across the microvessels probably represents a combination of lipid-mediated and paracellular transport. Airspace–capillary water permeability was reduced by ~30-fold double knockout mice lacking AQP1 and AQP5. AQP4 deletion by itself had little effect on water permeability, but deletion of AQP4 together with AQP1 resulted in ~30% decreased water permeability compared to deletion of AQP1 alone (Song *et al.*, 2000b). It was reasoned that the relatively low lung water permeability in AQP1 null mice might permit the detection of a small incremental effect of AQP4 deletion.

Interestingly, despite the marked decrease in airspace–capillary water permeability after deletion of AQP1 or AQP5, an important function of the lung alveolus, active alveolar fluid clearance, was not affected (Fig. 6). In addition, fluid clearance from the airspace compartment in the neonatal lung was not affected by aquaporin deletion (Song *et al.*, 2000b). In lungs from wild-type mice harvested at specified times after spontaneous delivery, lung water content (assessed by wet-to-dry weight ratios) decreased from 7.9 at birth to 5.3 at 24 h after birth with 50% reduction at ~25 min. At 45 min after birth, wet-to-dry weight ratios were similar for wild-type mice and mice lacking AQP1, AQP4, or AQP5. Thus a near-isosmolar

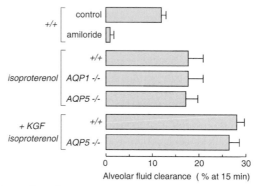

FIGURE 6 Alveolar fluid clearance in lungs of aquaporin knockout mice. Measurements were done at 37°C using an *in situ* perfused lung preparation in which the airspace was instilled with an isosmolar solution containing[131]I-albumin as a volume marker. Alveolar fluid clearance is expressed as percentage fluid absorption (mean ± SD, $n = 6$–10 mice) at 15 min. Where indicated, the instillate solution contained isoproterenol (0.1 m*M*) and/or measurements were done 72 h after treatment of mice with KGF. (Data taken from Bai *et al.*, 1999, and Ma *et al.*, 2000a.)

fluid transport mechanism does not necessarily require aquaporins. There are several possible explanations for this finding in lung that will require further study. The rate of airspace fluid absorption in lung is substantially less than the rate of active fluid absorption (per unit surface area) in proximal tubule and active fluid secretion in the pilocarpine-stimulated salivary gland acinus. In the latter systems, active fluid transport exceeds 0.5 μl/min per square micron of epithelial surface area. In mouse the lung maximum rate of fluid absorption is less than 0.02 μl/min per square micron. Slower rates of active fluid transport probably do not require as high a cell membrane water permeability. Another consideration is that in the alveolus, unlike the kidney proximal tubule and salivary gland, salt and water move primarily through different cells, type II and type I cells, respectively. Thus three-compartment models of solute–solvent coupling to accomplish isosmolar fluid transport probably do not apply to the alveolar epithelium. Another difference between active fluid transport in alveolus vs. kidney proximal tubule and salivary gland is that fluid is rapidly cleared on both sides of the epithelium in kidney (by lumen and capillary flow) and salivary gland (by capillary flow and saliva expulsion from the acinus), whereas fluid in the alveolar airspaces moves very slowly.

In addition, aquaporin deletion did not affect lung fluid accumulation in three established models of lung injury: hyperoxia, acid aspiration, and thiourea administration (Song *et al.*, 2000a). In mice exposed to >95% oxygen, mean survival was not affected by aquaporin deletion, nor was wet-to-dry ratio after 65 h of hyperoxia. In mice undergoing intratracheal acid instillation to produce epithelial injury, aquaporin deletion did not affect mean wet-to-dry weight ratios at 2 h. In mice

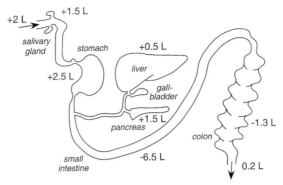

FIGURE 7   Schematic of the gastrointestinal tract showing amounts of fluid secretion and absorption in humans.

receiving intraperitoneal thiourea to produce endothelial lung injury, aquaporin deletion did not affect mean wet-to-dry weight ratios or the pleural fluid volume at 3 h. Together these data suggest that despite their major role in osmotically driven lung water transport, the known lung aquaporins have little importance in the physiological absorption of lung water or in the accumulation of fluid in the injured lung. Although not yet tested, the results so far would suggest that aquaporins are not important in the pathophysiology of other clinically relevant causes of acute lung injury such as pneumonia, sepsis, or the resolution of pulmonary edema from cardiac failure. Our results thus suggest that modulation of lung aquaporin expression and function by pharmacologic or genetic means would not have clinical utility.

G. Aquaporins and Gastrointestinal Function

Aquaporins are expressed throughout the gastrointestinal tract in salivary gland, stomach, intrahepatic cholangiocytes, gallbladder, pancreas, small intestine, and colon (for review, see Ma and Verkman, 1999). Next to kidney, the gastrointestinal tract transports the greatest amount of fluid, with fluid secretion by the salivary gland, stomach, and hepatobiliary tract, and fluid absorption by the small intestine and colon (Fig. 7). The results described in Section D above implicate a role for AQP5 in salivary gland fluid secretion. We describe below data on the role of AQP4 in colonic fluid absorption and gastric acid secretion, and AQP1 in dietary fat processing.

AQP4 is expressed at the basolateral membrane of surface epithelial cells in rat colon (Frigeri *et al.*, 1995a). AQP4 knockout mice were used to test the hypothesis that AQP4 is involved in colon water transport and fecal dehydration (Wang *et al.*,

2000b). AQP4 was immunolocalized to the basolateral membrane of colonic surface epithelium of wild-type mice. The transepithelial osmotic water permeability coefficient (P_f) of the *in vivo* perfused colon was measured using radiolabeled polyethylene glycol as a volume marker. P_f in wild-type mice was 0.016 cm/s and independent of osmotic gradient magnitude and direction, as well as the size of the solute used to induce osmosis and the lumenal perfusion rate. Also, P_f of proximal colon was 2-fold greater than that of distal colon. P_f was significantly reduced by 40–60% when measured in full-length colon and proximal colon of AQP4 null mice, but not in distal colon. There was no difference in water content of cecal stool from wild-type vs. AQP4 null mice (fraction water, 0.80 ± 0.01 vs. 0.81 ± 0.01, mean $\pm$ SE), but there was slightly higher water content in defecated stool in wild-type mice (0.68 ± 0.01 vs. 0.65 ± 0.01, P<0.05). Despite the differences in water permeability with AQP4 deletion, colonic secretory function was not impaired in a model of theophylline-induced secretion. The data suggested that transcellular water transport through AQP4 water channels in colonic epithelium facilitates transepithelial osmotic water permeability, but has little or no effect on colonic fluid secretion or fecal dehydration. Dehydration of feces in the distal colon may rely primarily on aquaporin-independent, cryptal absorption mechanisms. The colon also expresses AQP8, and possibly as yet unidentified aquaporins that may contribute to transepithelial fluid transport and active absorptive/secretory processes. Analysis of colonic fluid transport in AQP8 null mice and mice containing multiple aquaporin deletions may be useful in assessing the role of each aquaporin in colon physiology.

AQP4 has been proposed to play a role in gastric acidification based on its strong expression at the basolateral membrane of gastric parietal cells in rat (Frigeri *et al.*, 1995b). Similar localization was found in wild-type mice, which were used to test the hypothesis that AQP4 facilitates stomach acid secretion (Wang *et al.*, 2000a). The AQP4 null mice showed no differences in gastric morphology by light microscopy. Gastric acid secretion was measured in anesthetized mice in which the stomach was lumenally perfused at 0.3 ml/min with 0.9% NaCl containing radiolabeled polyethylene glycol as a volume marker. Collected effluent was assayed for titrable acid content and radioactivity. After a 45-min baseline perfusion, acid secretion was stimulated by high-dose intravenous pentagastrin for 1 h or intravenous histamine plus intralumenal carbachol. Baseline gastric acid secretion was 0.24 μEq/h. Pentagastrin stimulated acid secretion was 2.3–2.8 μEq/h and not affected by AQP1 deletion. Histamine/carbachol stimulated acid secretion was 28–32 μEq/h and also not affected by AQP4 deletion. In addition, there was no effect of AQP4 deletion on gastric fluid secretion, gastric pH, or fasting serum gastrin concentrations. These results provide direct evidence against a role of AQP4 in gastric acid secretion. It is unclear why AQP4 is strongly expressed at the basolateral membrane in gastric parietal cells. AQP4 may increase basolateral membrane water permeability so as to maintain constant parietal cell volume under conditions where the apical cell surface is exposed to fluids of very different osmolalities.

Alternatively, AQP4 expression in gastric parietal cells may represent a vestigial remnant of an earlier time in which high parietal cell water permeability was required.

AQP1 is expressed at sites involved in dietary fat processing, including intrahepatic cholangiocytes, gallbladder, pancreatic microvascular endothelium, and intestinal lacteals. To determine whether AQP1 has a role in dietary fat digestion and/or absorption, mice were placed on a diet containing 50% vegetable fat (Ma *et al.*, 2000c). Whereas wild-type mice gained 35% body weight in 8 days, AQP1 null mice lost weight; the weights became similar after return to a low fat diet for 6 days. The AQP1 null mice on a high fat diet acquired an oily appearance and developed steatorrhea as assessed by Sudan IV stool staining and stool lipid extraction. Addition of a pancreatic enzyme preparation containing lipase, amylase, and protease (Pancrease) to the 50% fat diet resulted in partial correction of the weight loss in AQP1 knockout mice. To assess the role of biliary and pancreatic secretions, fluid collections were carried out in anesthetized mice by ductal cannulations. Bile fluid volume flow and bile salt concentration in adult mice were not affected by AQP1 deletion or by intravenous secretin/CCK administration. Pancreatic fluid flow increased in response to secretin/CCK, but volumes and pH were not affected by AQP1 deletion. However, there was a small but significant decrease in secreted lipase and amylase. The role of defective biliary vs. pancreatic vs. intestinal function in the defective dietary fat processing phenotype will require further analysis.

H. Aquaporin Expression Does Not Imply Physiological Significance

Of the 10 mammal aquaporins, at least 7 have a wide tissue distribution in multiple organs and cell types. Some aquaporins are expressed in tissues that have no obvious role in fluid transport, such as AQP1 in erythrocytes, AQP3 in skin and urinary bladder, AQP7 in fat, and AQP9 in leukocytes. Based on the phenotypic defects described earlier, the question arises whether the expression of an aquaporin in specific tissue indicates importance in normal organ physiology or adaptation to pathophysiological stress. Although many clear-cut phenotypic abnormalities in aquaporin null mice are described above, there are several well-characterized examples where no phenotypic abnormalities are demonstrable. For example, as mentioned earlier, AQP4 deletion in gastric parietal cells does not affect gastric acidification. We have found no abnormalities in the physiology of erythrocytes from AQP1 null mice. Additional examples are described below.

It has been proposed that aquaporin-4 (AQP4), a water channel expressed at the plasmalemma of skeletal muscle cells, is important in normal muscle physiology and in the pathophysiology of Duchenne's muscular dystrophy. Rapid changes in muscle cell volume occur in response to muscle contraction, which is associated with the intracellular generation of osmotically active solutes from lactate production and metabolism of creatine phosphate. AQP4 is expressed in skeletal muscle, but much less or not at all in smooth and cardiac muscle (Frigeri *et al.*, 1995). Freeze-fracture electron microscopy of skeletal muscle plasmalemma has shown that AQP4-containing orthogonal arrays of particles (OAPs) and AQP4 protein are markedly reduced in hereditary muscular dystrophies (Ellisman *et al.*, 1976; Wakayama *et al.*, 1989). It was reported that AQP4 in rat skeletal muscle is expressed more in fast than slow twitch fibers, and that apparent water permeability is higher in fast twitch fibers (Frigeri *et al.*, 1998).

To determine whether AQP4 plays an important role in skeletal muscle physiology, muscle water permeability and function were compared in wild-type and AQP4 knockout mice (Yang *et al.*, 2000c). Immunofluorescence and freeze-fracture electron microscopy showed AQP4 protein expression in plasmalemma of fast twitch skeletal muscle fibers of wild-type mice. Osmotic water permeability was measured in microdissected muscle fibers from the extensor digitalis longus (EDL) and fractionated membrane vesicles from EDL homogenates. Using spatial-filtering microscopy to measure osmotically induced volume changes in EDL fibers, the rate of osmotic equilibration was not affected by AQP4 deletion. Stopped-flow light scattering measurements of osmotically induced volume changes in plasmalemma vesicles also showed no significant differences in water permeability. Similar water permeability yet ~90% decreased AQP4 protein expression was found in EDL from mdx mice, which lack dystrophin. Skeletal muscle function was measured by force generation in isolated EDL, treadmill performance time, and *in vivo* muscle swelling in response to water intoxication. Using an apparatus to measure the force generation of isolated skeletal muscle (Fig. 8A), no differences were found in EDL force generation after electrical stimulation (Figs. 8B and 8C). Further, AQP4 deletion did not affect treadmill performance time (22 vs. 26 min; 29 m/min, 13-degree incline) or muscle swelling (2.8 vs. 2.9 % increased water content at 90 min after intraperitoneal water infusion). Together these results provided evidence against a significant role of AQP4 in skeletal muscle physiology or in the pathophysiology of hereditary muscular dystropies.

AQP1 is expressed in microcapillary endothelial beds in many tissues such as secretory glands. However, AQP1 deletion did not affect pilocarpine-stimulated saliva secretion (Ma *et al.*, 1999). Tear secretion was not affected by deletion of AQP1 in lacrimal gland microvessels or aquaporins (AQP3, AQP4, AQP5) expressed in the lacrimal gland epithelia (Moore *et al.*, 2000). AQP1 deletion in microvessels in the peritoneal barrier reduced by 3-fold osmotically driven water transport, as occurs in peritoneal dialysis, but did not affect the slower active absorption of peritoneal fluid (Yang *et al.*, 1999a). AQP1 deletion in pleural microvessels and mesangiol cells also produced a significant reduction in osmotic water permeability, but did not affect physiologically-relevant mechanisms of pleural effusion formation on resolution (Song *et al.*, 2000c). These examples underscore the need to evaluate the organ-specific importance of aquaporins on a case-by-case basis and under appropriate *in vivo* conditions.

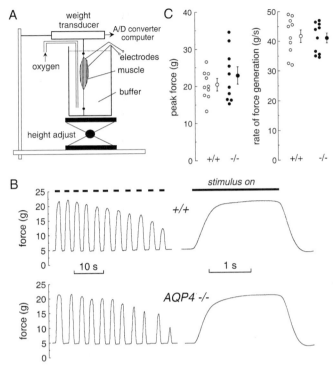

FIGURE 8 Evaluation of skeletal muscle function by electrically stimulated force generation. (A) Schematic of apparatus for measurement of muscle force generation in which a weight transducer records muscle contractions. (B) Time course of EDL muscle force generation. Where indicated by bars, EDL muscle was electrically stimulated by application of a pulse train. (C) Individual and averaged (SE) peak force generation and rate of force generation of skeletal muscle from wild-type vs. AQP4 knockout mice. Differences not significant. (Adapted from Yang *et al.*, 2000c.)

I. Perspective

Phenotype analysis of aquaporin knockout mice has been informative in defining the role of aquaporins in renal and extrarenal physiology. However, many questions remain to be addressed. In kidney, elucidation of the possible roles for AQP6 and AQP7 will require generation and phenotype analysis of knockout mice. The AQP2 knock-in mouse model of human nephrogenic diabetes insipidus should be useful in therapy development. There remains much to be learned about the role of aquaporins in organ physiology, particularly in the brain, eye, and gastrointestinal tract, where multiple aquaporins are expressed at sites of active fluid transport. In addition to these problems, our laboratory is actively studying the role of aquaporins in tumor angiogenesis (AQP1), skin hydration (AQP3), and lung humidification (AQP3 and AQP4).

The phenotype results described above suggest new therapeutic strategies for human disease. Aquaporin water channel inhibitors may be helpful as aquaretics in kidney, edema-inhibiting agents in brain, and possibly as regulators of intraocular and intracerebral pressures. AQP1 or AQP3 inhibitors, especially in combination with conventional salt transport-blocking diuretics, might be highly effective in inducing a diuresis in refractory states of edema associated with congestive heart failure and cirrhosis. The data in aquaporin knockout mice thus provides a rational basis to search for pharmacologically helpful aquaporin blockers. Although crystallographic resolution of aquaporin structure is at present inadequate for structure-based drug design, drug discovery by combinatorial chemistry and high-throughput screening (Tamarappoo *et al.*, 1999a) has considerable potential in identifying aquaporin inhibitors.

III. BIOPHYSICAL ANALYSIS OF AQUAPORIN FUNCTION

A. Overview

The identification of aquaporins and the generation of transgenic mouse models has mandated the need to develop new methods for measurement of water permeability in heterologous expression systems and in tissues derived from transgenic mouse models. The older goal of cataloging water transport parameters has in large part been replaced by hypothesis-driven research to determine whether aquaporin proteins mediate water transport across defined tissue barriers and to elucidate the molecular mechanisms of regulated water transport. In this section, the relevant biophysical principles of water transport are reviewed, and newer approaches to quantify water transport properties in cellular systems are described.

B. Biophysics of Water Transport and Water Pores

For a semipermeable membrane separating compartments 1 and 2, the volume flow (J_v, cm^3/s) across the membrane is defined by (Finkelstein, 1987)

$$J_v = P_f S v_w [(c_{i2} - c_{i1}) + \sigma_p (c_{p2} - c_{p1}) + (P_1 - P_2)/RT], \tag{1}$$

where P_f (cm/s) is the osmotic water permeability coefficient, S (cm^2) the membrane surface area, v_w (18 cm^3/mol) the partial molar volume of water, P the hydrostatic pressure (atm), σ_p the reflection coefficient of the permeant solute, c_{i1} and c_{i2} the osmolalities of impermeant solutes on sides 1 and 2, and c_{p1} and c_{p2} the osmolalities of permeant solutes on sides 1 and 2. It is assumed in Eq. (1) that the membrane barrier is homogenous with respect to permeability properties and that P_f, S, and σ_p are independent of osmotic gradient size and direction.

The osmotic water permeability coefficient P_f provides the most helpful single parameter characterizing the water transporting properties of a defined barrier. P_f relates net volume flux across a barrier to osmotic and hydrostatic driving forces. For simple membrane barriers like planar bilayers, liposomes, and cell plasma membranes, the absolute value of P_f provides a helpful index about whether water transport is facilitated by molecular pores such as aquaporin water channels. P_f greater than 0.01 cm/s (at 25–37°C) is considered to be high and suggests the involvement of molecular water channels, whereas P_f less than 0.005 cm/s suggests water diffusion through the lipid portion of a membrane. The interpretation of measured P_f in terms of molecular water channels assumes accurate definition of barrier surface area S and the absence of significant unstirred layer effects.

The Arrhenius activation energy (E_a, kcal/mol) is defined by the relation ln $P_f = E_a / RT + A$, where R is the gas constant, T is absolute temperature, and A is an entropic term. E_a is determined from the slope of an Arrhenius plot of ln P_f vs. $1/RT$. E_a provides a measure of the energy barrier to water movement across a membrane. E_a for water movement across a lipid membrane is generally high (>8– 10 kcal/mol) because of hydrogen bonding interactions between water and lipid molecules, as well as the lipid dynamics required for creation of a water pathway. For water movement through aqueous channels, E_a is generally low (3–6 kcal/mol). The low E_a associated with water pores is assumed to be related to the weak temperature dependence of water self-diffusion. However, a rigorous theoretical basis for a low E_a in water pores is lacking, because water movement through molecular water channels probably involves rate-limiting interactions between water molecules and the pore wall rather than water self-diffusion. E_a is also low when P_f is unstirred layer-limited because of the weak temperature dependence of solute diffusion in water. The independence of E_a on surface area (S) can be an advantage in measuring E_a rather than P_f because S may be difficult to determine accurately. However, for complex epithelial and multicellular barriers the utility of measuring E_a is questionable. Complex barriers often contain multiple serial and parallel water transporting pathways, each of which may be nonideal in terms of temperature-independent E_a and osmotic gradient-independent P_f. Also, temperature-dependent changes can occur in cell shape due to changes in solute transporter activities and cytoskeletal assembly. Comparative E_a measurements may be helpful in some complex systems, such as intact tissues or organs from control vs. aquaporin knockout mice.

The ratio of osmotic-to-diffusional water permeability (P_f/P_d) has been proposed as an independent parameter in providing information about the existence of water pores. The diffusional water permeability coefficient P_d (cm/s) is defined as the rate of water transport (exchange) across a membrane in the absence of an osmotic gradient. Experimental methods for measurement of P_d involving tritiated water flux, nuclear magnetic resonance, H_2O/D_2O exchange, and other approaches have been reviewed previously (Verkman, 1995). P_f/P_d should equal unity for simple lipid bilayer membranes that do not contain water channels. P_f/P_d can be greater than unity when water moves through a wide pore or narrow channel, or when measured P_d is less than true membrane P_d because of unstirred layers. Various equations have been derived that relate P_f/P_d to simple pore geometries such as smooth right cylindrical pores of fixed length, however it is unlikely that simplified pore equations can provide useful information about pore geometry. Further, there is probably little utility in attempting to measure P_d in systems more complex than liposomes or small suspended cells like erythrocytes since P_d is unstirred layer-limited in epithelia, oocytes, planar lipid membranes, and large cells, often by several orders of magnitude.

The solute reflection coefficient, σ_p, is defined by Eq. (1), and the equation for the flux (J_s, mol/s) of an uncharged permeant solute (c_p) is

$$J_s = P_s S(c_{p1} - c_{p2}) + J_v(1 - \sigma_p)\langle c_p \rangle, \tag{2}$$

where P_s (cm/s) is the solute permeability coefficient and $\langle c_p \rangle$ is the "mean solute concentration" in the pore. Equations (1) and (2) (Kedem and Katchalsky, 1958) are derived from nonequilibrium thermodynamics and so are formally valid for small osmotic and solute gradients. The second term in Eq. (2) defines the solvent drag of a solute with σ_p less than unity. σ_p has been measured experimentally by induced osmosis and solvent drag strategies as reviewed previously (Verkman *et al.*, 1996). Induced osmosis involves the measurement of J_v in response to osmotic gradients created by an impermeant vs. test solute as defined by Eq. (1). Solvent drag involves the measurement of J_s in the absence and presence of forced osmosis as defined by Eq. (2). The reflection coefficient has been used to evaluate whether a common water–solute transporting pathway exists. σ_p is defined by the relation, $\sigma_p = 1 = P_s v_s/P_f v_w - f$, where v_s is the partial molar volume of solute and f is a "frictional" term arising from water-solute interactions in a water pore. Values of σ_p less than unity ($f > 0$) have been taken as evidence for a common water–solute pathway, provided that $P_s v_s/P_f v_w$ is appropriately evaluated. However σ_p should be interpreted with caution in biological systems. σ_p determination can be exceedingly difficult technically because of the presence of coupled and uncoupled water and solute transport. Rapid solute diffusion causes an underestimation of σ_p in induced osmosis measurements because of osmotic gradient dissipation by solute diffusion. σ_p is also generally underestimated in solvent drag measurements because of solute movement by both diffusive and solvent drag mechanisms. There is an even greater conceptual concern in measuring and interpreting σ_p in biological systems. The σ_p concept and validation were done on artificial porous membranes containing very long narrow macroscopic water conduits containing many millions of water molecules. Because the concentration of water is $\sim$55 molar and solute concentration is generally <1 molar, it is unlikely that a molecular water pore, such as an aquaporin water channel, would contain even one solute molecule at any instant. For these technical and theoretical reasons, the determination of σ_p in complex biological systems probably has little or no value.

C. Water Transport across Cell Plasma Membranes

Osmotic water transport of a cell membrane is measured from the time course of cell volume change in response to an osmotic challenge. The problem of water permeability determination is thus a problem of cell volume measurement. Various volume-dependent physical parameters have been used to develop strategies for measurement of membrane water permeability. In unlabeled cells, increased cell volume is often accompanied by small changes in light scattering. Cytoplasmic dilution due to cell swelling also causes a decrease in cytoplasmic refractive index. Direct three-dimensional cell shape reconstruction by bright-field confocal microscopy is theoretically possible, but not generally practical because of poor contrast and scattering from intracellular structures. Additional measurement possibilities are offered by labeling cells with fluorescent dyes. Dilution of an aqueous-phase fluorescent dye distributed in the cytoplasmic compartment can be measured by total internal reflection fluorescence microscopy, confocal fluorescence microscopy, or wide-field fluorescence microscopy with appropriate optics to create partial confocality. It is also possible to exploit "cell volume-sensitive" fluorescent indicators in which integrated cell fluorescence provides a direct measure of cell volume. Further methodological details on these approaches with a discussion of their biological applications have been reviewed by Verkman (2000a). Discussion of newer biophysical measurement methods is provided below.

1. Total Internal Reflection Fluorescence Microscopy

Total internal reflection fluorescence microscopy (TIRFM) provides another approach to measure the concentration of an aqueous-phase fluorophore in cytoplasm. TIRFM involves the excitation of fluorophores in membrane-adjacent cytosol near a high-to-low refractive index interface (Ölveczky et al., 1997). Fluorescence excitation is generally accomplished using a laser source and glass prism to illuminate the sample at a subcritical angle at a glass–aqueous interface (Fig. 9A). Cells are loaded with an aqueous-phase dye. Cell swelling in response to an osmotic gradient results in cytosolic fluorophore dilution and decreased fluorescence signal (Farinas et al., 1995). Cell shrinkage produces increased TIRFM fluorescence. Because the effective depth of TIRFM illumination (<150–200 nm) is much smaller than cell thickness (and much smaller than is possible with confocal microscopy), TIRFM provides a quantitative and relatively simple approach to monitor relative cell volume in adherent cells of arbitrary shape and size. Fluorophore photobleaching in TIRFM is minimal because a small fraction of total cell volume is illuminated and fluorophores diffuse rapidly in cytoplasm. TIRFM has been applied to measure osmotic water permeability in cells transfected with the vasopressin-regulated water channel AQP2 (Katsura et al., 1995; Valenti et al., 1996). Addition of cAMP agonists produces exocytic plasma membrane insertion of functional AQP2 water channels and increased osmotic water permeability. TIRFM was used to measure water permeability in individual microdissected skeletal muscle fibers that were

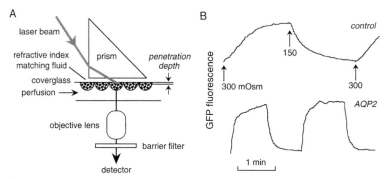

FIGURE 9 Measurement of osmotic water permeability in adherent cells by total internal reflection fluorescence microscopy. (A) Cells are loaded with a membrane impermeant volume marker. A thin (500–200 nm) layer of cytosol (labeled "penetration depth") is illuminated by a laser beam directed through a glass prism at a subcritical illumination angle. As the cell shrinks in response to an osmotic gradient, fluorophore concentration in the illuminated region increases, producing an increase in detected signal. (B) Time course of TIRFM fluorescence in CHO cells after transient transfection with cytoplasmic GFP alone (control) or together with cDNA encoding AQP2 water channels. GFP provides a volume marker in the transfected cells. Perfusate osmolality was changed as indicated to create osmotic gradients. (Adapted from Tamarappoo and Verkman, 1998.)

loaded with a fluorescent dye and immobilized on a polylysine-coated coverglass (Frigeri *et al.*, 1998). TIRFM can also be used to measure water permeability in cells expressing cytoplasmic green fluorescent protein (GFP) in the aqueous-phase cytoplasm. Water permeability was measured in transiently transfected cells coexpressing GFP and wild-type vs. mutant AQP2 water channels (Tamarappoo and Verkman, 1998). Even with imperfect transfection efficiency, water permeability could be measured selectively in the fluorescent cells coexpressing AQP2 (Fig. 9B). The ability to target GFP to selected cell types suggests a number of interesting possibilities, such as water permeability measurements in GFP-targeted tissues in transgenic mice.

2. Spatial Filtering Microscopy Approaches

Cell swelling causes dilution of cytoplasmic proteins and solutes producing a small change in intracellular refractive index. Doubling of cell volume in cultured mammalian cells decreased the refractive index from 1.367 to 1.405; cell shrinking by 50% increased refractive index to 1.347 (Farinas and Verkman, 1996). Measurement of intracellular refractive index provides a direct index of relative cell volume in unlabeled cell layers on solid or transparent supports, or epithelial tissues such as amphibian urinary bladder. An initial proof-of-principle study was done utilizing laser interferometry (Farinas and Verkman, 1996) in which changes in intracellular refractive index were quantified by measurement of the small changes in optical pathlength of a laser beam passing through a cell layer. Plasma

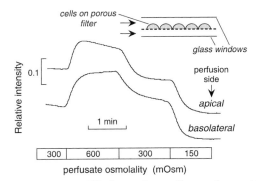

FIGURE 10 Osmotic water permeability in monolayers of human airway epithelial cells measured by spatial filtering microscopy. Cells were cultured on porous transparent supports and mounted in a perfusion chamber with separate apical and basolateral perfusion (*inset*). Cell volume was followed in response to indicated changes in the osmolalities of solutions perfusing the apical or basolateral surfaces. The osmolality of solution perfusing the contralateral surface was maintained at 300 mOsm. (Adapted from Farinas *et al.*, 1997.)

membrane P_f was measured in cultured epithelial cells and intact toad urinary bladder. A simple method to exploit the dependence of intracellular refractive index on cell volume was subsequently developed utilizing spatial filtering Fourier optics. Spatial filtering is the optical principle responsible for contrast generation in phase contrast and dark-field microscopes. Interference of zero-order (nonscattered) and phase-shifted first-order (scattered) beams produces image contrast. A rigorous mathematical basis for this phenomenon as applied to cells is reported in Farinas *et al.* (1997). With the appropriate arrangement of optics, the intensity of transmitted monochromatic light provides a semiquantitative measure of cell volume. The spatial filtering method was used to measure water permeability in aquaporin transfected cells grown on solid supports, apical vs. basolateral membrane water permeabilities in epithelial cells, and intact toad urinary bladder (Farinas *et al.*, 1997). Figure 10 shows the time course of transmitted light in a phase contrast microscope for primary cultures of human tracheal epithelial cells grown on porous supports. Apical and basolateral membrane water permeabilities were deduced from light transmittance signal changes (proportional to volume changes) produced by changes in osmolality of the perfusate bathing the apical or basolateral surfaces. The spatial filtering method provides excellent quality data in unlabeled cell layers during one-sided or two-sided perfusion.

3. Volume-Sensitive Fluorescent Dyes

Direct cell volume readout using a volume-sensitive fluorescent indicator provides a simple approach to measure water permeability in heterogeneous and polarized cell layers. We originally introduced a fluorescence quenching method to measure water permeability in liposomes and vesicles in which the readout was volume-sensitive fluorescence quenching (Verkman *et al.*, 1988). However, available fluorescent indicators cannot be loaded into living cells at high enough concentrations to make fluorescence quenching possible. We evaluated a series of cell-loadable fluorophores whose fluorescence quantum yield is sensitive to the concentration of endogenous cytoplasmic contents. The chloride-sensitive indicator SPQ was found to be quenched strongly by cytoplasmic organic anions and proteins by a collisional mechanism that conferred volume-dependent fluorescence (reviewed in Mansoura *et al.*, 1999). Cell SPQ fluorescence thus provides a semiquantitative measure of volume, with a caveat that fluorescence is also dependent on chloride concentration. Recently, a bright cell trappable anion indicator was introduced, LZQ (Jayaraman *et al.*, 1999), whose fluorescence is quenched by exogenously added iodide. LZQ has been used to measure water permeability in various cells after iodide loading. Water permeability measurements using volume-sensitive fluorescent indicators can be carried out in complex tissue geometries.

D. Water Transport across Epithelia and Complex Tissues

The methods described above for determination of cell membrane water permeability involve transient measurements of cell volume in response to rapidly imposed osmotic gradients. Often it is necessary to measure net water permeability across a tissue barrier, such as in cultured epithelia, kidney tubules, and bladder sacs. Measurement of steady-state osmotic water transport across a tissue barrier generally involves determination of the volume that has moved across the barrier in response to a continuously imposed osmotic gradient. Several approaches are applicable depending on barrier geometry, required time resolution, and water permeability. The classical gravimetric method is useful for large tissues sacs such as toad urinary bladder, where serial measurements of sac weight provide a quantitative index of time-integrated water movement across the limiting barrier. Imaging methods to quantify sac volume are suitable for P_f measurements in smaller tissue sacs such as mouse gallbladder. For planar bilayers mounted in Ussing-type chambers, dye dilution and capacitance strategies are applicable to deduce osmotically induced water flux from changes in volume on either side of the barrier. These classical methods are reviewed in Verkman (2000a).

Because of the many possible ways to assemble multiple cells types in a living tissue or whole organ, there are no clear-cut prescriptions to perform and analyze water permeability measurements. The principal challenges in complex tissues are defining the geometry and effective surface area of the barrier(s) to water transport, and determining the effects of unstirred layers on measured water fluxes. One type of complex barrier is that found in multilayer tissues such as cornea or epidermis. The degree to which water moves between cells in this type of barrier can have a major effect on water permeability properties. Another type of barrier is an epithelium–interstitium–endothelium as in lung or organs carrying out fluid secretion or absorption. Water movement involves at least three distinct compartments separated by serial barriers with different permeabilities and permselectivity properties. Using the lung as an example, two general approaches, surface fluorescence and gravimetry, are described which provide quantitative permeability data. Other general approaches to quantify water movement in complex tissues include wet-to-dry weight determination, as used recently in transgenic mice for measurements of lung fluid accumulation (Song *et al.*, 2000a) and brain edema (Manley *et al.*, 2000), and volume marker dilution in fluid cavities, as used recently in transgenic mice to measure osmotically induced water transport across the peritoneal barrier (Yang *et al.*, 1999a).

1. Water Transport in Perfused Cylindrical Tubules

For cylindrical epithelial cell layers such as kidney tubules, P_f has been measured by *in vitro* microperfusion using a membrane-impermeant volume marker perfused through the lumen. In the presence of a bath-to-lumen osmotic gradient, transepithelial P_f is determined from perfused vs. collected concentrations of the volume marker, lumen flow, lumen and bath osmolalities, and tubule length and surface area. A fluorescence method to measure marker concentration (without the need for collection of luminal fluid) was developed in which a membrane-impermeant fluorescent indicator such as FITC-dextran was used as the luminal volume marker (Kuwahara *et al.*, 1988). Steady-state osmotic water movement out of the lumen produces a progressive increase in fluorophore concentration along the lumen axis that is measurable by wide-field fluorescence microscopy. In addition to kidney tubules, this method has been used to measure water permeability in isolated microperfused distal airways (Folkesson *et al.*, 1996).

2. Water Permeability in Lung Measured by Pleural Surface Fluorescence

For studies of lung water permeability in small animals, where rapid airspace fluid instillation and sampling is not practical, a pleural surface fluorescence method was developed in which airspace osmolality is deduced from the fluorescence of an indicator dissolved in the airspace fluid (Carter *et al.*, 1996). The principle of the method is shown in Fig. 11A. The airspace is filled with fluid containing a membrane-impermeant fluorophore and the pulmonary artery is perfused with solutions of specified osmolalities. Because of the finite penetration depth of the excitation light (as in TIRFM), only lung tissue within 100–200 μm of the pleural surface is illuminated. Under these conditions, the surface fluorescence signal is directly proportional to the airspace fluorophore concentration. In response to an osmotic gradient, water flows between the airspace and perfusate compartments, resulting in a change in fluorophore concentration and thus pleural surface fluorescence. This approach has the advantage of excellent time resolution without the need for invasive sampling. In response to doubling of perfusate osmolality from

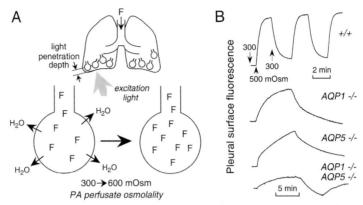

FIGURE 11   Measurement of osmotic water permeability in intact lung by a pleural surface fluorescence method. (A) To measure airspace–capillary water permeability, the airspace is filled with saline
containing a membrane-impermeant fluorescent probe (F) and pleural surface fluorescence is monitored
by epifluorescence microscopy ("excitation light"). In response to an increase in osmolality in fluid
perfused into the pulmonary artery, water moves out of the airspace and concentrates the fluorophore.
Surface fluorescence provide a quantitative measure of intra-alveolar fluorophore concentration.
(B) Representative time course data shown for lungs of mice of indicated genotypes. Osmolalities
were switched between 300 and 500 mOsm. (Adapted from Carter *et al.,* 1996; Song *et al.,* 2000b; and
Ma *et al.,* 2000a.)

300 to 600 mOsm, the pleural surface fluorescence signal approximately doubles
as predicted theoretically. P_f in mouse lung was 0.017 cm/s at 23°C, independent
of the solute used to induce osmosis, independent of osmotic gradient size and direction, weakly temperature dependent, and inhibited by $HgCl_2$. Figure 11B shows
an ~10-fold decrease in airspace-capillary water permeability in lungs from mice
lacking AQP1 and AQP5.

   A related strategy was developed to measure microvascular endothelial water
permeability in intact lung (Carter *et al.,* 1998). The airspace is filled with an inert,
water-insoluble perfluorocarbon to restrict lung water to two compartments, the
interstitium and capillaries, and thus establish a single rate-limiting permeability
barrier—the capillary endothelium. The pulmonary artery is perfused with solutions of specified osmolalities containing equal concentrations of high molecular
weight fluorescein-dextran. In response to a change of perfusate osmolality, water
is osmotically driven into or out of the capillaries, resulting in fluorophore dilution or concentration, respectively. The prompt change (decrease for fluorophore
dilution) in pleural surface fluorescence provides a quantitative measure of microvascular osmotic water permeability. Utilizing a three-compartment model to
compute capillary P_f from the fluorescence data, microvascular P_f was found to
be ~0.03 cm/s, weakly temperature dependent, and inhibited by mercurials. These
general strategies should be applicable for measuring P_f in various other organs.

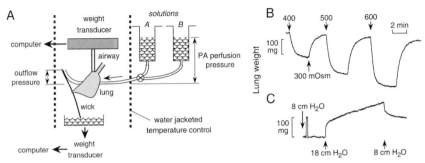

FIGURE 12 Gravimetric measurement of lung water permeability. (A) Apparatus for continuous measurement of lung weight during perfusion. The airspace compartment is filled with air, saline, or perfluorocarbon, and the pulmonary artery perfused with solutions of specified osmolality (solutions A and B) at specified pulmonary artery (PA) perfusion and outflow pressures. Lung weight is measured continuously by a gravimetric transducer and perfusate exit via the outflow catheter is measured by a second transducer. (B) Measurement of osmotic water permeability. Lung weight was recorded continuously in response to indicated changes in perfusate osmolality. The airspace was filled with isosmolar saline and perfusion pressure was 20 cmH$_2$O. (C) Measurement of hydrostatic "filtration". Venous outflow pressure was set at 5 cmH$_2$O and pulmonary artery perfusate pressure was increased from 8 to 18 cmH$_2$O without change in perfusate osmolality. (Adapted from Song *et al.*, 2000b.)

3. Capillary Filtration in Intact Organs

There is considerable interest in water permeability across microvascular endothelial beds in intact organs, both for osmotically driven and hydrostatically driven (filtration) water transport. Capillary filtration has been measured for many years in lung and muscle of dogs and other large animals by gravimetric determination of organ weight changes in response to changes in hydrostatic pressures. Lung weight includes the sum of fluids contained in the airspace, interstitial, and capillary compartments. We recently adapted this classical gravimetric method to measure osmotic water permeability and filtration in mouse lungs (Song *et al.*, 2000b). Figure 12A shows the apparatus in which lung weight is measured continuously during pulmonary artery perfusion. When the airspace compartment is filled with isosmolar saline, changes in perfusate osmolality result in water movement across endothelial and epithelial barriers, which produces changes in the water content of both the interstitial and airspace compartments observed as changes in lung weight. Figure 12B shows representative gravimetric data in which decreasing perfusate osmolality from 300 to 400, 500 and 600 mOsm produced decreases in lung weight as water is driven out of the extravascular spaces of the lung. Figure 12C shows a filtration study in which the pulmonary artery was continuously perfused with saline and the pulmonary artery pressure was increased from 8 to 18 cmH$_2$O at a constant left atrial pressure of 5 cmH$_2$O. The increased pressure produced a prompt increase in lung weight due to vascular engorgement, followed by a further approximately linear increase in lung weight due to fluid filtration. Gravimetry is a simple method that appears to be applicable for measurement of microvascular water permeability properties in other organs and tissues.

IV. AQUAPORIN STRUCTURE AND FUNCTION

The basic features of aquaporin structure have been defined using mutagenesis, epitope-tagging, spectroscopic, and freeze-fracture electron microscopy methods. Aquaporins appear to assemble in membranes as homotetramers in which each monomer, consisting of six-membrane-spanning α-helical domains with cytoplasmically oriented amino and carboxy termini, contains a distinct water pore. As described below, medium-resolution structural analysis by electron cryocrystallography indicated that the six tilted helical segments form a barrel surrounding a central porelike region that contains additional protein density. Several of the mammalian aquaporins (e.g., AQPs 1, 2, 4, and 5) appear to be highly selective for the passage of water, whereas others (called *aquaglyceroporins*) also transport glycerol (e.g., AQPs 3 and 7) and possibly ions (AQP6) and larger solutes (AQP9). The subject of aquaporin structure and function was reviewed recently (Verkman and Mitra, 2000). We describe below recent studies from our laboratory on this subject.

A. AQP1 and Carbon Dioxide Transport

AQP1 has been thought to function primarily as a water channel that facilitates the transmembrane transport of water in response to osmotic gradients and hydrostatic pressure differences. As reviewed by Verkman and Mitra (2000), proteoliposomes reconstituted with AQP1 are highly water permeable compared to control liposomes, but do not have increased permeability to protons, urea, and other small molecules. Expression of AQP1 in *Xenopus* oocytes confers increased water permeability without detectable increases in ion conductance and solute permeabilities. Although earlier reports ascribed cAMP-stimulated cation conductance and glycerol permeability AQP1, these findings were not subsequently confirmed by several laboratories. Nakhoul *et al.* (1998) reported that oocytes expressing AQP1 had 30–40% increased CO_2 permeability compared to control oocytes. Using oocytes microinjected with carbonic anhydrase and pH-sensitive microelectrodes, they found an increased rate of acidification in AQP1-expressing oocytes in response to external addition of CO_2. Follow-up work from the same group showed $HgCl_2$ inhibition of CO_2 transport by AQP1 in oocytes, and another report demonstrated that AQP1 reconstitution into proteoliposomes conferred a small increase in liposome CO_2 permeability, which was inhibited by $HgCl_2$ (Prasad *et al.*, 1998). The transport of CO_2 by AQP1 suggested by these studies is a potentially important observation because of its physiological implications, particularly for gas exchange in lung where AQP1 is strongly expressed in erythrocyte plasma membranes and alveolar capillary endothelia.

We recently examined the physiological consequences of CO_2 transport by AQP1 by comparing CO_2 permeabilities in erythrocytes and intact lung of wild-type and AQP1 null mice (Yang *et al.,* 2000a). Erythrocytes from wild-type mice strongly expressed AQP1 protein and had 7-fold greater osmotic water permeability than erythrocytes from null mice. CO_2 permeability was measured from the rate of intracellular acidification in response to addition of CO_2/HCO_3^- in a stopped-flow fluorimeter using BCECF as a cytoplasmic pH indicator. In erythrocytes from wild-type mice, acidification was rapid ($t_{1/2}$ 7.3 ms) and blocked by acetazolamide and raising external pH (to decrease CO_2/HCO_3^- ratio). Apparent CO_2 permeability (P_{CO_2}) was not different in erythrocytes from wild-type vs. AQP1 null mice (Fig. 13A). Lung CO_2 transport was measured in anesthetized ventilated mice subjected to a decrease in inspired CO_2 content from 5% to zero, producing an average decrease in arterial blood pCO_2 from 77 to 39 mm Hg with a $t_{1/2}$ of 1.4 min. The pCO_2 values and kinetics of decreasing pCO_2 were not different in wild-type vs. null mice (Fig. 13B). Because AQP1 deletion did not affect CO_2 transport in erythrocytes and lung, we measured CO_2 permeability in AQP1-reconstituted liposomes containing carbonic anhydrase (CA) and a fluorescent pH indicator. Whereas osmotic water permeability in AQP1-reconstituted liposomes was >100-fold greater than control liposomes, apparent P_{CO_2} ($\sim 10^{-3}$ cm/s) did not differ. Measurements using different CA concentrations and $HgCl_2$ indicated that liposome P_{CO_2} is unstirred-layer limited, and that $HgCl_2$ slows acidification because of inhibition of CA rather than AQP1. These results provided direct evidence against physiologically significant AQP1-mediated CO_2 transport, and established an upper limit to the CO_2 permeability through single AQP1 water channels. However, we cannot rule out the transport of small amounts of CO_2 through AQP1 that are below the detection sensitivity of the methods used here. It would be interesting to determine whether the non-water-selective aquaporins (AQP3, AQP7, AQP9) have significant CO_2 permeability. From steric considerations, it follows that the non-water-selective aquaporins might have substantially increased CO_2 permeability compared to AQP1, and thus provide a sensitive experimental test of whether aquaporins can conduct CO_2. However the intrinsic CO_2 permeability of an aquaporin would need to be exceptionally high to significantly increment the already very high CO_2 permeability of biological membranes.

B. Crystallographic Determination of AQP1 Structure

Because of the ease of AQP1 purification, reconstitution and formation of highly ordered two-dimensional crystals, several laboratories have analyzed AQP1

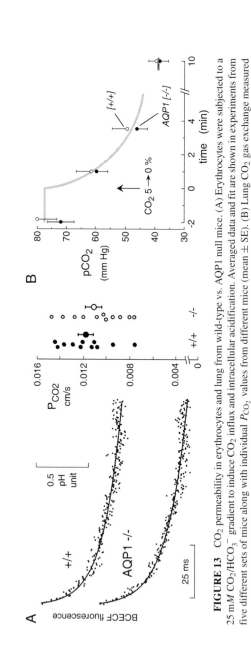

FIGURE 13 CO_2 permeability in erythrocytes and lung from wild-type *vs.* AQP1 null mice. (A) Erythrocytes were subjected to a 25 m*M* CO_2/HCO_3^- gradient to induce CO_2 influx and intracellular acidification. Averaged data and fit are shown in experiments from five different sets of mice along with individual P_{CO_2} values from different mice (mean ± SE). (B) Lung CO_2 gas exchange measured in *in vivo* anesthetized and ventilated mice. Mice underwent tracheostomy and were ventilated at a rate of 65–70/min and tidal volume of 6–7 ml/kg. Serial arterial blood samples were withdrawn from a carotid catheter for blood gas measurements. The CO_2 content of the inspired gas was set to 5% (with 95% O_2) for 15 min and then decreased to zero where indicated without change in ventilation rate. Data are the mean ± SE for mice of indicated genotypes. (Adapted from Yang *et al.*, 2000a.)

structure by electron crystallography. An historical account of work in this subject is provided in Verkman and Mitra (2000). In 1997, our group reported the unperturbed 7-Å resolution 3D structure of frozen-hydrated crystals of AQP1 deduced by analysis of minimal dose images and electron diffraction patterns recorded from crystals tilted up to 45 degrees (Cheng *et al.*, 1997). Similar 3D density maps were reported by the other two groups (Walz *et al.*, 1997; Li *et al.*, 1997) using specimens preserved in trehalose with the tilted projections comprised of images and electron diffraction patterns or only images. All three maps reveal a barrel formed by six helices surrounding additional density; the helices pack with a right-handed twist in the models of Cheng *et al.* and Walz *et al.* but with a left-handed twist in the model of Li *et al.* Walz *et al.* suggested a possible positional assignment of the six helices based on their AFM results (Walz *et al.*, 1996) and attributed some of the density features in the 3D map to interhelix loops.

The Cheng *et al.* (1997) structure shows four AQP1 monomers arranged symmetrically around a 4-fold axis oriented perpendicular to the bilayer (Fig. 14A). The overall density for each monomer is approximately cylindrical ($\sim$30 Å in diameter and $\sim$60 Å in height). The prominent feature in a monomer is a barrel formed by six, approximately cylindrical, tilted (18–30 Å) rods of density (A–F $\sim$36–44 Å length) representing the six transmembrane α helices. All six helices show some degree of curvature, especially helix D on the cytoplasmic side and helix F, which displays a distinct kink and could be thought of as being composed of two short helices. The bends in the helices could be due to the presence of several glycine and/or proline residues located within the putative membrane-spanning regions of the polypeptide chain. Within a monomer, interactions among the six helices are elaborated in three tightly packed two-helix bundles. Also, near the 4-fold axis, interactions that may contribute to the stability of a tetramer are suggested by strong overlap of density between helices B, C and D, E belonging to adjacent monomers. The six-helix barrel encloses a central block of density, which appears to be linked to the surrounding protein barrel. Two sets of bridges of density suggest linkages of this central block to the surrounding barrel. One, located on the proximal side as viewed in Fig. 14B, is directed toward helices F and D, and the other, located on the distal side, is directed toward helices A and C. These bridges and the block of density reside within the interior of the protein and have been attributed to predominantly hydrophobic segments containing the conserved NPA sequences in loops connecting helices 2, 3 and 5, 6 on the opposite sides of the bilayer.

The 3D structure of AQP1 shows evidence of an internal symmetry related to the homology of the first and second halves of the aquaporin amino acid sequence. Examination of our 3D density map revealed a symmetric disposition of protein density in the hydrophobic core of each AQP1 monomer. Thus planes of 3D density, parallel to the bilayer and equidistant above and below a particular plane (located $\sim$3 Å toward the extracellular side from the center of mass), can be superposed by

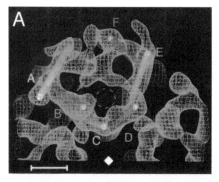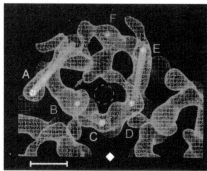

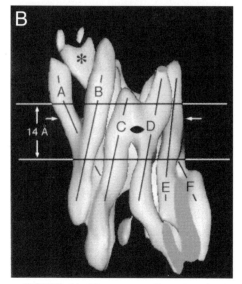

FIGURE 14 Electron crystallography of AQP1 crystals in reconstituted proteoliposomes. (A) Stereo pair of 3D density maps of frozen-hydrated AQP1 viewed approximately perpendicular to the bilayer. The rods trace the approximate paths of the centers of six tilted helices packed with a right-handed twist. These helices surround a vestibular region which narrows to a diameter of ~8 Å (*indicated by dashed circle*). The arrow indicates density assigned to the vertically apposed NPA sequences. (B) Surface-shaded representation of the 6-helix barrel viewed parallel to the bilayer. In-plane molecular pseudo-two-fold symmetry (*black ellipse*) is strongest within the demarcated transmembrane region. (Adapted from Cheng *et al.*, 1997.) (See Color Plate.)

a 180-degree rotation around an axis inclined by ~10 degrees to a lattice edge. This (2-fold) "pseudo"-symmetry is strongest within a span of ~14 Å near the center of an AQP1 molecule. The 2-fold symmetry in the membrane plane and the tandem repeats in the AQP1 sequence provide a simple explanation for bidirectional water transport through aquaporins. Although the current resolution of the map does not allow unique positional assignment of helices, symmetry and other considerations (connectivity, mercurial binding site) have suggested a specific assignment of helices (Verkman and Mitra, 2000) that will require atomic resolution data (<2.5 Å in all dimensions) for verification. Our collaborator Michael Wiener has made considerable progress in the 3D crystallization and x-ray structural analysis of human erythrocyte AQP1 (Wiener *et al.*, 2000), although generation of an atomic resolution map will require considerable further effort.

C. Cellular Biogenesis of Aquaporins

Like most eukaryotic multispanning membrane proteins, aquaporins are synthesized and assembled in the rough endoplasmic reticulum (ER) (Shi *et al.*, 1995; Skach *et al.*, 1994). During this process, AQP topology is established as transmembrane (TM) segments are properly oriented in the ER membrane through a series of translocation and membrane integration events. It is predicted that these early events of AQP biogenesis are facilitated by interactions between the Sec61 translocation machinery and topogenic sequence determinants, such as signal anchor and stop-transfer sequences, encoded within the nascent polypeptide (Bibi, 1998; Blobel, 1980; Hegde and Lingappa, 1999). Consistent with this hypothesis, functional topogenic determinants have been identified in AQP1 (Skach *et al.*, 1994) and AQP4 (Shi *et al.*, 1995). AQP1 and AQP4 cDNAs were sequentially truncated and ligated to a passive translocation reporter. Topogenic activities of AQP peptide regions were then characterized based on their ability to initiate and/or terminate polypeptide translocation in microinjected *Xenopus* oocytes and in a cell-free translation system. These early studies confirmed that AQP topology is established at the ER membrane and began to define the individual translocation events that generate each of the six predicted TM helices.

Analysis of AQP1 and AQP4 biogenesis also revealed that despite their similar structures and functional properties, AQP proteins utilize different folding pathways to acquire their transmembrane topology. During AQP4 biogenesis, the nascent polypeptide is cotranslationally translocated into the ER lumen by the action of signal sequences encoded within TM1, TM3, and TM5 (Shi *et al.*, 1995). Stop-transfer sequences encoded within TM2, TM4, and TM6 function in concert to terminate translocation and establish the orientation of cytoplasmic and extracytoplasmic peptide loops (Fig. 15). The topology of each TM segment in AQP4 is thus established cotranslationally as it emerges from the ribosome during protein synthesis. In contrast, AQP1 utilizes a more complex folding pathway. First, AQP1 encodes only two signal sequences and two stop-transfer sequences, and these determinants direct translocation events that establish only four of the six TM segments (Skach *et al.*, 1994). TM2 and TM4 are initially positioned on

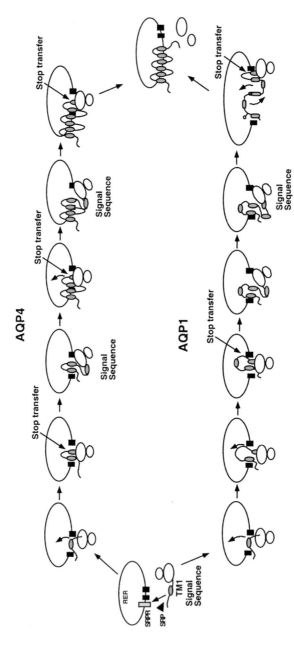

FIGURE 15 Schematic of aquaporin biogenesis at the endoplasmic reticulum. Different pathways for AQP1 and AQP4 biogenesis. AQP4 TM segments are serially inserted into the ER membrane to give the final topology consisting of six-membrane-spanning helical domains (*shown at right*). The biogenesis pathway for AQP1 is more complex in which a four-spanning topology is converted to a six-spanning topology by reorientation of a TM segment. See text for details.

lumenal and cytosolic faces of the membrane, respectively. Using a variation on the C-terminus translocation reporter technique, analysis of AQP1 fusion proteins containing two distinct translocation reporters recently demonstrated that this four-spanning structure represents a folding intermediate in the AQP1 biosynthetic pathway (Lu *et al.*, 2000). During and following AQP1 synthesis, TM3 undergoes a topologic reorientation in which its N-terminus flanking residues are repositioned from the ER lumen to the cytosol, and its C-terminus residues are repositioned from the cytosol to the ER lumen. This 180-degree rotation of TM3 also orients TM2 and TM4 into their proper mature topology (Fig. 15). Active areas of investigation are the identification of the driving force for AQP1 reorientation and the determination of whether reorientation occurs during association of the nascent chain with the Sec61 translocation channel or after release into the lipid bilayer.

A detailed examination of AQP1/AQP4 chimeric proteins has recently identified two peptide regions responsible for different AQP folding pathways (Foster *et al.*, 2000). One region is located at the N terminus of TM2 where hydrophilic residues in AQP1 (N49 and K51) interfere with stop-transfer activity and prevent TM2 from terminating translocation and cotranslationally spanning the membrane. A second region was identified in the TM3–4 peptide loop that influences the ability of TM3 to reinitiate translocation. By appropriate substitution of these two regions it was possible to convert AQP1 into a cotranslational mode of biogenesis similar to that observed for AQP4. Our studies of AQP biogenesis have provided insight on how complex topologies of polytopic transmembrane proteins can be established. Whereas the final topologies of AQP1 and AQP4 are similar, the translocation and integration events that establish this topology are quite different. This unexpected level of complexity demonstrates not only that related proteins may utilize alternative biogenesis folding pathways, but also that different folding pathways may result from relatively minor sequence variations.

D. Diffusion and Cellular Processing of GFP-Aquaporin Chimeras

We have used fluorescence recovery after photobleaching to address questions about the intramembrane mobility of aquaporins, the existence of AQP2-protein interactions, and the interactions between mutant AQP2 (causing nephrogenic diabetes insipidus) and the ER-quality control machinery. The strategy is to measure the mobility of GFP-tagged aquaporin chimeras in mammalian cells. Clones of stably transfected LLC-PK1 cells were isolated with plasma membrane expression of GFP-AQP1 and GFP-AQP2, in which GFP was fused upstream and in-frame to each aquaporin (Umenishi *et al.*, 2000) (Fig. 16A). The GFP fusion did not affect aquaporin tetrameric association or water permeability. By spot photobleaching, fluorescence in GFP-AQP1 expressing cells recovered to >96% of its initial level

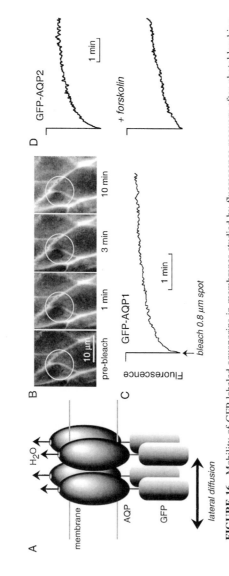

FIGURE 16 Mobility of GFP-labeled aquaporins in membranes studied by fluorescence recovery after photobleaching. LLC-PK1 cells were stably transfected with GFP-AQP1 and GFP-AQP2 chimeras containing the green fluorescent protein at the aquaporin N termini. (A) Cartoon showing a GFP-AQP tetramer in a membrane containing GFP labeled AQP monomers. (B) GFP fluorescence of GFP-AQP1 transfected LLC-PK1 cells showing a membrane expression pattern. Fluorescence was bleached by a brief laser pulse within the area demarcated by the white circle. Images shown before bleach (prebleach) and at indicated times after bleach. (C) Recovery curve measured using an 0.8-µm-diameter focused laser spot. (D) Recovery curve for GFP-AQP2 transfected cells before (*top*) and after (*bottom*) cAMP stimulation by forskolin. (Adapted from Umenishi *et al.*, 2000.)

with a diffusion coefficient of 5.3×10^{-11} cm^2/s. Figure 16B shows a series of images in which a large spot on the cell (in white circle) was bleached; Fig. 16C shows quantitative spot photobleaching using a small, 0.8-μm-diameter laser spot. GFP-AQP1 diffusion was abolished by paraformaldehyde fixation, slowed >50-fold by the cholesterol-binding agent filipin, but not affected by cAMP agonists. Similar diffusional rates were found for cells expressing GFP-AQP2, except that cAMP agonist forskolin significantly slowed GFP-AQP2 mobility significantly (Fig. 16D). The cAMP slowing was blocked by actin filament disruption with cytochalasin D, by K$^+$ depletion in combination with hypotonic shock, and by mutation of the protein kinase A phosphorylation consensus site (S256A) at the AQP2 C-terminus. These results indicate unregulated diffusion of AQP1 in membranes, but regulated AQP2 diffusion that was dependent on phosphorylation at serine 256, and an intact actin cytoskeleton and clathrin coated pit. The cAMP-induced immobilization of phosphorylated AQP2 provides evidence for AQP2–protein interactions that may be important for retention of AQP2 in specialized membrane domains for efficient membrane recycling. More recently, a series of GFP-AQP2 chimeras containing mutations in AQP2 were generated and expressed in mammalian cells (Umenishi and Verkman, 1999). The GFP tag did not affect the mistrafficking pattern of mutant AQP2s. Mobility measurements of the GFP-AQP2 mutants at the ER provided biophysical evidence for an interaction between misfolded AQP2 mutants with resident ER protein(s). The photobleaching approach permits real-time measurements of AQP diffusion and membrane protein interactions than cannot be accomplished by classical biochemical methods.

V. NEW DIRECTIONS IN AQUAPORIN PHYSIOLOGY AND BIOPHYSICS

Although research in aquaporin structure and function has advanced rapidly in part because of the relative ease of aquaporin purification (in the case of AQP1), membrane crystallization, mutagenesis, and functional analysis, major issues remain unresolved. Little is known about intramolecular dynamics in aquaporin proteins—it is not known whether aquaporins undergo conformational changes or gating. Substantially improved resolution in electron crystallographic analysis and/or high-resolution X-ray crystallography will be needed to unambiguously identify the water pathway. Purification and high-resolution crystallographic analysis of other aquaporins is needed, particularly the aquaglyceroporins, to identify the small solute pathway and to determine whether all aquaporins have similar basic structure. As nonmercurial water channel inhibitors are identified, structural analysis may provide key information about inhibition mechanisms and facilitate the design of compounds with greater inhibitory potencies. There remain many questions about aquaporin function. The possibility that aquaporins transport small gases and other solutes requires further investigation, as does the possibility that intrinsic aquaporin function may be subject to biochemical regulation. The mechanism and role of aquaporin-mediated small solute transport remain unknown.

We believe that the most exciting unresolved issues in aquaporin research involve the role of aquaporins in mammalian physiology and the development of clinically useful therapies based on aquaporin biology. Although phenotype studies of transgenic mouse models have begun to define the role of aquaporins in kidney, brain, lung, and secretory glands, much remains to be done in defining the role of aquaporins in gastrointestinal, neuromuscular, eye, skin, and reproductive function. Genetic deletion and replacement of aquaporins in mice will continue to provide important insights into the role of aquaporins in physiology and disease. The development of agents to correct AQP2 cellular misprocessing in some forms of NDI is particularly intriguing. The discovery of nonmercurial aquaporin inhibitors by high-throughput screening approaches is an important area of investigation. Such inhibitors would provide fundamental information about the physiological role of aquaporins, and might be useful in humans as aquaretics, agents that inhibit brain swelling, regulators of intraocular pressure, and so on. Finally, genetic models of aquaporin gene knockout and replacement may provide hints about novel roles of aquaporins in disease, possibly in embryonic development, tumor angiogenesis, atherogenesis, and immunocompetence.

Acknowledgments

We thank Drs. Dennis Brown, Charles Epstein, Hans Folkesson, Mark Knepper, Michael Matthay, Thomas Pallone, Jurgen Schnermann, Alfred van Hoek, Jean-Marc Verbavatz, and Kasper Wang for collaborations on the generation and phenotype analysis of aquaporin null mice. This work was supported by NIH grants DK35124, HL59198, HL51854, HL60288 and DK43840, and grant R613 from the National Cystic Fibrosis Foundation.

References

Bai, C., Biwersi, J., Verkman, A. S., and Matthay, M. A. (1998). A mouse model to test the *in vivo* efficacy of chemical chaperones. *J. Pharm. Toxicol. Methods* **40**, 39–45.

Bai, C., Fukuda, N., Song, Y., Ma, T., Matthay, M. A., and Verkman, A. S. (1999). Lung fluid transport in aquaporin-1 and aquaporin-4 knockout mice. *J. Clin. Invest.* **103**, 555–561.

Bibi, E. (1998). The role of the ribosome–translocon complex in translation and assembly of polytopic membrane proteins. *Trends Biochem. Sci.* **23**, 51–55.

Bichet, D. G. (1998). Nephrogenic diabetes insipidus. *Am. J. Med.* **47**, 1344–1347.

Blobel, G. (1980). Intracellular protein topogenesis. *Proc. Natl. Acad. Sci. USA* **77**, 1496–1500.

Carter, E. P., Matthay, M. A., Farinas, J., and Verkman, A. S. (1996). Transalveolar osmotic and diffusional water permeability in intact mouse lung measured by a novel surface fluorescence method. *J. Gen. Physiol.* **108**, 133–142.

Carter, E. P., Umenishi, F., Matthay, M. A., and Verkman, A. S. (1997). Developmental changes in alveolar water permeability in perinatal rabbit lung. *J. Clin. Invest.* **100**, 1071–1078.

Carter, E. P., Ölveczky, B. P., Matthay, M. A., and Verkman, A. S. (1998). High microvascular endothelial water permeability in mouse lung measured by a pleural surface fluorescence method. *Biophys. J.* **74**, 2121–2128.

Cheng, A., van Hoek, A. N., Yaeger, M., Verkman, A. S., and Mitra, A. K. (1997). Three-dimensional organization of a human water channel. *Nature* **387,** 627–630.

Chou, C. L., Ma, T., Yang, B., Knepper, M. A., and Verkman, A. S. (1998). Four-fold reduction in water permeability in inner medullary collecting duct of aquaporin-4 knockout mice. *Am. J. Physiol.* **274,** C549–C554.

Chou, C. L., Knepper, M. A., Van Hoek, A. N., Brown, D., Ma, T., and Verkman, A. S. (1999). Reduced water permeability and altered ultrastructure in thin descending limb of Henle in aquaporin-1 null mice. *J. Clin. Invest.* **103,** 491–496.

Deen, P. M., Verkijk, M. A., Knoers, N. V., Wieringa, B., Monnens, L. A., van Os, C. H., and van Oost, B. A. (1994). Requirement of human renal water channel aquaporin-2 for vasopressin-dependent concentration of urine. *Science* **264,** 92–95.

Deen, P. M., Croes, H., van Aubel, R. A., Ginsel, L. A., and van Os, C. H. (1997). Water channels encoded by mutant aquaporin-2 genes in nephrogenic diabetes insipidus are impaired in their cellular routing. *J. Clin. Invest.* **9,** 2291–2296.

Dobbs, L., Gonzalez, R., Matthay, M. A., Carter, E. P., Allen, L., and Verkman, A. S. (1998). Highly water-permeable type I alveolar epithelial cells confer high water permeability between the airspace and vasculature in rat lung. *Proc. Natl. Acad. Sci. USA* **95,** 2991–2996.

Ecelbarger, C. A., Terris, J., Frindt, G., Echevarria, M., Marples, D., Nielsen, S., and Knepper, M. A. (1995). Aquaporin-3 water channel localization and regulation in rat kidney. *Am. J. Physiol.* **269,** F663–F672.

Ellisman, M. H., Rash, J. E., Staehelin, L. A., and Porter, K. R. (1976). Studies of excitable membranes. II. A comparison of specializations at neuromuscular junctions and nonjunctional sarcolemmas of mammalian fast and slow twitch muscle fibers. *J. Cell Biol.* **68,** 752–774.

Farinas, J., and Verkman, A. S. (1996). Measurement of cell volume and water permeability in epithelial cell layers by interferometry. *Biophys. J.* **71,** 3511–3522.

Farinas, J., Simenak, V., and Verkman, A. S. (1995). Cell volume measured in adherent cells by total internal reflection microfluorimetry: Application to permeability in cells transfected with water channel homologs. *Biophys. J.* **68,** 1613–1620.

Farinas, J., Kneen, M., Moore, M., and Verkman, A. S. (1997). Plasma membrane water permeability of cultured cells and epithelia measured by light microscopy with spatial filtering. *J. Gen. Physiol.* **110,** 283–296.

Finkelstein, A. (1987). "Water Movement through Lipid Bilayers, Pores, and Plasma Membranes: Theory and Reality." Wiley & Sons, New York.

Folkesson, H. G., Matthay, M. A., Hasegawa, H., Kheradmand, F., and Verkman, A. S. (1994). Transcellular water transport in lung alveolar epithelium through mercurial-sensitive water channels. *Proc. Natl. Acad. Sci. USA* **91,** 4970–4974.

Folkesson, H., Matthay, M. A., Frigeri, A., and Verkman, A. S. (1996). High transepithelial water permeability in microperfused distal airways: Evidence for channel-mediated water transport. *J. Clin. Invest.* **97,** 664–671.

Foster, W., Helm, A., Turnbull, I., Gulati, H., Yang, B., Verkman, A. S., and Skach, W. R. (2000). Identification of sequence determinants that direct different intracellular folding pathways for aquaporin-1 and aquaporin-4. *J. Biol. Chem.* **275,** 34157–34165.

Frigeri, A., Gropper, M., Turck, C. W., and Verkman, A. S. (1995a). Immunolocalization of the mercurial-insensitive water channel and glycerol intrinsic protein in epithelial cell plasma membranes. *Proc. Natl. Acad. Sci. USA* **92,** 4328–4331.

Frigeri, A., Gropper, M., Umenishi, F., Kawashima, M., Brown, D., and Verkman, A. S. (1995b). Localization of MIWC and GLIP water channel homologs in neuromuscular, epithelial and glandular tissues. *J. Cell Sci.* **108,** 2993–3002.

Frigeri, A., Nicchia, G. P., Verbavatz, J. M., Valenti, G., and Svelto, M. (1998). Expression of aquaporin-4 in fast-twitch fibers of mammalian skeletal muscle. *J. Clin. Invest.* **102,** 695–703.

Hegde, R., and Lingappa, V. (1999). Regulation of protein biogenesis at the endoplasmic reticulum membrane. *Trends Cell Biol.* **9,** 132–137.

Jayaraman, S., Teitler, L., Skalski, B., and Verkman, A. S. (1999). Long-wavelength iodide-sensitive fluorescent indicators for measurement of functional CFTR expression in cells. *Am. J. Physiol.* **277,** C1008–C1018.

Katsura, T., Verbavatz, J. M., Farinas, J., Ma, T., Ausiello, D. A., Verkman, A. S., and Brown, D. (1995). Constitutive and regulated membrane expression of aquaporin-CHIP and aquaporin-2 water channels in stably transfected LLC-PK1 cells. *Proc. Natl. Acad. Sci. USA* **92,** 7212–7216.

Kedem, O., and Katchalsky, A. (1958). Thermodynamic analysis of the permeability of biological membranes to nonelectrolytes. *Biochim. Biophys. Acta* **27,** 229–246.

King, L. S., Nielsen, S., and Agre, P. (1996). Aquaporin-1 water channel protein in lung-ontogeny, steroid-induced expression, and distribution in rat. *J. Clin. Invest.* **97,** 2183–2191.

Kuwahara, M., Berry, C. A., and Verkman, A. S. (1988). Rapid development of vasopressin-induced hydroosmosis in kidney collecting tubules measured by a new fluorescence technique. *Biophys. J.* **54,** 595–602.

Li, H., Lee, S., and Jap, B. K. (1997). Molecular design of aquaporin-1 water channel as revealed by electron crystallography. *Nat. Struct. Biol.* **4,** 263–265.

Lu, Y., Turnbull, I. R., Bragin, A., Carveth, K., Verkman, A. S., and Skach, W. R. (2000). Reorientation of aquaporin-1 topology during maturation in the endoplasmic reticulum. *Mol. Biol. Cell* **11,** 2973–2985.

Ma, T., and Verkman, A. S. (1999). Aquaporin water channels in gastrointestinal physiology. *J. Physiol.* **517,** 317–306.

Ma, T., Yang, B., Gillespie, A., Carlson, E. J., Epstein, C. J., and Verkman, A. S. (1997). Generation and phenotype of a transgenic knock-out mouse lacking the mercurial-insensitive water channel aquaporin-4. *J. Clin. Invest.* **100,** 957–962.

Ma, T., Yang, B., Gillespie, A., Carlson, E. J., Epstein, C. J., and Verkman, A. S. (1998). Severely impaired urinary concentrating ability in transgenic mice lacking aquaporin-1 water channels. *J. Biol. Chem.* **273,** 4296–4299.

Ma, T., Song, Y., Gillespie, A., Carlson, E. J., Epstein, C. J., and Verkman, A. S. (1999). Defective secretion of saliva in transgenic mice lacking aquaporin-5 water channels. *J. Biol. Chem.* **274,** 20071–20074.

Ma, T., Fukuda, N., Song, Y., Matthay, M. A., and Verkman, A. S. (2000a). Lung fluid transport in aquaporin-5 knockout mice. *J. Clin. Invest.* **105,** 87–100.

Ma, T., Song, Y., Yang, B., Gillespie, A., Carlson, E. J., Epstein, C. J., and Verkman, A. S. (2000b). Nephrogenic diabetes insipidus in mice deficient in aquaporin-3 water channels. *Proc. Natl. Acad. Sci. USA* **97,** 4386–4391.

Ma, T., Jayaraman, S., Wang, K. S., Song, Y., Bastidas, J. A., and Verkman, A. S. (2000c). Defective dietary fat processing in transgenic mice lacking aquaporin-1 water channels. *Am. J. Physiol.* In press.

Maeda, Y., Smith, B. L., Agre, P., and Knepper, M. A. (1995). Quantification of aquaporin-CHIP water channel protein in mirodissected renal tubules by fluorescence-based ELISA. *J. Clin. Invest.* **950,** 422–428.

Manley, G. T., Fujimura, M., Ma, T., Feliz, F., Bollen, A., Chan, P., and Verkman, A. S. (2000). Aquaporin-4 deletion in mice reduces brain edema following acute water intoxication and ischemic stroke. *Nature Med.* **6,** 159–163.

Mansoura, M., Biwersi, J., Ashlock, M., and Verkman, A. S. (1999). Fluorescent chloride indicators to assess the efficacy of CFTR cDNA delivery. *Hum. Gene Ther.* **10,** 861–875.

Matthay, M. A., Folkesson, H., and Verkman, A. S. (1996). Salt and water transport across alveolar and distal airway epithelia in the adult lung. *Am. J. Physiol.* **270,** L487–L503.

Moore, M., Ma, T., Yang, B., and Verkman, A. S. (2000). Tear secretion by lacrimal glands in mice is not affected by deletion of water channels AQP1, AQP3, AQP4 or AQP5. *Exp. Eye Res.* **70**, 557–562.

Nakhoul, N. L., Davis, B. A., Romero, M. F., and Boron, W. F. (1998). Effect of expressing the water channel aquaporin-1 on the CO_2 permeability of *Xenopus* oocytes. *Am. J. Physiol.* **274**, C543–C548.

Nielsen, S., Fror, J., and Knepper, M. A. (1998). Renal aquaporins: Key roles in water balance and water balance disorders. *Curr. Opin. Nephrol. Hypertens.* **7**, 509–516.

Ölveczky, B. P., Periasamy, N., and Verkman, A. S. (1997). Mapping fluorophore distributions in three dimensions by quantitative multiple angle-total internal reflection fluorescence microscopy. *Biophys. J.* **73**, 2836–2847.

Pallone, T. L., Edwards, A., Ma, T., Silldorff, E., and Verkman, A. S. (2000). Requirement of aquaporin-1 for NaCl driven water transport across descending vasa recta. *J. Clin. Invest.* **105**, 215–222.

Prasad, G. V., Coury, L. A., Finn, F., and Zeidel, M. L. (1998). Reconstituted aquaporin 1 water channels transport CO_2 across membranes. *J. Biol. Chem.* **273**, 33123–33126.

Preston, G. M., Smith, B. L., Zeidel, M. L., Moulds, J. J., and Agre, P. (1994). Mutations in aquaporin-1 in phenotypically normal humans without functional CHIP water channels. *Science* **265**, 1585–1587.

Rao, S., and Verkman, A. S. (2000). Analysis of functional organ physiology in transgenic mice. *Am. J. Physiol.* **279**, C1–C18.

Schnermann, J., Chou, C. L., Ma, T., Traynor, T., Knepper, M. A., and Verkman, A. S. (1998). Defective proximal tubular fluid reabsorption in transgenic aquaporin-1 null mice. *Proc. Natl. Acad. Sci. USA* **95**, 9660–9664.

Shi, L. B., Fushimi, K., and Verkman, A. S. (1991). Solvent drag measurement of transcellular and basolateral membrane NaCl reflection coefficient in mammalian proximal tubule. *J. Gen. Physiol.* **98**, 379–398.

Shi, L. B., Skach, W. R., and Verkman, A. S. (1994). Functional independence of monomeric CHIP28 water channels revealed by expression of wild type-mutant heterodimers. *J. Biol. Chem.* **269**, 10417–10422.

Shi, L. B., Skach, W. R., Ma, T., and Verkman, A. S. (1995). Distinct biogenesis mechanisms for water channels MIWC and CHIP28 at the endoplasmic reticulum. *Biochemistry* **34**, 8250–8256.

Skach, W., Shi, L. B., Calayag, M. C., Frigeri, A., Lingappa, V., and Verkman, A. S. (1994). Biogenesis and transmembrane topology of the CHIP28 water channel in the endoplasmic reticulum. *J. Cell Biol.* **125**, 803–815.

Song, Y., Fukuda, N., Bai, C., Ma, T., Matthay, M. A., and Verkman, A. S. (2000a). Role of aquaporins in alveolar fluid clearance in neonatal and adult lung, and in edema formation following lung injury. *J. Physiol.* **525**, 771–779.

Song, Y., Ma, T., Matthay, M. A., and Verkman, A. S. (2000b). Role of aquaporin-4 in airspace-to-capillary water permeability in intact mouse lung measured by a novel gravimetric method. *J. Gen. Physiol.* **115**, 17–27.

Song, Y., Yang, B., Matthay, M. A., Ma, T., and Verkman, A. S. (2000c). Role of aquaporin water channels in pleural fluid dynamics. *Am. J. Physiol.* In press.

Spring, K. R. (1998). Routes and mechanisms of fluid transport by epithelia. *Annu. Rev. Physiol.* **60**, 105–119.

Tamarappoo, B. K., and Verkman, A. S. (1998). Defective trafficking of AQP2 water channels in nephrogenic diabetes insipidus and correction by chemical chaperones. *J. Clin. Invest.* **101**, 2257–226.

Tamarappoo, B. K., Koyama, N., Gabbert, M., and Verkman, A. S. (1999a). High-throughput screening of combinatorial drug libraries to identify non-mercurial aquaporin inhibitors. *J. Am. Soc. Nephrol.* **10**, 25A.

Tamarappoo, B. K., Yang, B., and Verkman, A. S. (1999b). Misfolding of mutant aquaporin-2 water channels in nephrogenic diabetes insipidus. *J. Biol. Chem.* **274**, 34825–34831.

Terris, J., Ecelbarger, C. A., Marples, D., Knepper, M. A., and Nielsen, S. (1995). Distribution of aquaporin-4 water channel expression within rat kidney. *Am. J. Physiol.* **269**, F775–F785.

Umenishi, F., and Verkman, A. S. (1999). Cellular targeting and diffusional mobility of green fluorescent protein-labeled aquaporin-2 mutants causing human nephrogenic diabetes insipidus. *J. Am. Soc. Nephrol.* **10**, 25A.

Umenishi, F., Carter, E. P., Yang, B., Oliver, B., Matthay, M. A., and Verkman, A. S. (1996). Sharp increase in rat lung water channel expression in the perinatal period. *Am. J. Resp. Cell Mol. Biol.* **15**, 673–679.

Umenishi, F., Verbavatz, J. M., and Verkman, A. S. (2000). cAMP regulated membrane diffusion of a green fluorescent protein-aquaporin-2 chimera measured by photobleaching recovery. *Biophys. J.* **78**, 1024–1035.

Valenti, G., Frigeri, A., Ronco, P. M., D'Ettorre, C., and Svelto, M. (1996). Expression and functional analysis of water channels in stably AQP2-transfected human collecting duct cell line. *J. Biol. Chem.* **271**, 24365–24370.

Van Hoek, A. N., Ma, T., Yang, B., Verkman, A. S., and Brown, D. (2000). Aquaporin-4 is expressed in basolateral membranes of juxtamedullary proximal tubules in mouse kidney. *Am. J. Physiol.* **278**, F310–F316.

Verbavatz, J. M., Brown, D., Sabolic, I., Valenti, G., Ausiello, D. A., Van Hoek, A. N., Ma, T., and Verkman, A. S. (1993). Tetrameric assembly of CHIP28 water channels in liposomes and cell membranes. A freeze-fracture study. *J. Cell Biol.* **123**, 605–618.

Verbavatz, J. M., Ma, T., Gobin, R., and Verkman, A. S. (1997). Absence of orthogonal arrays in kidney, brain and muscle from transgenic knockout mice lacking water channel aquaporin-4. *J. Cell. Sci.* **110**, 2855–2860.

Verkman, A. S. (1995). Optical methods to measure membrane transport processes. *J. Membrane Biol.* **148**, 99–110.

Verkman, A. S. (1998). Water transporting mechanisms in airways and lung. In "Pulmonary Edema" (M. Matthay and D. Ingbar, eds.), pp. 525–547. Marcel Dekker, New York.

Verkman, A. S. (2000a). Measurement of water permeability in living cells and complex tissues. *J. Membrane Biol.* **173**, 73–87.

Verkman, A. S. (2000b). Physiological importance of aquaporins: Lessons from knockout mice. *Curr. Opin. Nephrol. Hypertens.* **9**, 517–522.

Verkman, A. S., and Mitra, A. S. (2000). Structure and function of aquaporin water channels. *Am. J. Physiol.* **278**, F13–F28.

Verkman, A. S., Lencer, W., Brown, D., and Ausiello, D. A. (1988). Endosomes from kidney collecting tubule contain the vasopressin-sensitive water channel. *Nature* **333**, 268–269.

Verkman, A. S., Van Hoek, A. N., Ma, T., Frigeri, A., Skach, W. R., Mitra, A. K., Tamarappoo, B. K., and Farinas, J. (1996). Water transport across mammalion cell membranes. *Am. J. Physiol.* **270**, C12–C30.

Verkman, A. S., Matthay, M. A., and Song, Y. (2000). Aquaporin water channels and lung physiology. *Am. J. Physiol.* **278**, L867–L879.

Wakayama, Y., Jimi, T., Misugi, N., Kumagai, T., Miyake, S., Shibuya, S., and Miike, T. (1989). Dystrophin immunostaining and freeze-fracture studies of muscles of patients with early stage amyotrophic lateral sclerosis and Duchenne muscular dystrophy. *J. Neurol. Sci.* **91**, 191–205.

Walz, T., Tittmann, P., Fuchs, K. H., Muller, D. J., Smith, B. L., Agre, P., Gross, H., and Engel, A. (1996). Surface topographies at subnanometer-resolution reveal asymmetry and sidedness of aquaporin-1. *J. Mol. Biol.* **264**, 907–918.

Walz, T., Hirai, T., Murata, K., Heymann, J. B., Mitsuoka, K., Fujiyoshi, Y., Smith, B. L., Agre, P., and Engel, A. (1997). The three-dimensional structure of aquaporin-1. *Nature* **387**, 624–627.

Wang, K. S., Komar, A. R., Ma, T., Filiz, F., McLeroy, J., Hoda, K., Verkman, A. S., and Bastidas, J. A. (2000a). Gastric acid secretion in aquaporin-4 knock-out mice. *Am. J. Physiol.* **279**, G448–G453.

Wang, K. S., Ma, T., Feliz, F., Verkman, A. S., and Bastidas, J. A. (2000b). Colonic water transport in transgenic mice lacking aquaporin-4 water channels. *Am. J. Physiol.* **279**, G463–G470.

Wiener, M. C., Verkman, A. S., Stroud, R. M., and van Hoek, A. N. (2000). Mesotropic surfactant organization and membrane protein crystallization. *Protein Sci.* **9**, 1407–1409.

Yamamoto, T., and Sasaki, S. (1998). Aquaporins in the kidney: Emerging new aspects. *Kid. Int.* **54**, 1041–1051.

Yang, B., Brown, D., and Verkman, A. S. (1996). The mercurial-insensitive water channel (AQP-4) forms orthogonal arrays in stably transfected CHO cells. *J. Biol. Chem.* **271**, 4577–4580.

Yang, B., Folkesson, H. G., Yang, J., Matthay, M. A., Ma, T., and Verkman, A. S. (1999a). Reduced water permeability of the peritoneal barrier in aquaporin-1 knockout mice. *Am. J. Physiol.* **276**, C76–C81.

Yang, B., Gillespie, A., Carlson, E. J., Epstein, C. J., and Verkman, A. S. (2000d). Neonatal mortality in an aquaporin-2 knock-in mouse model of nephrogenic diabetes insipidus. *J. Biol. Chem.* In press.

Yang, B., Ma, T., Xu, Z., and Verkman, A. S. (1999b). cDNA and genomic cloning of mouse aquaporin-2: Functional analysis of an orthologous mutant causing nephrogenic diabetes insipidus. *Genomics* **57**, 79–83.

Yang, B., Fukuda, N., van Hoek, A. N., Matthay, M. A., Ma, T., and Verkman, A. S. (2000a). Carbon dioxide permeability of aquaporin-1 measured in erythrocytes and lung of aquaporin-1 null mice and in reconstituted proteoliposomes. *J. Biol. Chem.* **275**, 2686–2692.

Yang, B., Ma, T., Dong, J. Y., and Verkman, A. S. (2000b). Partial correction of the urinary concentrating defect in aquaporin-1 null mice by adenovirus-mediated gene delivery. *Hum. Gene Ther.* **11**, 567–575.

Yang, B., Verbavatz, J. M., Song, Y., Manley, G., Vetrivel, L., Kao, W. M., Ma, T., and Verkman, A. S. (2000c). Skeletal muscle function and water transport in aquaporin-4 deficient mice. *Am. J. Physiol.* **278**, C1108–C1115.

Yasui, M., Serlachius, E., Lofgren, M., Belusa, R., Nielsen, S., and Aperia, A. (1997). Perinatal changes in expression of aquaporin-4 and other water and ion transporters in rat lung. *J. Physiol.* **505**, 3–11.

CHAPTER 6

Trafficking of Native and Mutant Mammalian MIP Proteins

Peter M. T. Deen* and **Dennis Brown**[†]

*Department of Cell Physiology, University Medical Center of Nijmegen, 6500 HC, Nijmegen, The Netherlands; [†]Renal Unit and Program in Membrane Biology, Harvard Medical School, Massachusetts General Hospital East, Charlestown, Massachusetts 02129

I. NORMAL ROUTING OF MIP PROTEINS

At present, the subcellular localization of the different mammalian members of the MIP family of proteins is well established, but the identity of the targeting and trafficking signals that sort different AQPs to distinct plasma membrane domains is largely unknown. Different aquaporins are located on apical membranes, basolateral membranes, or on both membrane domains of epithelial cells. It is clear that aquaporin targeting, like that of other membrane proteins, is a complex process that is not only protein specific, but also cell-type specific, as discussed below for both AQP1 and AQP2. AQP1 is often constitutively located on both apical and basolateral membranes of expressing cells, for example, in the proximal tubule and efferent ducts of the male reproductive tract (Nielsen *et al.*, 1993a, 1993b; Sabolic *et al.*, 1992). However, it is found only on the apical plasma membrane in the choroid plexus (Nielsen *et al.*, 1993a), and it is inserted into the plasma membrane of bile duct cholangiocytes upon cAMP elevation by secretin stimulation (Marinelli *et al.*, 1997). This cell-specific subcellular localization and trafficking regulation complicates the search for specific targeting motifs because they may be interpreted differently by distinct cell types. Possibly because of its implications in human diseases, most routing information has been obtained for AQP2, which is therefore the main subject of this chapter.

A. AQP2 Recycling–The "Shuttle Hypothesis"

Long before the discovery of AQP2, a large amount of morphological and functional data led to the hypothesis that water channels [originally visualized as integral membrane protein (IMP) aggregates by freeze–fracture electron microscopy] are stored on intracellular vesicles, and that they move to and fuse with the plasma membrane after vasopressin stimulation. Several reviews summarize this early literature (Brown *et al.*, 1990; Hays, 1983), much of which was based on work carried out using amphibian epithelia (urinary bladder and epidermis). The idea that vasopressin stimulates vesicle exocytosis was originally proposed by Masur *et al.* (1971) and the concept of a vasopressin-induced vesicle "shuttling" mechanism was formulated by Wade *et al.* (1981). However, in addition to this regulated exocytotic pathway, water channels are continually retrieved from the plasma membrane by endocytosis, even whereas vasopressin is present, although an enhanced rate of apical membrane internalization can be detected after vasopressin washout (Harris *et al.*, 1986; Rapaport *et al.*, 1997; Strange *et al.*, 1988). This internalization process is believed to occur via clathrin-coated pits (Brown and Orci, 1983; Rapaport *et al.*, 1997; Strange *et al.*, 1988).

Endocytosis of water channels results in a reduction of overall membrane permeability and in the formation of endosomes that are highly water permeable. Direct measurements on renal papillary and toad bladder endosomes that had internalized volume-sensitive fluorescent probes during vasopressin action revealed the very high, mercurial-sensitive water permeability of these endosomes

(Harris *et al.*, 1990; Shi and Verkman, 1989; Verkman *et al.*, 1988). Internalized water channels are then recycled back to the cell surface upon subsequent vasopressin stimulation, a process that received support from combined freeze–fracture and HRP tracer studies in toad bladder (Coleman *et al.*, 1987) and more recently was directly demonstrated in cell cultures transfected with AQP2 (Katsura *et al.*, 1996). Following the cloning and sequencing of the aquaporin family of water channels, the shuttle hypothesis was tested directly using specific antibodies and was found to be essentially correct, as detailed below.

B. AQP2 Recycling in Collecting Duct Principal Cells

Rat AQP2 was first identified and sequenced by Fushimi *et al.* (1993). Specific antibodies were used by several groups to demonstrate that AQP2 is abundantly expressed in the apical plasma membrane of collecting duct principal cells, as predicted from earlier studies (Hayashi *et al.*, 1994; Marples *et al.*, 1995; Nielsen *et al.*, 1993c; Sabolic *et al.*, 1995; Fig. 1). AQP2 was also detected in small intracellular vesicles, as expected based on the proposed "shuttle hypothesis" of

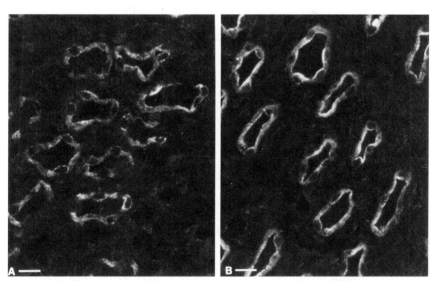

FIGURE 1   Immunofluorescence staining of AQP2 in papillary collecting ducts from (A) a control Brattleboro rat and (B) a vasopressin-treated Brattleboro rat. Note that vasopressin induced a marked shift in staining from the cytoplasm to the apical region of principal cells. Bar = 20 μm. (Originally published in Sabolic *et al.*, 1995.)

vasopressin action described earlier. Different groups carried out studies both *in vivo* and *in vitro* to examine the effect of vasopressin stimulation and vasopressin withdrawal on the distribution of AQP2 in principal cells. In all cases, vasopressin stimulation caused a decrease in intracellular AQP2 staining and an increase in apical plasma membrane staining for AQP2 (Hayashi *et al.*, 1994; Marples *et al.*, 1995; Yamamoto *et al.*, 1995). This localization was reversed by vasopressin washout in the *in vitro* studies (Nielsen *et al.*, 1995a). Vasopressin or dDAVP treatment of vasopressin-deficient Brattleboro rats or normal rats induced a striking redistribution of AQP2 from intracellular vesicles to the apical plasma membrane (Sabolic *et al.*, 1995), whereas a rapid internalization of AQP2 was induced by water loading rats or by infusion of a vasopressin V_2 receptor antagonist (Christensen *et al.*, 1998). Thus, direct support for a reversible vasopressin-induced shuttling of AQP2 from intracellular vesicles to the apical plasma membrane was obtained.

C. Aquaporin Expression in Xenopus Oocytes

Xenopus oocytes were used to elucidate the function of CHIP28, now renamed AQP1—the first water channel to be identified (Preston and Agre, 1991; Preston *et al.*, 1992). The altered membrane permeability of oocytes resulting from the translation of injected AQP mRNA, followed by membrane insertion of the translated protein, is easily measured by computer-assisted analysis of oocyte swelling in response to a hypotonic buffer (Zhang *et al.*, 1990). This assay is helpful in comparing the water permeabilities of membrane proteins, as well as examining the properties of mutated aquaporins and modifiers of aquaporin permeability. However, as with all expression systems, some important caveats must be considered. For example, measurements of aquaporin water permeability must be accompanied by some quantitative measurement of the amount of protein that reaches the oocyte plasma membrane. Only in this way can statements concerning the relative water permeabilities of aquaporins and aquaporin mutations be made. However, the oocyte sorting machinery seems to recognize most of the aquaporins as constitutive membrane proteins, so that even AQP2 is inserted into the plasma membrane without the need for any apparent exogenous stimulatory factor, such as cAMP (Kamsteeg *et al.*, 1999a, Fig. 8). Nevertheless, when interpreted appropriately, oocyte expression is a powerful experimental tool and has been used to identify several mutations in AQP2 that cause an accumulation of the protein within the oocyte, probably due to folding defects that lead to retention in the rough endoplasmic reticulum (ER) and/or Golgi apparatus (Deen *et al.*, 1994a; Mulders *et al.*, 1998), as discussed in more detail in Section II of this chapter.

D. Expression of AQPs in Nonpolarized Cells

Expression of aquaporins in nonepithelial cells, such as Chinese hamster ovary (CHO) cells, usually results in the constitutive appearance of the protein at the cell surface. As is the case for oocytes, these cells are not suitable for studies on the polarized trafficking and membrane insertion of these proteins. However, non-polarized cells have been very helpful for morphological and functional studies on different exogenously expressed aquaporins. For example, the tetrameric structure of AQP1 was directly visualized in freeze–fracture studies on transfected CHO cells (Verbavatz *et al.*, 1993), and transfection of CHO cells with AQP4 cDNA showed that this protein forms a characteristic pattern of orthogonal IMP arrays of proteins (OAPs) (Yang *et al.*, 1996) that had been previously reported in several cell types, including collecting duct principal cells (on the basolateral plasma membrane) (Orci *et al.*, 1981).

Finally, transfection of CHO cells with AQP2 did not result in the appearance of IMP clusters or aggregates on the plasma membrane, despite the presence of AQP2 at the cell surface (Van Hoek *et al.*, 1998a). This suggests that vasopressin-induced IMP clusters (previously described on collecting duct principal cells) are formed during the water channel recycling process and do not result from an intrinsic tendency of the AQP2 protein to form clusters. Indeed, recent label-fracture data on AQP2-transfected LLC-PK$_1$ cells show that although AQP2 is located in IMP clusters during part of the membrane recycling process, the IMP clusters can exist at the cell surface in the absence of AQP2 (Van Hoek *et al.*, 1998b). Thus, it is possible that the clusters of IMPs, previously shown to be associated with the vasopressin stimulation of water channel recycling, are hallmarks of increased endocytotic activity (i.e., sites of clathrin-coated pits), but that the IMPs within each cluster represent several unidentified membrane proteins, in addition to AQP2.

E. Expression of AQPs in Epithelial Cells–Studies on AQP2 Recycling

With the availability of cDNAs encoding the various aquaporins, several stable cell lines expressing AQP2 have been generated. cAMP-dependent translocation of AQP2 has been reconstituted in LLC-PK$_1$ cells (Katsura *et al.*, 1995), transfected rabbit collecting duct epithelial cells (Valenti *et al.*, 1996), Madin–Darby canine kidney (MDCK) cells (Deen *et al.*, 1997a), and primary cultures of inner medullary collecting duct cells (Maric *et al.*, 1998). Originally, two lines of stably transfected LLC-PK$_1$ and MDCK renal epithelial cells were produced that retained constitutive (AQP1) and regulated (AQP2) membrane localization of AQP1 and AQP2, respectively (Deen *et al.*, 1997a, 1997b; Katsura *et al.*, 1995).

AQP1-transfected LLC-PK$_1$ and MDCK cells showed constitutive plasma membrane expression of the protein, whereas AQP2-transfected cells had intracellular vesicular labeling that relocated to the plasma membrane only after vasopressin stimulation, or after increasing cytosolic cAMP levels with forskolin. Functional studies showed that AQP1 transfected cells had a high constitutive water permeability, whereas AQP2-transfected cells acquired the same degree of permeability only after stimulation. Incubation of the transfected LLC-PK$_1$ cells with FITC-dextran, followed by vasopressin treatment, resulted in a rate of internalization of the fluorescent marker that was 6-fold greater in AQP2-transfected cells than in AQP1 cells or nontransfected cells. Thus, the expression of AQP2, but not AQP1, can increase the vasopressin-induced endocytotic activity, implying that in LLC-PK$_1$ cells, the information for active recycling is encoded in the sequence of the AQP2 molecule. However, vasopressin treatment does not result in an increased endocytosis in AQP2-expressing over nontransfected MDCK cells (Deen, P.M.T., unpublished results), implying that this phenomenon is also cell-type dependent.

F. Polarity of Insertion of AQP2 – Apical versus Basolateral

In the original study in LLC-PK$_1$ cells (Katsura *et al.*, 1995), AQP2 was unexpectedly inserted into the basolateral plasma membrane upon vasopressin stimulation (Fig. 2A). In MDCK cells and rabbit collecting duct cells, apical insertion of AQP2 was reported, more closely resembling the polarity of insertion in principal cells *in vivo* (Deen *et al.*, 1997a; Valenti *et al.*, 1996; Fig. 2C). However, in primary cultures of IMCD cells, AQP2 was inserted both apically and basolaterally (Maric *et al.*, 1998). This is intriguing because AQP2 is, in addition to the apical membrane, also expressed in the basolateral membrane of collecting duct principal cells *in vivo*, especially in the inner medulla (Marples *et al.*, 1995). This basolateral location of AQP2 is less evident in the inner stripe. Furthermore, the amount of AQP2 on the basolateral cell surface is increased by prior cold-treatment of tissue slices prior to fixation (Breton and Brown, 1998). Thus, trafficking of AQP2 to different membrane domains occurs in a variety of transfected cell lines, as well as *in vivo*. These unique targeting properties can be used to examine how different cell types interpret polarity signals on proteins and how the intracellular transport machinery translates them.

G. Recycling of AQP2 in Transfected Epithelial Cells

The use of transfected cells allows the AQP2 recycling process to be dissected in a way that is difficult to achieve using the intact kidney, or even in isolated tissue slices or perfused tubules. As mentioned above, vasopressin stimulation of

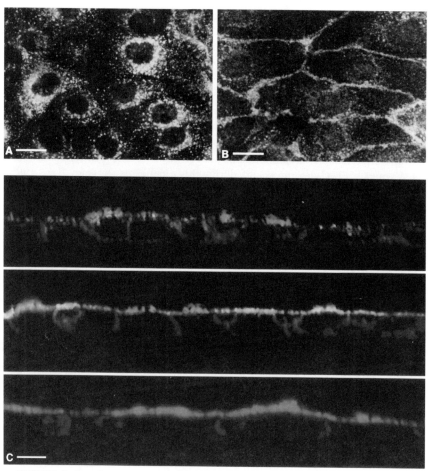

FIGURE 2   Localization of AQP2 in transfected LLC-PK₁ and MDCK cells. LLC-PK₁ cells
stably expressing an AQP2-c-*myc* fusion protein were treated (A) without or (B) with vasopressin,
fixed and stained for c-*myc*. Without stimulation, AQP2 is mainly located on intracellular, perinuclear
vesicles, whereas after vasopressin stimulation, AQP2 is mainly located on the plasma membrane.
(C) Transfected MDCK cells stably expressing AQP2 (WT10 cells; upper, middle panel) or native
MDCK cells (lower panel) grown on coverslips were treated without (upper panel) or with forskolin
(middle, lower panel), fixed, and apically labeled with biotin LC-hydrazide. After permeabilization,
cells were incubated with affinity-purified rabbit anti-AQP2 antibodies, rat anti-mouse E-cadherin, and
TRITC-coupled extravidin, and after washing, with affinity-purified goat anti-rabbit IgG coupled to
FITC and Cy-5-coupled goat anti-rat IgG. Following washing, dehydration and embedding, *x-z* axes
images were obtained with a Bio-Rad MRC-1000 laser scanning confocal imaging system. E-cadherin
(blue), apical glycoproteins (red) and AQP2 (green) are indicated. Note the redistribution of AQP2
from intracellular vesicles to the apical membrane upon forskolin treatment. (Originally published in
Deen *et al.*, 1997a.) (See Color Plate.)

AQP2-transfected LLC-PK$_1$ cells increases not only exocytosis, but also endo-cytosis (Katsura *et al.,* 1995). This implies that a vigorous membrane recycling process occurs in the presence of vasopressin. However, the initial burst of endo-cytosis that follows vasopressin stimulation is probably a homeostatic phenomenon that regulates cell membrane area in these cells. This so-called exocytosis-induced or compensatory endocytosis had been described in other cell types as a way of preventing a large increase in membrane surface area subsequent to a burst of exocytotic membrane insertion (Engisch and Nowycky, 1998; Smith and Neher, 1997). In LLC-PK$_1$ cells, most of the AQP2 in the cell is located at the cell surface 10 min after vasopressin stimulation, yet considerable endocytosis of the fluid-phase marker FITC-dextran occurs during this time frame (Katsura *et al.,* 1995). This implies that AQP2 is not concentrated in the FITC-labeled endosomes that are formed during the initial stages of vasopressin stimulation. Only at later time points, 10–30 min after vasopressin withdrawal, do AQP2-containing vesicles reappear in the cytoplasm. Most of the cell surface AQP2 is recovered into intracellular vesicles 60 min after vasopressin withdrawal.

Based on the altered transcellular water transport permeabilities at different time points following addition or removal of vasopressin, apical cell surface expression and removal of AQP2 occurs at a similar speed in transfected MDCK cells (Deen *et al.,* 1997a). In LLC-PK$_1$ cells, the internalized AQP2 can be redelivered to the cell surface by repeated stimulations, even in the presence of the protein synthe-sis inhibitor cycloheximide, indicating that the same cohort of AQP2 can indeed recycle between the cell surface and intracellular vesicles (Katsura *et al.,* 1996). However, the actual residence time of the AQP2 protein at the cell surface, after exocytosis, is unknown.

H. Targeting of AQP2 in Epithelial Cells

As in renal proximal tubules, AQP1 is located in both membranes of transfected MDCK cells, and this localization is not changed by activation of the PKA or PKC signaling cascade (Deen *et al.,* 1997b). In AQP2-transfected MDCK cells, however, AQP2 is located in intracellular vesicles without stimulation and is translocated to the apical membrane by vasopressin, which is similar to AQP2 in collecting duct cells (Deen *et al.,* 1997a). Because the transfected cells are the same, the do-mains determining the different targeting of AQP1 and AQP2 in these cells must be intrinsic to the proteins. In an attempt to dissect targeting domains, these cells were transfected with expression constructs encoding a hybrid aquaporin contain-ing the AQP1 sequence with its C terminus replaced by 52 or 42 or amino acids from the AQP2 molecule (Deen *et al.,* 1999). With or without vasopressin stim-ulation, both hybrids were expressed in the apical membrane, whereas a chimera of placental alkaline phosphatase coupled to a vesicular stomatitis virus G-protein transmembrane domain and the last 48 amino acids of AQP2 were retained in the cell. These results indicated that the AQP2 C terminus was necessary, but not sufficient, for targeting to the apical membrane. In addition, the C tail might be necessary but is not sufficient for recycling between the apical membrane and intracellular vesicles.

Also LLC-PK$_1$ cells expressing a hybrid aquaporin consisting of AQP1 with its C terminus replaced by the last 41 amino acids of AQP2 were generated (Toriano *et al.,* 1998). Stimulation of these cells with forskolin resulted in an increase in water permeability compared to controls, but that increase was less than that of cells transfected with AQP2 alone. Data from both cell types thus indicate that domains other than the C terminus are involved in regulating membrane insertion or retrieval of AQP2. A similar situation exists for the GLUT4 glucose transporter, where both N- and C-terminal motifs are known to be necessary for the normal trafficking of this protein (Garippa *et al.,* 1994; Verhey and Birnbaum, 1994).

I. Intracellular Pathways of AQP2 Recycling

Although it is clear that AQP2 recycles between cytoplasmic vesicles and the cell surface, the pathway followed by AQP2 during its intracellular transit period is only now beginning to be understood. Using a 20°C block of intracellular trafficking, AQP2 can be induced to accumulate in a clathrin-positive, Golgi-associated compartment within LLC-PK$_1$ cells (Gustafson *et al.,* 2000). A similar block is also induced by the H$^+$-ATPase inhibitor, bafilomycin, implying that AQP2 exit from this compartment requires a vesicle or luminal acidification step (Gustafson *et al.,* 2000; Fig. 3). The compartment in which blocked AQP2 accumulates is not labeled with antibodies against Golgi cisternae, such as giantin, and it is not stained for β-COP, a marker of intra-Golgi transport vesicles as well as Golgi cisternae (Waters *et al.,* 1991). Because the compartment is clathrin positive, it is likely to be the trans-Golgi network (TGN) from which clathrin-coated vesicles are known to bud (Griffiths *et al.,* 1985). However, part of the "recycling endosome," which is closely related to the TGN, can also have clathrin-coated domains (Futter *et al.,* 1998), so that the exact nature of this Golgi-related recycling compartment for AQP2 remains to be established. Another recycling protein, the GLUT4 glucose transporter, was reported to be at least partially located in a very specialized recycling endosomal compartment during its intracellular transit, although some was also present in the TGN (Martin *et al.,* 1999).

Whatever the precise nature of this recycling compartment is for AQP2 in LLC-PK$_1$ cells, it now appears that after endocytosis, AQP2 is delivered to a perinuclear, Golgi-related compartment where the recycled protein is repackaged along with newly synthesised AQP2 into a new vesicle that can eventually move to and fuse with the cell surface.

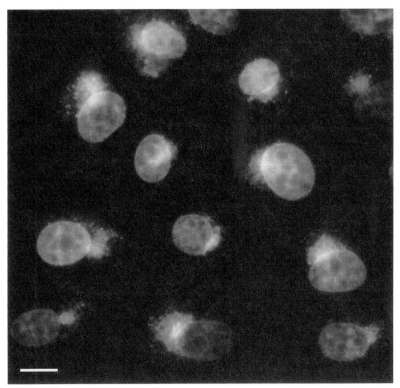

FIGURE 3 Localization of AQP2 in LLC-PK$_1$ cells on treatment with bafilomycin. Treatment of AQP2-expressing LLC-PK$_1$ cells with the H$^+$-ATPase-inhibitor bafilomycin for 2 h causes an accumulation of AQP2 (orange/red) in a dense perinuclear patch, coresponding at least partially to the trans-Golgi network. Nuclei are counterstained with DAPI (blue). Very few AQP2-positive vesicles are detectable in the rest of the cytoplasm after bafilomycin treatment, indicating that virtually all of the intracellular AQP2, both newly synthesized and recycling, are trapped in this perinuclear patch. A similar perinuclear accumulation is seen after exposure of cells to low temperature (20°C) for 2 h. Bar = 5 μm (Courtesy of D. Brown.) (See Color Plate.)

J. Microtubules and Actin Microfilaments

1. Microtubules

Most, if not all, vesicle trafficking processes in cells require the involvement of microtubules, microfilaments, or both (Brown and Stow, 1996; Fig. 4). Microtubule-depolymerizing agents such as colchicine and nocodazole markedly inhibit the vasopressin-induced permeability increase in target epithelia (Brown and Stow, 1996). Colchicine disrupts the apical localization of AQP2 in principal cells and

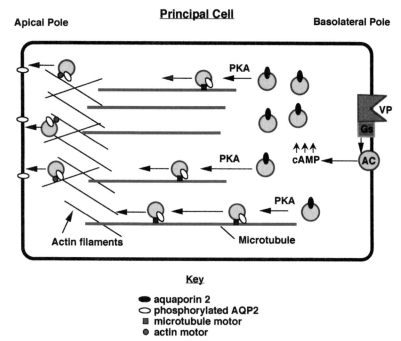

FIGURE 4 Involvement of microtubules and actin cytoskeleton in AQP2 redistribution from vesicles to the apical membrane. Vasopressin binding to the V2 receptor on the basolateral membrane of principal cells initiates a cascade of events that results in aquaporin-2 (AQP2) insertion into the apical membrane. Stimulation of adenylate cyclase (AC) by the heterotrimeric G-protein Gs increases intracellular cAMP, and AQP2 is phosphorylated at serine 256 by protein kinase A (PKA). AQP2-containing vesicles move toward the plasma membrane, with which they fuse by exocytosis. Vesicle movement toward the membrane involves both microtubules and actin filaments. Microtubule motors are required for long-range vesicle movement along microtubules, whereas actin-based motors such as myosin I might be required for the final "short-range" delivery of vesicles to the plasma membrane. The mechanisms by which PKA phosphorylation of AQP2 facilitates the cell-surface accumulation of AQP2 is unknown. [Originally published in Brown, D. (2000). Targeting of membrane transporters in renal epithelia: When cell biology meets physiology. *Am. J. Physiol. Renal Physiol.* **278,** F192–F201.]

causes it to be scattered on vesicles throughout the cytoplasm. *In vivo* and *in vitro* studies have shown that vesicles move along microtubules in an ATP-dependent manner, and that this requires the involvement of other proteins known as microtubule motors (Allan and Schroer, 1999). Motor proteins can be divided into two groups. The minus end-directed motor, cytoplasmic dynein, will transport vesicles toward the microtubule organizing center, whereas the plus end-directed motor, kinesin, will transport vesicles in the opposite direction.

     Microtubules in epithelial cells are organized with the minus-ends projecting toward the cell periphery (Bacallao *et al.,* 1989). Marples *et al.* (1998) found that dynactin, a protein complex that links dynein to microtubules and vesicles, is associated with AQP2 bearing vesicles, consistent with the view that microtubule motor proteins are involved in the vasopressin-regulated trafficking of AQP2-bearing vesicles. In transfected LLC-PK$_1$ cells, microtubules are required for AQP2 insertion into the basolateral plasma membrane (Brown, D., unpublished observations), suggesting that the basolateral transport of AQP2 also occurs along microtubule tracks. One current theme of intensive research in this area is to understand how cargo molecules interact with microtubules and microtubule motors, and how the selectivity of vesicle association with the microtubular transport mechanism is regulated to allow bidirectional vesicle movement during vesicle recycling.

2. Actin Cytoskeleton

Cytochalasins, which disrupt actin filaments, markedly inhibit the vasopressin response in target epithelia (Kachadorian *et al.,* 1979; Pearl and Taylor, 1985; Phillips and Taylor, 1989). Vasopressin administration causes a primary modification of the cortical actin network in toad bladder (Davis *et al.,* 1978; Hartwig *et al.,* 1987) and collecting duct principal cells (Hays *et al.,* 1993). The role of actin rearrangements in vasopressin action is unclear, but one possibility is that the actin cytoskeleton may be involved in controlling the access of vesicles to the underside of the plasma membrane by forming a reversible physical barrier. Principal cells of rat and mouse kidney contain high levels of the actin-severing protein adseverin, which may have a role in remodeling the actin cytoskeleton in these cells (Lueck *et al.,* 1998). Based on recent findings in other systems, AQP2 transport vesicle movement along actin filaments close to the plasma membrane is also a possibility (Rogers and Gelfand, 1998). Long-range microtubule-dependent transport may bring vesicles close to the membrane, and actin and associated motor proteins such as the myosins might then take over for the final short-range steps in the transport process. It is not known how actin interacts with AQP-containing vesicles, but preliminary data from transfected LLC-PK$_1$ cells suggest that AQP2 contains a putative actin-binding domain at its N terminus and that actin can be co-immunoprecipitated with AQP2 (Brown *et al.,* 1996). This work, coupled with the finding that myosin I is also associated with AQP2 containing vesicles (Marples *et al.,* 1997), supports the notion that actin, as well as tubulin-based vesicle movement, may contribute to vasopressin-induced AQP2 recruitment to the plasma membrane.

K. AQP2 Fusion Machinery–The SNARE Hypothesis

As for all intracellular transport vesicles, specificity and selectivity of intracellular transport and membrane fusion must be ensured by the cellular sorting machinery. The docking step for vasopressin-induced exocytosis of AQP2-containing vesicles is also likely to be mediated by vesicle targeting proteins (Hays *et al.*, 1994; Mandon *et al.*, 1996, 1997; Nielsen *et al.*, 1995b). This involves an interaction between integral membrane "SNARE" proteins present on the vesicles (v-SNAREs) and the target membrane (t-SNAREs) (Sollner *et al.*, 1993). The formation of this complex is thought to be important for the eventual vesicle fusion process, although the precise role of the SNARE proteins in vesicle fusion is highly controversial at present (Valenti *et al.*, 1998). In the collecting duct principal cell, VAMP-2 is present in AQP2-containing vesicles (Nielsen *et al.*, 1995b), and the t-SNAREs syntaxin-4 (Mandon *et al.*, 1996) and SNAP23 (Mandon *et al.*, 1997) are present in the apical plasma membrane. The presence of SNARE receptors in collecting duct principal cells and on AQP2-transporting vesicles indicates a potential role of these components in AQP2 trafficking, although functional experiments are needed both *in vivo* and *in vitro* to test this hypothesis.

Whereas the SNARE hypothesis is likely to apply also to AQP2 exocytosis and vesicle fusion, the overall mechanism involved is much more complex, requiring modulatory input from other components of the cellular regulatory machinery. For example, data from transfected rabbit IMCD cells have shown that, as in some other exocytotic pathways, heterotrimeric GTP-binding proteins are involved in AQP2 delivery to the cell surface (Valenti *et al.*, 1998). Using permeabilized cells to which specific peptide fragments corresponding to domains of GTP-protein α subunits were added, a role for the G-protein $G_{\alpha i3}$ in vasopressin-induced AQP2 exocytosis was demonstrated. AQP2 exocytosis was also inhibited by pertussis toxin, which inhibits the activity of many of the Gi-family of heterotrimeric GTP-binding proteins.

L. Role of AQP2 Phosphorylation

When intracellular cAMP is elevated, AQP2 is phosphorylated by protein kinase A (PKA) on a serine residue at position 256 on the cytoplasmic C terminus. Phosphorylation could modulate the water permeability of AQP2 in the plasma membrane or it could be required for the regulated trafficking and plasma membrane insertion of vesicles containing AQP2. An initial study showed that AQP2 phosphorylation is required to increase the water permeability of oocytes expressing AQP2 (Kuwahara *et al.*, 1995), but other studies on isolated kidney papillary vesicles containing AQP2 showed that water permeability was not dependent on the phosphorylation state of AQP2 (Lande *et al.*, 1996). However, the permeability of AQP4 (Han *et al.*, 1998) and of the plant water channels TIP and PM28A (Johansson *et al.*, 1998; Maurel *et al.*, 1995) appears to be regulated directly by phosphorylation, and the ion channel properties of AQP0 (MIP26) are gated by calcium and calmodulin (Peracchia and Girsch, 1989).

The role of AQP2 phosphorylation was addressed using LLC-PK$_1$ cells stably transfected with an AQP2 construct bearing a point mutation that converted the S256 residue to an alanine (AQP2-S256A). This AQP2 mutant was located on perinuclear intracellular vesicles in the basal state, and it did not move to the cell surface after stimulation of the cells with either vasopressin or forskolin (Fushimi *et al.,* 1997; Katsura *et al.,* 1997). This supports the idea that AQP2 phosphorylation is involved in the regulated redistribution of AQP2 from intracellular vesicles to the plasma membrane. Interesting in this respect is that, using antibodies specifically recognizing PKA-phosphorylated AQP2 (p-AQP2), p-AQP2 is already present in the intracellular vesicles of collecting ducts (Christensen *et al.,* 2000). Because AQP2 is expressed in membranes as homotetramers (Kamsteeg *et al.,* 1999b), the number of phosphorylated AQP2 monomers in a tetramer might determine whether translocation occurs. Indeed, expression studies in oocytes revealed that in an AQP2 tetramer at least three AQP2-S256D monomers (mimicking phosphorylated AQP2) were necessary for expression in the plasma membrane (Kamsteeg *et al.,* 1999a). These results suggest that in principal cells, three out of four monomers need to be phosphorylated for expression of an AQP2 tetramer in the apical membrane of collecting ducts.

Whether phosphorylation induces an increased interaction of vesicles with the cytoskeleton via microtubule and/or microtubule motors is currently under study. In support of this is the recent finding that PKA and several PKA anchoring proteins (AKAPs) were enriched in AQP2-immunopurified vesicles from IMCD cells. The observed specific inhibition of forskolin-induced AQP2 translocation with a peptide that prevents PKA–AKAP interaction demonstrated that, in addition to its activity, tethering of PKA to subcellular compartments is essential for AQP2 translocation (Klussmann *et al.,* 1999). Alternatively, phosphorylation could inhibit the endocytotic step of AQP2 recycling, leading to accumulation of exocytosed AQP2 at the cell surface. In this regard, studies on transfected LLC-PK$_1$ cells have shown that AQP2 can recycle constitutively between the plasma membrane and intracellular vesicles, even in the absence of increased intracellular cAMP (Gustafson *et al.,* 2000). However, accumulation of AQP2 at the cell surface does not usually occur under baseline conditions. Thus, regulation of the amount of AQP2 on the plasma membrane could be achieved by either increasing the rate of exocytosis, inhibiting endocytosis, or both.

M. Other Aquaporins

None of the other aquaporins have been shown to be regulated acutely by vesicle trafficking, although dehydration increases AQP3 expression in kidney principal cells (Ishibashi *et al.,* 1997). This constitutive membrane expression is characterized by the apparent paucity of intracellular vesicles containing these aquaporins.

Dissection of targeting domains on aquaporins will clearly require a considerable investment of time and effort but will be facilitated by the availability of expression systems described earlier in this section.

II. DISTURBED TRAFFICKING OF MIP PROTEINS

A. General Aspects of Mutations

1. Classification of Gene Mutations

The synthesis, maturation, and regulation of routing of plasma membrane proteins are extremely complex processes that require specific interactions between many different intracellular components. It is not surprising, therefore, that flaws in these processes are responsible for many pathophysiological conditions. Diseases can be divided into congenital and acquired forms. An acquired form of a disease is usually not present at birth and is often caused by external factors, such as medications or complications in other parts of the body. Such acquired diseases may result from perturbations of synthesis, maturation, or regulation of routing of a wild-type (wt) protein, and identification of the mechanisms responsible for inducing the disease often provides important information on the processing, trafficking, and regulation of the protein involved. With respect to aquaporins, extensive evidence has been obtained during the past five years that the expression and subcellular localization of AQP2 is affected in disorders such as congestive heart failure, liver cirrhosis, preeclampsia, polycystic kidney disease, and acquired nephrogenic diabetes insipidus (Devuyst *et al.*, 1996; Kuwahara *et al.*, 1995; Nielsen *et al.*, 1999; Schrier *et al.*, 1998). These subjects are discussed in detail in the chapter by Knepper *et al.*

The focus of this section is on the misrouting of aquaporins caused by mutations in aquaporin genes. Such a mutation may lead to a congenital form of a disease, which is, in contrast to acquired disorders, present from birth onward. Cosegregation of a mutation and a disease within a family, and subsequent identification that the encoded mutant is nonfunctional, provides direct proof for the involvement of the protein in a particular physiological process. In the last two decades, numerous mutations have been identified in genes encoding proteins involved in many diseases (http://www2.ebi.ac.uk/mutation; http://ariel.ucs.unimelb.edu.au/~cotton/dblist.htm; http://www.uwcm.ac.uk/uwcm/mg/hgmd0.htm). Most of the reported mutations are point mutations involving one or a few nucleotides. In cystic fibrosis (CF), a lethal disease caused by mutations in the cystic fibrosis transmembrane conductance regulator (CFTR) chloride channel gene, more than 800 unique mutations have been described (http://www.genet.sickkids.on.ca/cftr/), which are distributed as follows: 40% missense, 18% nonsense, 18% splice-site, 22% frameshift, and 2% other (promoter, in-frame deletions, etc). Missense and nonsense mutations

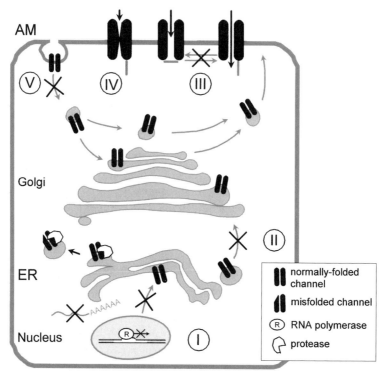

FIGURE 5 Schematic representation of the cell-biological outcome of the five different classes of gene mutations in diseases. The apical membrane (AM) and endoplasmic reticulum (ER) are indicated. For details see text. (Courtesy of P. M. T. Deen.)

change one amino acid into another, or result in a premature termination site, respectively. Splice-site mutations lead to skipping of exons or the introduction of a novel exon. Frameshift mutations, nucleotide insertions, or deletions change the translational reading frame, which usually results in a truncated protein, but also often affects the stability of its mRNA. These genetic errors may result in abnormal synthesis of proteins, abnormal folding, processing, or routing of proteins, or changes in their functional properties. Based on the cell-biological outcome, genotypes can be assigned to five different classes (Fig. 5):

1. Class I mutations lead to defects in the synthesis of stable mRNA, which precludes the formation of the protein. Many mutations, including promoter alterations, exon skipping, aberrant mRNA splicing, frameshifts, and premature translation termination, are thought to fall into this category. With such mutations in the CFTR gene, it is thought that a reduction in mRNA levels to less than 10%

of normal results in CF (Zeitlin, 1999). In a patient with leprechaunism, a genetic syndrome associated with intrauterine growth retardation and extreme insulin resistance, a similar reduction in insulin receptor mRNA levels has been found that results from the introduction of a nonsense mutation (Kadowaki *et al.*, 1990). Also, several nonsense mutations and deletions of parts of the low-density lipoprotein (LDL) receptor gene, found in patients with familial hypercholesterolemia (FH), have been assigned to this category (Hobbs *et al.*, 1990).

2. With class II mutations, the translation of the protein is completed, but the abnormal protein fails to be exported from the ER (for a review, see Kuznetsov and Nigam, 1998). Consequently, little or no protein reaches its final destination. In the ER, the maturation of secretory and membrane proteins occurs. This process, which includes several folding steps, glycosylation and trimming of sugar groups, and oligomerization, is facilitated by molecular chaperones and catalytic enzymes, such as the immunoglobulin binding protein (BiP), the glucose-regulated protein of 94 kDa (grp94), calreticulin, calnexin, peptidyl-prolyl isomerase (PPI), and protein disulfide isomerase (PDI; Brooks, 1997; Kuznetsov and Nigam, 1998). Depending on the stage of maturation, different chaperones associate and dissociate independently or cooperatively with the folding protein. These ATP-consuming interactions are usually short lived. When the structure of a protein is altered as a result of a mutation (or stress), the interaction of the misfolded protein with chaperones may be prolonged, which results in an impaired transport from the ER, often inducing increased chaperone expression levels (Kim and Arvan, 1998). Alternative models for ER retention of mutant proteins, for which less evidence has been obtained, are that mutant proteins are unable to present themselves to cargo receptors mediating ER exit or that mutant proteins form aggregates, which sterically preclude the proteins from entering transport vesicles (for review, see Kim and Arvan, 1998). In many cases, misfolding is caused by missense mutations (Deen and Knoers, 1998). In fact, more than 50% of the mutations in the LDL receptor fall into this category (Hobbs *et al.*, 1990). As such, the ER serves as a quality control organelle for proper processing and assembly of secretory and membrane proteins.

Retention in the ER is usually followed by degradation of the protein (Pind *et al.*, 1994). The intracellular location of the proteolytic enzymes involved, however, is still controversial. Recent studies revealed that misfolded membrane proteins can be extruded from the ER into the cytosol, become attached to ubiquitin molecules, and subsequently be degraded by the proteasome (for a review, see Hochstrasser, 1996). This mechanism has been proposed for the most common mutation in CFTR, the deletion of phenylalanine 508 (CFTRΔF508; Jensen *et al.*, 1995; Ward *et al.*, 1995), mutant alpha$_1$-antitrypsin (Qu *et al.*, 1996), and apolipoprotein B-100 (Fisher *et al.*, 1997). Alternatively, proteolytic enzymes might be located in the ER (Gardner *et al.*, 1993). Prompt disposal of the misfolded proteins may be necessary to maintain the balance of key regulatory proteins and to prevent formation of intracellular aggregates, which could be deleterious to cells. Indeed, evidence has been obtained that the inability to remove toxic aggregates of misfolded vasopressin-neurophysin precursor protein underlies dominant central diabetes insipidus (Ito and Jameson, 1997; Repaske *et al.,* 1996; Rutishauser *et al.,* 1996).

Because the majority of ER-retarded mutants are degraded to undetectable levels *in vivo,* most of the above-mentioned data have been obtained from *in vitro* studies, using transfected cells or immortalized cells expressing the mutant protein. The available data from *in vivo* studies, however, are generally in line with the *in vitro* results (Dray-Charier *et al.,* 1999; Fransen *et al.,* 1991; Lloyd and Olsen, 1987; Lorenzsonn *et al.,* 1993; Semenza and Auricchio, 1989). The study of Kalin *et al.* (1999), however, in which processing and localization of CFTR-ΔF508 in respiratory and intestinal tissues of CF patients was not different from wt-CFTR in healthy subjects, whereas expression of CFTR-ΔF508 in sweat glands of CF patients was absent, indicates that processing and targeting of a mutant protein might be tissue specific.

3. Class III mutations disrupt the activation and regulation of the protein at the plasma membrane, but biosynthesis, processing, and trafficking are undisturbed. The protein may be defective with respect to, for example, ATP binding or hydrolysis, or phosphorylation. In one patient suffering from nephrogenic diabetes insipidus (NDI), a disease in which the kidney is unable to concentrate urine in response to vasopressin, an R137H substitution has been identified in the vasopressin V_2 receptor (V_2R), which does not affect binding of vasopressin but specifically blocks stimulation of the G_s-adenylyl cyclase system (Rosenthal *et al.,* 1993). In CF, several mutations have been identified in the CFTR chloride channel that interfere with its activation by PKA or with its ATP binding and hydrolysis (for a review, see Zeitlin, 1999).

4. Class IV mutations affect the conductance or gating (open probability) of a channel (or hormone binding by a receptor) but have no effect on processing or routing of the protein. Several mutations of this class, resulting in the absence of agonist binding, have been identified in the V_2R of NDI patients (Oksche *et al.,* 1996a; Tsukaguchi *et al.,* 1995; for a review, see Deen and Knoers, 1998). Also, mutations have been found in the CFTR protein that specifically affect the rate of chloride flow (R347P) or reduce the channel open time (R117H, P574H; Zeitlin, 1999, and references therein).

5. Class V mutations do not cause misfolding of the protein but lead to other disturbances in the routing of the protein (e.g., endocytosis or exocytosis). A few of these rare mutations have been found in FH, in which LDL receptors that contained a missense mutation in the C-terminal tail (Y807C; Davis *et al.,* 1986) or lacked almost the entire tail (Lehrman *et al.,* 1985) failed to concentrate in coated pits and were not internalized. Subsequent analysis of the C tail of the LDL receptor revealed that these mutants were not endocytosed because they lacked the internalization motif NPXY (Chen *et al.,* 1990). A similar phenomenon has been identified

as the cause of Liddle's syndrome (Schild *et al.*, 1996; Shimkets *et al.*, 1994; Staub *et al.*, 1996). In this autosomal dominant disease, which is characterized by a hereditary form of systemic hypertension, C-terminal deletions and missense mutations in the β and γ subunits of the epithelial sodium channel ENaC have been identified that interfere with endocytosis of this pentameric protein complex via a PPXY domain. Because these mutants also exert an increased open probability (Kellenberger *et al.*, 1998), the mutations also belong to the class IV category. Another mutation in this class is found in hyperoxaluria (Purdue *et al.*, 1991). Here, a mutation in the alanine-glyoxalate aminotransferase gene results in a leucine for proline 11 substitution, which unmasks a cryptic site that targets the mutant enzyme to mitochondria instead of peroxisomes.

2. Candidate Screening

Because identification of mutations in a gene in a disease proves the critical involvement of the encoded protein in a given physiological process, researchers often "search" for a disease that fits with a phenotype predicted to result from a nonfunctional or constitutively activated target protein. For this so-called candidate screening, it is essential to know:

- The chromosomal localization of the human gene, which can be obtained by *in situ* chromosomal hybridization using the gene or cDNA as a probe
- The tissue-specificity of expression, which is usually obtained by immunohistochemistry and immunoblotting using specific antibodies
- The function (and selectivity) of the protein, which is often obtained from analysis in expression systems.

Using these data, the OMIM database can be screened for the presence of a disease with the expected phenotype around the gene's localization (http://www3.ncbi. nlm.nih.gov/Omim/). This database reveals the localization of many diseases, which have been narrowed down to "small" chromosomal areas on the basis of segregation analysis and positional cloning. Because many mutant phenotypes are known in mice, determination of the chromosomal localization of the mouse gene and searching the mouse genome might also come up with a putative hit. Alternatively, a region in the mouse genome that is homologous to the location of the human gene can be analyzed. Upon identification of a disease that fits with the expected phenotype (with or without use of the databases) and other available data, the gene or cDNA of the target protein of affected individuals needs to be analyzed for mutations.

3. Screening for Aquaporin Diseases

This risky approach has been applied several times in the aquaporin field, but in many cases the target AQP appeared not to be involved in the disease in question.

Segregation analysis of (congenital) primary nocturnal enuresis (PNE; nightly bed-wetting) revealed that a subset of patients had a positive two-point LOD score of 4.2 close to the chromosomal localization of AQP2 (Arnell *et al.*, 1997). We speculated that in these patients an activating mutation in AQP2 could be involved in PNE. Our hypothesis was that such a mutation would keep AQP2 in the apical membrane of the cell, resulting in continuous water reabsorption. As a consequence, the vasopressin release from the pituitary would be reduced, which would lower the AQP2 expression level and hence total water reabsorption. The usual diurnal need for urine concentration that accompanies the lack of drinking during sleep would thus require new synthesis of AQP2, which might take several hours. This time lag in the recovery of urine concentrating ability could then be the cause of PNE. This theory is consistent with the fact that in these PNE patients, the vasopressin analog dDAVP can ameliorate PNE (Hogg and Husmann, 1993). However, amplification of exons 1–4 of the AQP2 gene of several affected individuals in these families and subsequent sequence analysis revealed no mutation that might cause PNE, which clearly indicated that AQP2 was not involved in PNE in these patients.

Because AQP4 is abundantly expressed in brain and the localization of the mouse AQP4 gene is close to the ataxia phenotype Turtzo *et al.* (1997) speculated that AQP4 might be involved in this autosomal recessive neurological syndrome, which is characterized by progressive tremor, ataxia, weakness, and ultimately paralysis of the limbs. Detailed analysis, however, revealed no difference in the AQP4 cDNA sequence, and mRNA or protein expression between healthy and ataxia mice. Therefore, AQP4 was excluded as a candidate gene in ataxia. The recent generation of AQP4 knockout mice revealed only a minor renal phenotype (Chou *et al.*, 1998), which indicates that, at least in mice, AQP4 is not essential for physiological processes in lung or brain. Although these knockout mice reveal the redundancy of AQP4, an important function for AQP4 in brain cannot be excluded. In fact, the abundance of this protein in brain predicts a physiological role in water metabolism within the central nervous system (Nielsen *et al.*, 1997).

In contrast to the negative examples above, application of the candidate screening approach in aquaporin research also had its successes. These are discussed in detail in the following subsections.

B. MIP26 (Aquaporin-0)

The major intrinsic protein of 26 kDa (MIP26), or aquaporin-0, is the founder member of the MIP family, to which all aquaporins belong, and is expressed specifically and in high abundance in lens fibers of the vertebrate eye. In 1993, the mouse MIP gene was known to be localized to the distal end of chromosome 10 (Griffin and Shiels, 1992). By analogy to its family members AQP1 and AQP2, MIP26 was additionally expected to facilitate transmembrane water permeability

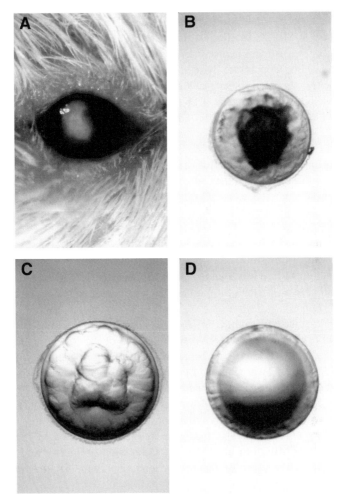

FIGURE 6 Phenotype of the lens of a CAT mouse. (A) the eye of a mouse homozygous for the CAT mutation in the MIP protein. (B–D) dissected lenses from homozygote (B) or heterozygote (C) CAT mice or a wild-type mouse (D). (Courtesy of A. Shiels, Department of Opthalmology, Washington University School of Medicine, St. Louis, Missouri.) (See Color Plate.)

(Shiels and Griffin, 1993), which was confirmed later (Kushmerick *et al.*, 1995; Mulders *et al.*, 1995). Because a mutation causing cataract in the homozygous CAT^{Fr} mouse (Fig. 6) had been assigned to the distal end of chromosome 10 (Muggleton-Harris *et al.*, 1987) and one of the first signs of the opacification process was the anterior swelling of the embryonic lens fiber cells (Sakuragawa *et al.*, 1975), Shiels and Griffin realized that the CAT^{Fr} mutation might be located in the

MIP26 gene. Indeed, immunocytochemical analysis of the CAT^{Fr} lens revealed no expression of MIP26, in contrast to the lens of healthy mice, and a smaller MIP26 transcript was identified in CAT^{Fr} mice by Northern blot analysis (Shiels and Griffin, 1993). Conclusive evidence was obtained with the analysis of the MIP26 gene (Shiels and Bassnett, 1996). The CAT^{Fr} phenotype was caused by an insertion of a long terminal repeat (LTR) of an embryo transposon at the boundary of intron 3 and the last MIP26 exon. Therefore, the CAT^{Fr} MIP26 gene encoded an MIP26-LTR fusion protein, in which an LTR replaced the sixth transmembrane domain and C-terminal tail of MIP26. The LTR insertion is a category I mutation because it results in a 40% reduction of MIP26 mRNA.

In the Lop mouse, in which the cataract phenotype was also assigned to the distal end of chromosome 10, a G-to-C nucleotide change in the MIP gene, resulting in an amino acid change of alanine to proline at position 51, caused the disease (Shiels and Bassnett, 1996). Immunohistochemical staining for MIP26 in lenses of Lop mice revealed a diffuse pattern in which MIP was localized to intracellular membranes. Superimposition with the staining pattern of the ER constituent PDI revealed that Lop MIP26 was retained in the ER. Because Lop MIP26 mRNA was as stable as that of wt-MIP26, the Lop MIP26 gene mutation falls into the second category.

Recently, the cause of cataract in *Hfi* mice has been shown to be caused by a 76-bp deletion, containing the consensus sequence of the MIP26 exon 2 splice donor site (Sidjanin, D. J., personal communication). Consequently, with MIP26 mRNA splicing, exon 2 was skipped. Because exon 1 ends and exon 3 starts with a full codon, this results in an in-frame deletion of amino acids 121–175. The 22-kDa protein product lacks the predicted fourth to fifth transmembrane segments of MIP26. No MIP26 immunoreactivity was observed in lenses of homozygous *Hfi* mice, and immunoblot analysis revealed a marked decrease of MIP22 expression, compared to MIP26 in normal mice. Expression in oocytes and HEK293 cells showed that the mutant MIP protein was retained in the ER and was rapidly degraded (category II mutation).

The Cat^{Fr}, Lop, and *Hfi* mutations have been described as semidominant, because the original heterozygous mutants presented with a milder phenotype than the inbred homozygotes (Shiels and Bassnett, 1996; Sidjanin *et al.*, 2000). In most cases, dominant-negative inheritance of a disease is caused by interference of the wild type by the mutant protein or by haplotype insufficiency (i.e., the reduced expression of the wt protein in heterozygotes is not enough to prevent the disease). Because MIP26 has a limited capacity to transport water (Mulders *et al.*, 1995; Verbavatz *et al.*, 1994), constitutes more than 50% of lens membrane proteins, and appears to be concentrated in crystalline tetrameric arrays in fiber cell membranes (Konig *et al.*, 1997), Shiels and Bassnett (1996) reasoned that MIP26 might primarily serve a structural function in maintaining the refraction and accommodation properties of the lens. Mutations in genes for structural proteins are frequently dominant, whereas mutations in genes for nonstructural proteins (e.g., AQP1, AQP2, CFTR, and LDL receptor) are usually recessive. This is in line with their notion that mutations in the crystalline gene family, which encode the major cytosolic structural proteins of the lens, also confer dominant cataract phenotypes in mammals (Shiels and Bassnett, 1996). These observations support the idea that cataract is caused by haploinsufficiency. The finding that increased cleavage of MIP26 by a cysteine protease in lenses of transgenic mice accelerates the opacification of the lens is also in agreement with MIP26 being a structural protein (Mitton *et al.*, 1996).

Further evidence corroborating haplotype insufficiency as a cause for cataract formation in mice with MIP26 mutations comes from data on AQP1 in Colton-negative individuals (Section II.C) and AQP2 mutants in NDI (Section II.D). Aquaporin-1 and AQP2 are highly related to MIP26 but are nonstructural proteins. All mutations in the AQP1 or AQP2 gene encoding truncated or ER-retarded mutants are inherited recessively, which indicates that the mutants are not able to form an interaction with the wild-type protein. Haplotype insufficiency does not occur because the parents of the Colton-null individuals or patients with recessive NDI, who encode a mutant and wild-type protein, are positive for a Colton antigen (AQP1) or are healthy (AQP2), respectively. Also, in the rare case of an AQP2 mutation resulting in a dominant-inheritance of NDI, the phenotype shows a full penetration (i.e., the NDI phenotype in the patients is as severe as seen in patients with the recessive form), and the mutant is able to heterotetramerize with wt-AQP2. The expression and subcellular localization of the MIP mutants resembles that of recessively inherited AQP1 and AQP2 mutants, but the phenotype of the heterozygotes with MIP mutations (mild) is between that of heterozygotes encoding recessive (co-positive; healthy) and dominant AQP2 mutations (severe and full penetration). Analyses of whether the MIP mutants are able to heterotetramerize with wt-MIP could provide conclusive evidence of whether cataract is caused by haplotype insufficiency or a dominant-negative effect of the MIP mutants.

C. Aquaporin-1

The archetypal water channel, aquaporin-1, is abundantly expressed in red cells, renal proximal tubules, descending thin limbs of Henle, and several other water-permeable epithelia. Therefore, it was predicted that mutations in AQP1 would result in a severe or lethal phenotype. Surprisingly, therefore, three individuals who were negative for Colton blood group antigens ($Co^{a-}Co^{b-}$), who essentially are AQP1 knockout individuals, had no apparent clinical symptoms (Preston *et al.*, 1994). In a previous study of the same group, it had become clear that the Colton blood group antigens, which were assigned to the same chromosomal location as the AQP1 gene, were in fact an Ala/Val polymorphism at position 45 in AQP1 (Smith *et al.*, 1994). Genomic analyses revealed that one individual lacked most of exon 1, which encodes half of AQP1. The second person had a single base insertion at position 307, producing a frameshift starting after Gly 104, located in the third transmembrane region. In the third proband, a nucleotide change resulted in a substitution of a leucine for proline 38, which is located immediately after the first transmembrane domain. Immunoblot analysis of red cell membranes and urine samples revealed that AQP1 immunoreactivity was undetectable in the first two probands, whereas a weak AQP1 signal was obtained in red cells from the third person. Presumably, the mutations in probands 1 and 2 lead to unstable AQP1 mRNA (class I). Proline or glycine at position 38 is well conserved in most aquaporins (Reizer *et al.*, 1993). The low level of AQP1-P38L expression in red cells and the instability of AQP1-P38L upon expression in oocytes are consistent with a structural requirement for Pro or Gly at position 38 (Preston *et al.*, 1994) and suggest that this mutant is retained in the ER and subsequently degraded (category II).

Recently, a different mutation was identified in another Colton-negative individual, resulting in an N192K amino acid substitution (Chretien and Catron, 1999). Because N192 is part of the highly conserved second NPA box and expression of AQP1-N192D or AQP1-N192Q in oocytes results in reduced water permeability and decreased plasma membrane localization (Mathai and Agre, 1999), it was anticipated that N192 would be essential for channel function (Chretien and Catron, 1999). All Colton-negative individuals were homozygous for their respective mutations (Chretien and Catron, 1999; Preston *et al.*, 1994). Red cells from the first proband's mother and daughter gave weakened reactions with anti-Co[a] and no reaction with anti-Co[b] antibodies. Red cells of these heterozygotes had a reduced AQP1 expression level and water permeability (Lacey *et al.*, 1987; Mathai *et al.*, 1996). Because all of the tested relatives of the Colton-negative individuals were Colton a or b positive (see above; personal communications with Jean-Pierre Cartron, Institut National de la Transfusion Sanguine, France), it can be concluded that all identified AQP1 mutants do not interfere with plasma membrane expression of wild-type AQP1 and, therefore, that the Colton-negative phenotype in these families is inherited recessively. This is in line with most AQP2 mutations in NDI (Section II.D).

The identification of only five Colton-negative individuals among millions of blood donors and transfusion recipients shows that this condition is extremely rare and suggests that these particular individuals might in some way compensate for the loss of AQP1 (King and Agre, 1996). Indeed, AQP1 knockout mice have a severely impaired urinary concentrating ability after water deprivation (Ma *et al.*, 1998), but they can concentrate their urine mildly when water is given *ad libitum* because salt transporters are functional, and collecting duct water reabsorption still occurs. Therefore, AQP1-deficient humans, who are not subject to water deprivation, might not exert any abnormality (Schnermann *et al.*, 1998). Careful water deprivation tests may be needed to uncover renal defects in humans lacking AQP1.

D. Aquaporin-2

Originally, aquaporin-2 was only found to be expressed in the renal collecting duct (Fushimi *et al.*, 1993), but recently it has also been found in epithelial cells of the vas deferens (Nelson *et al.*, 1998; Stevens *et al.*, 2000). It is the only aquaporin known to be involved in a human disease. This rare disease, nephrogenic diabetes insipidus, occurs in four out of a million newborns (Bichet *et al.*, 1992) and is characterized by the inability of the kidney to concentrate urine in response to normal or elevated levels of the antidiuretic hormone arginine-vasopressin (AVP). In 90% of NDI families, the disease is caused by mutations in the vasopressin V_2 receptor, which resides at q28 of the X chromosome (for a review, see Bichet *et al.*, 1998; Deen and Knoers, 1998). However, well before identification of the involvement of AQP2, it had been reported that NDI could also be inherited as an autosomal recessive trait (Langley *et al.*, 1991). Aquaporin-2 was a strong candidate for the autosomal inheritance of NDI for these reasons:

- In the kidney, AQP2 is expressed only in the collecting ducts (Fushimi *et al.*, 1993).
- It is expressed in the apical membrane, the site at which regulation of urine concentration by AVP is believed to occur (Fushimi *et al.*, 1993).
- Dehydration (i.e., increased AVP levels) significantly increases renal AQP2 mRNA (Ma *et al.*, 1994; van Os *et al.*, 1994; Yamamoto *et al.*, 1995) and protein levels (Hayashi *et al.*, 1994; Nielsen *et al.*, 1993c).
- The AQP2 gene is localized at position q13 on (the autosomal) chromosome 12 (Deen *et al.*, 1994b; Sasaki *et al.*, 1994).

Indeed, analysis of the AQP2 gene of an NDI patient, who had no mutation in his V_2R gene, revealed that he was a compound heterozygote for missense mutations in the third and fourth exon of AQP2 (Deen *et al.*, 1994a). Expression analysis in oocytes revealed that, in contrast to wt-AQP2, the encoded AQP2-R187C and AQP2-S216P mutants were nonfunctional and impaired in their export from the ER (Deen *et al.*, 1994a, 1995). Coexpression of these mutants with wt-AQP2 showed no functional consequences for wt-AQP2, which is in keeping with the autosomal recessive inheritance of NDI observed in this family (Deen *et al.*, 1994a).

1. Routing of AQP2 Mutants in Recessive NDI

At present, 18 different families have been described in which recessive NDI is caused by mutations in the AQP2 gene (Deen *et al.*, 1994a; Goji *et al.*, 1998; Hochberg *et al.*, 1997; Mulders *et al.*, 1997, 1998; Oksche *et al.*, 1996b; Tamarappoo and Verkman, 1998; van Lieburg *et al.*, 1994; Vargas-Poussou *et al.*, 1998). In four of these families, the patients were compound heterozygotes for AQP2 mutations, whereas the other patients had identical mutations in both AQP2

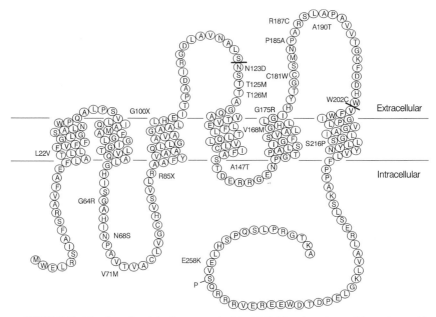

FIGURE 7 Mutations found in the aquaporin-2 in NDI. Mutations detected in patients with autosomal recessive or dominant NDI are indicated by text (missense/nonsense) or bars (nucleotide insertions or deletions resulting in a frameshift). For details see text. (Courtesy of P. M. T. Deen.)

alleles, mostly because they were from consanguineous matings. In these families, 19 different mutations were identified, of which three mutations (two nonsense and one nucleotide deletion) encode a truncated AQP2 protein (Hochberg *et al.,* 1997; van Lieburg *et al.,* 1994; Vargas-Poussou *et al.,* 1998), one might lead to aberrant splicing of intron 3 (Oksche *et al.,* 1996b), and the other 15 are missense mutations (Fig. 7). Of the missense mutations, 10 have been analyzed in cells. In oocytes, AQP2 proteins with G64R, N68S, T126M, A147T, R187C, or S216P substitutions were all impaired in their export from the ER. Although all cRNAs were stable, some of these mutants (A147T, S216P) were unstable compared to wt-AQP2 (Deen *et al.,* 1995; Mulders *et al.,* 1997; Tamarappoo and Verkman, 1998). Retardation in the ER was concluded from the finding that immunoblotting of oocyte lysates expressing these mutants revealed a 32-kDa band, which was not present in the lane of wt-AQP2 expressing oocytes, and which was reduced to the size of unglycosylated AQP2 of 29 kDa after digestion with endoglycosidase H (Deen *et al.,* 1995). This enzyme removes core glycosylation groups, which are added in the ER to asparagines in the glycosylation consensus sequence N-X-S/T. In the trans-Golgi complex, these core groups are exchanged for a complex glycosylation form that is insensitive to endo H and increases the size of AQP2 to 40–45 kDa.

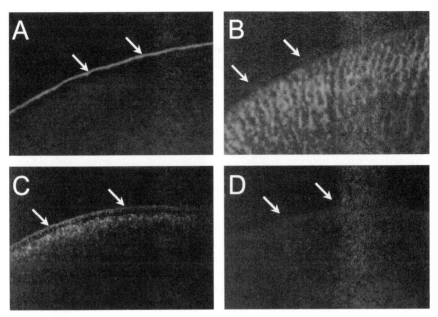

FIGURE 8 Immunocytochemistry of AQP2 proteins expressed in *Xenopus* oocytes. Oocytes were injected with low amounts of cRNAs coding for (A) wild-type AQP2, (B) an AQP2 mutant in recessive NDI (AQP2-R187C), (C) an AQP2 mutant in dominant NDI (AQP2-E258K), or (D) were not injected. Two days after injection, oocytes were sectioned and the AQP2 proteins were visualized using AQP2-specific rabbit antibodies and Alexa-594 coupled goat anti-rabbit antibodies. The plasma membranes are indicated by arrows. (Courtesy of P. M. T. Deen.)

In addition, immunocytochemistry of oocytes expressing these mutants showed the characteristic dispersed pattern of ER-retarded proteins, whereas wt-AQP2 was localized only at the plasma membrane (Fig. 8). These results were confirmed for AQP2-T126M, AQP2-A147T, and AQP2-R187C in transfected mammalian cells (Tamarappoo and Verkman, 1998). In these cells, the three mutants also appeared to be less stable than wt-AQP2.

The reduced stability is in agreement with *in vivo* studies (Deen *et al.*, 1996). In addition to AQP1, AQP2 can also be detected in antidiuretic, but not diuretic urine, of healthy individuals (Deen *et al.*, 1996; Kanno *et al.*, 1995). Because the osmolality of urine of NDI patients is in the same range as that of the diuretic urine of controls, the failure to detect AQP2 in the urine of NDI patients was anticipated. However, NDI patients may not be comparable to control individuals in diuresis because they have a physiological need to concentrate their urine. Indeed, in the urine of a female NDI patient, whose V_2R alleles encode a wild-type and a truncated V_2R, an AQP2 signal was obtained that was similar to that in the antidiuretic urine of a healthy control. This indicated that an NDI patient expressing only a limited amount of functional V_2R synthesizes and excretes enough wt-AQP2 to be clearly detectable in urine. Therefore, the absence of AQP2 in urine of the NDI patients homozygous for AQP2-G64R, AQP2-A147T, or AQP2-R187C is a clear indication that *in vivo* the AQP2 mutants have a reduced stability.

On the basis of these results, the encoding mutations can be placed in the class II category. Another mutant in NDI, AQP2-C181W, might also belong to this category, because its stability was also reduced compared to wt-AQP2 upon expression in oocytes (Tamarappoo and Verkman, 1998). Unfortunately, this could not be confirmed, because AQP2-C181W expression could not be detected with immunoblotting, and immunocytochemical analysis was not performed.

Another mutation, which encodes AQP2-L22V in the same patient, seems to behave differently from the mutants above (Tamarappoo and Verkman, 1998). In expression studies, this mutant appeared to be as stable and water permeable as wt-AQP2 and was retained in the ER, but was, in contrast to the other mutants, only detected as a 29-kDa (wt) protein on blots. Possibly, AQP2-L22V is not retained in the ER as efficiently as other AQP2 mutants in NDI. An induced expression of this functional mutant would also provide an explanation for the observed increased antidiuresis in the patient upon dDAVP admission (Canfield *et al.,* 1997).

Recently, Goji *et al.* (1998) described two mutations in NDI, coding for AQP2-T125M or AQP2-G175R mutants, which they speculated were intrinsically inactive water channels (class IV mutations). Upon injection of cRNAs into oocytes, the water permeability for both mutants was not different from control oocytes and significantly less than that of AQP2-T126M (which was taken as a control), whereas the plasma membrane expression of both mutants was significantly higher than that of AQP2-T126M. Also, immunocytochemistry showed no intracellular AQP2 staining in oocytes expressing either mutant. Because T125 is part of the N-glycosylation consensus sequence, it is not surprising that no ER-glycosylated T125M was detected on blots. However, and in contrast to others (Mulders *et al.,* 1997; Tamarappoo and Verkman, 1998), an ER-glycosylated band was also absent for AQP2-T126M (as well as AQP2-G175R), and a dispersed ER staining was not visible in oocytes expressing AQP2-T126M. As the authors state themselves, the ER-retained/glycosylated AQP2 might not be detectable by their antibody, or differences between batches of oocytes might explain this result. Alternatively, it is known that expression of high amounts of protein in oocytes and mammalian cells results in the accumulation of the protein in its normal organelle as well as at other locations. For example, on injection of 3 ng cRNA encoding AQP2-S256A into oocytes, the mutant accumulates at the plasma membrane to the extent that the oocyte's water permeability as well as AQP2-S256A expression level in the plasma membrane is not different from wt-AQP2 (Mulders *et al.,* 1998). However, at low levels of injection (0.3 ng of cRNAs), all AQP2-S256A is retained just below the surface of the plasma membrane and confers no water permeability (Kamsteeg *et al.,* 1999a). Therefore, overexpression might obscure the real molecular basis of these mutants, causing NDI.

2. Routing of AQP2 Mutants in Dominant NDI

Recently, a mutation has been identified in a family in which NDI is inherited in a dominant mode. This mutation substitutes lysine for glutamic acid 258, which is located in the C-terminal tail of AQP2 (Mulders *et al.,* 1998; Fig. 7). Analysis in oocytes revealed that AQP2-E258K was a stable, functional water channel that was, in contrast to mutants in recessive NDI, not retained in the ER, but rather in the Golgi complex (Fig. 8). Coexpression with wt-AQP2 indeed revealed a dominant-negative effect of AQP2-E258K, but not of AQP2-R187C, on the function of wt-AQP2. Dominant inheritance of a disease suggests that the protein involved is part of a multimeric complex, whereas in recessively inherited diseases this does not need to be the case. It has been shown that AQP1 forms homotetramers (Jung *et al.,* 1994), and recent freeze–fracture data are compatible with a tetrameric membrane structure for other aquaporins, including AQP2 (Van Hoek *et al.,* 1998a). Because it has also been shown that the AQP1 monomer is the functional unit (Preston *et al.,* 1993; Van Hoek *et al.,* 1991), we speculated that (1) AQP2 would also form homotetramers, (2) an AQP2 monomer would also be a functional unit, and (3) that the AQP2 mutant in dominant NDI would exert its effect by impairing the routing of wt-AQP2 after oligomerization.

In contrast, a mutant in recessive NDI was expected not to be able to oligomerize and therefore not to inhibit the routing of wt-AQP2. Indeed, sucrose gradient analyses of solubilized membranes of rat and human kidney revealed that AQP2 sedimented as a complex, consistent with a homotetramer (Kamsteeg *et al.,* 1999b). When expressed in oocytes, wt-AQP2 and AQP2-E258K also appeared to sediment as a homotetramer, whereas AQP2-R187C was in monomeric form. Subsequent analysis of oocytes coexpressing differently tagged wt-AQP2 and mutant forms (E258K and R187C) indeed revealed that AQP2-E258K, but not AQP2-R187C, forms heterotetramers. The heterotetramerization of AQP2-E258K with wt-AQP2 and inhibition of further routing of this complex to the plasma membrane explains dominant NDI in this particular family.

Although close to S256, of which phosphorylation by PKA has been shown to be essential for AQP2 shuttling to the plasma membrane (Section I; Fushimi *et al.,* 1997; Katsura *et al.,* 1997), the E258K mutation did not appear to affect its phosphorylation (Mulders *et al.,* 1998). The fact that AQP2-R253*, which lacks the AQP2 tail, was only slightly impaired in its transport to the plasma membrane indicates that the E258K mutation presumably introduces a Golgi retention signal instead of changing an existing AQP2 routing signal. Studies are under way to define the underlying mechanism.

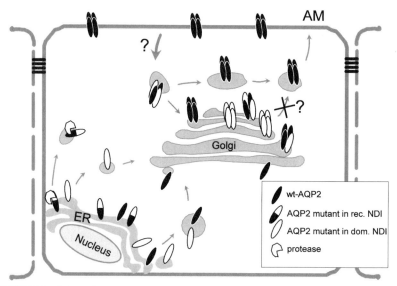

FIGURE 9 Schematic representation of the postulated molecular mechanisms underlying recessive and dominant NDI caused by AQP2 mutants. Indicated are recessive NDI (rec. NDI), dominant NDI (dom. NDI), the endoplasmic reticulum (ER), and the apical membrane (AM). For details see text. (Courtesy of P. M. T. Deen.)

On the basis of present data, it seems that the subcellular localization of the respective AQP2 mutants may reveal the mechanisms explaining recessive and dominant NDI (Fig. 9). The misfolded AQP2 mutants in recessive NDI are trapped in the ER (class II) and, possibly by compartmentalization, are not able to oligomerize with wt-AQP2. Consequently, in the healthy parents of the patients with a recessive form of NDI, whose genome encodes a mutant and a wt-AQP2, export of wt-AQP2 from the ER is not impaired. It can form homotetramers and fulfill its role in the concentration of urine. The genome of the patient with dominant NDI also encodes a mutant (AQP2-E258K) and a wt-AQP2 protein. However, the AQP2-E258K mutant is not misfolded and is able to heterotetramerize with wt-AQP2 in the ER or Golgi. Because this mutant is retained in the Golgi, the heterotetramer is entrapped in this cell organelle. This precludes routing of enough wt-AQP2 protein to the apical membrane of collecting duct cells, resulting in diuretic urine. On the basis of the proposed mechanism, intrinsically inactive AQP2 mutants that are not disturbed in their routing (class IV) can also explain recessive NDI. In NDI patients encoding these mutants, no functional water channels are produced. In the healthy parents, the AQP2 mutants would heterotetramerize with wt-AQP2, but this complex would not be affected in its routing or function. One would expect, therefore, that stable AQP2 mutant protein would be detectable in the urine of these patients. It would be interesting to see whether the AQP2-T125M and AQP2-G175R (Goji *et al.,* 1998) fit these criteria and would represent the first mutants of this class in cases of recessive NDI.

3. Structure–Function Information from AQP2 Mutants in Recessive NDI

In addition to the identification of the molecular basis by which these mutants cause NDI, functional testing of the mutants upon (over)expression in cell systems provides information on their structure–function relationships. AQP2-T126M appeared to be functional (Mulders *et al.,* 1997). Because many mutants that did not confer water permeability to oocytes had a stronger AQP2 staining in the plasma membrane than the functional AQP2-A147T, it was speculated that AQP2 mutants with G64R, N68S, R187C substitutions were nonfunctional. Using a new technique to efficiently isolate nearly pure plasma membranes it indeed appeared that AQP2-N68S, -R187C, and -S216P are nonfunctional, but that AQP2-G64R was weakly permeable for water (Marr *et al.,* 1999). These results were confirmed and extended in that AQP2-L22V was also a functional water channel and was also found to have, together with AQP2-T126M and AQP2-A147T, a single-channel water permeability identical to wt-AQP2 (Tamarappoo and Verkman, 1998). Other AQP2 mutants (C181W, T125M and G175R) were also nonfunctional.

Together, these data, which were obtained from expression studies in oocytes and mammalian cells, indicate that most mutations in the B-loop (N68S) and E-loop (G175R, R187C, C181W) destroy the water pore, whereas mutations in the first transmembrane domain (L22V), C-loop (T126M; except for T125M), and D-loop (A147T) do not have profound effects on the intrinsic water permeability of the AQP2 molecule. These data are in line with a mutational analysis study in AQP1, from which it was deduced that loops B and E are the pore-forming loops (Jung *et al.,* 1994).

Of the remaining five missense mutations found in NDI that have not been analyzed in cell systems, three exchanged amino acids are located in loops B or E, and might also be important for the formation of the water pore (V71M, P185A, A190T; Bichet *et al.,* 1995). Valine 71 and P185 are well conserved among aquaporins, which indicates their importance in the structure or function of these proteins. Another mutation (V168M) is located in and might disturb the formation of the fifth transmembrane α helix. Contrasting results were obtained from AQP2 proteins expressed in yeast vesicles (Shinbo *et al.,* 1999). With equal expression levels, stopped-flow Pf measurements and mercury inhibition examinations suggested that AQP2-L22V was fully functional, whereas mutants N68S, R187C, and S216P were partially functional. In contrast, mutants N123D, T125M, T126M, A147T, and C181W had very low water permeability. In contrast to oocytes and mammalian cells, the AQP2 mutants did not seem to be impaired in their trafficking in yeast because a high-mannose form of the AQP2 mutants was not detected. This might indicate that in yeast, AQP2 and its mutants are folded differently and have a different tertiary structure, resulting in other permeability characteristics.

Of interest, all AQP2 mutations in recessive NDI, which lead to misfolded, ER retained mutants or possibly to intrinsically inactive but properly folded mutants, are found between the first and the last transmembrane region. However, a mutation in dominant NDI that leads to properly folded mutants that are disturbed in their routing later than the ER is found in the C-terminal tail. Another mutation in the C tail (S256A) has been shown to impair shuttling of the AQP2 mutant to the plasma membrane (Fushimi et al., 1997; Kamsteeg et al., 1999a; Katsura et al., 1997). Therefore, mutations in the C-terminal tail might only affect the (regulation of) routing of properly folded AQP2 protein, whereas mutations in the remaining part of the protein may affect proper folding and/or proper formation of the water pore. Future studies on the routing of naturally occurring AQP2 mutants will show whether this hypothesis holds and will provide more insight as to the molecular causes of NDI.

E. Future Perspectives

The tissue-specific expression and subcellular localization of the 10 identified mammalian aquaporins is now well established, but their specific functions and detailed information on (the regulation of) their routing are still largely unknown. In the coming years, it is expected that analysis of aquaporin chimeras expressed in polarized cell lines will reveal which parts of the molecule contain important routing information. In addition, promising techniques such as yeast two-hybrid screening might unravel the identity of aquaporin-interacting proteins, which might shed light on the mechanisms underlying routing and shuttling of aquaporins, their interaction with the cytoskeleton, or even processes that are activated in parallel with water permeation.

The identification of AQP2 mutations in NDI directly showed the role and importance of the protein in vasopressin-regulated water reabsorption in the collecting duct. Although AQP0 is essential in the eye, its role did not become clear with the identification of AQP0 mutants in cataract, nor did the identification of AQP1 mutants in Colton-negative individuals shed light on its function. Because the inactivated gene might be redundant in knockout models, the role of different aquaporins in mammals might only become clear by knocking out all independent aquaporins and subjecting AQP-expressing tissues of these animals and intercrosses to careful functional examination. Data on some AQP knockout animals are described in Chapter 5 by Verkman et al.

The determination of the molecular nature of the mutations causing NDI and other diseases provides an essential basis for disease treatment. As discussed earlier, misfolded mutants are thought to be retained in the ER because of prolonged binding to chaperones. Therefore it was anticipated that modulation of chaperone binding to functional, ER-retained mutants by influencing the expression of chaperones and/or treatment with chemical chaperones could enhance proper folding of the mutant. Indeed, plasma membrane expression of the functional, but misfolded, CFTR-ΔF508 chloride channel appeared to be partially rescued *in vitro* by growing expressing cells at reduced temperatures and by treatment with chemical chaperones, such as dimethylsulfoxide (DMSO), trimethylamine oxide (TMAO), or sodium 4-phenylbutyrate (4PBA). Certain chemicals are already being used in clinical trials (Denning *et al.,* 1992; Rubenstein *et al.,* 1997; Zeitlin, 1999). In fact, treatment of mammalian cells expressing the AQP2 mutants T126M, A147T, or R187C with glycerol, TMAO, or DMSO partially relieved their ER retention (Tamarappoo and Verkman, 1998). Alternatively, gene therapy might provide a way to cure NDI.

In conclusion, we have come a long way in our understanding of water balance in mammals, but many questions remain to be addressed. Undoubtedly, many exciting discoveries will be made in the coming years that will broaden our insight on (the regulation of) the routing of aquaporins in health and disease.

References

Allan, V. J., and Schroer, T. A. (1999). Membrane motors. *Curr. Opin. Cell Biol.* **11,** 476–482.

Arnell, H., Hjalmas, K., Jagervall, M., Lackgren, G., Stenberg, A., Bengtsson, B., Wassen, C., Emahazion, T., Anneren, G., Pettersson, U., Sundvall, M., and Dahl, N. (1997). The genetics of primary nocturnal enuresis: Inheritance and suggestion of a second major gene on chromosome 12q. *J. Med. Genet.* **34,** 360–365.

Bacallao, R., Antony, C., Dotti, C., Karsenti, E., Stelzer, E. H., and Simons, K. (1989). The subcellular organization of Madin–Darby canine kidney cells during the formation of a polarized epithelium. *J. Cell Biol.* **109,** 2817–2832.

Bichet, D. G., Hendy, G. N., Lonergan, M., Arthus, M.-F., Ligier, S., Pausova, Z., Kluge, R., Zingg, H., Saenger, P., Oppenheimer, E. *et al.* (1992). X-linked nephrogenic diabetes insipidus: From the ship *Hopewell* to RFLP studies. *Am. J. Hum. Genet.* **51,** 1089–1102.

Bichet, D. G., Arthus, M.-F., Lonergan, M., Balfe, W., Skorechi, K., Nivet, H., Robertson, G., Oksche, A., Rosenthal, W., Fujiwara, M., Morgan, K., and Sasaki, S. (1995). Autosomal dominant and autosomal recessive nephrogenic diabetes insipidus: Novel mutations in the AQP2 gene [abstract]. *J. Am. Soc. Nephrol.* **6,** 717.

Bichet, D. G., Turner, M., and Morin, D. (1998). Vasopressin receptor mutations causing nephrogenic diabetes insipidus. *Proc. Assoc. Am. Physicians* **110,** 387–394.

Breton, S., and Brown, D. (1998). Cold-induced microtubule disruption and relocalization of membrane proteins in kidney epithelial cells. *J. Am. Soc. Nephrol.* **9,** 155–166.

Brooks, D. A. (1997). Protein processing: A role in the pathophysiology of genetic disease. *FEBS Lett.* **409,** 115–120.

Brown, D., and Orci, L. (1983). Vasopressin stimulates formation of coated pits in rat kidney collecting ducts. *Nature* **302,** 253–255.

Brown, D., and Stow, J. L. (1996). Protein trafficking and polarity in kidney epithelium: From cell biology to physiology. *Physiol. Rev.* **76,** 245–297.

Brown, D., Grosso, A., and DeSousa, R. C. (1990). Membrane architecture and water transport in epithelial cell membranes. *In* "Advances in Membrane Fluidity" (R. C. Aloia, Ed.) pp. 103–132. Alan Liss, New York.

Brown, D., Cunningham, C., Hartwig, J., McLaughlin, M., and Katsura, T. (1996). Association of AQP2 with actin in transfected LLC-PK1 cells and rat papilla [abstract]. *J. Am. Soc. Nephrol.* **7,** 1265A.

Canfield, M. C., Tamarappoo, B. K., Moses, A. M., Verkman, A. S., and Holtzman, E. J. (1997). Identification and characterization of aquaporin-2 water channel mutations causing nephrogenic diabetes insipidus with partial vasopressin response. *Hum. Mol. Genet.* **6,** 1865–1871.

Chen, W. J., Goldstein, J. L., and Brown, M. S. (1990). NPXY, a sequence often found in cytoplasmic tails, is required for coated pit-mediated internalization of the low density lipoprotein receptor. *J. Biol. Chem.* **265,** 3116–3123.

Chou, C. L., Ma, T. H., Yang, B. X., Knepper, M. A., and Verkman, S. (1998). Fourfold reduction of water permeability in inner medullary collecting duct of aquaporin-4 knockout mice. *Am. J. Physiol.* **43,** C549–C554.

Chretien, S., and Catron, J. P. (1999). A single mutation inside the NPA motif of aquaporin-1 found in a Colton-null phenotype [letter]. *Blood* **93,** 4021–4023.

Christensen, B. M., Marples, D., Jensen, U. B., Frokiaer, J., Sheikh-Hamad, D., Knepper, M. A., and Nielsen, S. (1998). Acute effects of vasopressin V_2-receptor antagonist on kidney AQP2 expression and subcellular distribution. *Am. J. Physiol.* **275,** F285–F297.

Christensen, B. M., Zelenina, M., Aperia, A., and Nielsen, S. (2000). Localization and regulation of PKA-phosphorylated AQP2 in response to V_2-receptor agonist/antagonist treatment. *Am. J. Physiol.* **278,** F29–F42.

Coleman, R. A., Harris, H. W., Jr., and Wade, J. B. (1987). Visualization of endocytosed markers in freeze-fracture studies of toad urinary bladder. *J. Histochem. Cytochem.* **35,** 1405–1414.

Davis, C. G., Lehrman, M. A., Russell, D. W., Anderson, R. G., Brown, M. S., and Goldstein, J. L. (1986). The J.D. mutation in familial hypercholesterolemia: Amino acid substitution in cytoplasmic domain impedes internalization of LDL receptors. *Cell* **45,** 15–24.

Davis, W. L., Goodman, D. B., Jones, R. G., and Rasmussen, H. (1978). The effects of cytochalasin B on the surface morphology of the toad urinary bladder epithelium: A scanning electron microscopic study. *Tissue Cell* **10,** 451–462.

Deen, P. M. T., Verdijk, M. A. J., Knoers, N. V. A. M., Wieringa, B., Monnens, L. A. H., van Os, C. H., and van Oost, B. A. (1994a). Requirement of human renal water channel aquaporin-2 for vasopressin-dependent concentration of urine. *Science* **264,** 92–95.

Deen, P. M. T., Weghuis, D. O., Sinke, R. J., Geurts van Kessel, A., Wieringa, B., and van Os, C. H. (1994b). Assignment of the human gene for the water channel of renal collecting duct aquaporin 2 (AQP2) to chromosome 12 region q12→q13. *Cytogenet. Cell Genet.* **66,** 260–262.

Deen, P. M. T., Croes, H., van Aubel, R. A., Ginsel, L. A., and van Os, C. H. (1995). Water channels encoded by mutant aquaporin-2 genes in nephrogenic diabetes insipidus are impaired in their cellular routing. *J. Clin. Invest.* **95,** 2291–2296.

Deen, P. M. T., van Aubel, R. A., van Lieburg, A. F., and van Os, C. H. (1996). Urinary content of aquaporin 1 and 2 in nephrogenic diabetes insipidus. *J. Am. Soc. Nephrol.* **7,** 836–841.

Deen, P. M. T., Rijss, J. P. L., Mulders, S. M., Errington, R. J., van Baal, J., and van Os, C. H. (1997a). Aquaporin-2 transfection of Madin–Darby canine kidney cells reconstitutes vasopressin-regulated transcellular osmotic water transport. *J. Am. Soc. Nephrol.* **8,** 1493–1501.

Deen, P. M. T., Nielsen, S., Bindels, R. J. M., and van Os, C. H. (1997b). Apical and basolateral expression of aquaporin-1 in transfected MDCK and LLC-PK cells and functional evaluation of their transcellular osmotic water permeabilities. *Pflugers Arch.* **433,** 780–787.

Deen, P. M. T., and Knoers, N. V. A. M. (1998). Vasopressin type-2 receptor and aquaporin-2 water channel mutants in nephrogenic diabetes insipidus. *Am. J. Med. Sci.* **316,** 300–309.

Deen, P. M. T., Van Balkom, B., Kamsteeg, E. J., Van Raak, M., Rajendran, V., and Caplan, M. J. (1999). The C-terminus of aquaporin-2 is necessary, but not sufficient, for routing of AQP2 to the apical membrane [abstract]. *J. Am. Soc. Nephrol.* **10,** 62A.

Denning, G. M., Anderson, M. P., Amara, J. F., Marshall, J., Smith, A. E., and Welsh, M. J. (1992). Processing of mutant cystic fibrosis transmembrane conductance regulator is temperature-sensitive. *Nature* **358**, 761–764.

Devuyst, O., Burrow, C. R., Smith, B. L., Agre, P., Knepper, M. A., and Wilson, P. D. (1996). Expression of aquaporins-1 and -2 during nephrogenesis and in autosomal dominant polycystic kidney disease. *Am. J. Physiol.* **40**, F169–F183.

Dray-Charier, N., Paul, A., Scoazec, J. Y., Veissiere, D., Mergey, M., Capeau, J., Soubrane, O., and Housset, C. (1999). Expression of delta F508 cystic fibrosis transmembrane conductance regulator protein and related chloride transport properties in the gallbladder epithelium from cystic fibrosis patients. *Hepatol.* **29**, 1624–1634.

Engisch, K. L., and Nowycky, M. C. (1998). Compensatory and excess retrieval: Two types of endocytosis following single step depolarizations in bovine adrenal chromaffin cells. *J. Physiol. (Lond.)* **506 (Pt 3)**, 591–608.

Fisher, E. A., Zhou, M., Mitchell, D. M., Wu, X., Omura, S., Wang, H., Goldberg, A. L., and Ginsberg, H. N. (1997). The degradation of apolipoprotein B100 is mediated by the ubiquitin-proteasome pathway and involves heat shock protein 70. *J. Biol. Chem.* **272**, 20427–20434.

Fransen, J. A., Hauri, H. P., Ginsel, L. A., and Naim, H. Y. (1991). Naturally occurring mutations in intestinal sucrase-isomaltase provide evidence for the existence of an intracellular sorting signal in the isomaltase subunit [published erratum appears in *J. Cell Biol.* (1991) **115**(5), following 1473]. *J. Cell Biol.* **115**, 45–57.

Fushimi, K., Uchida, S., Hara, Y., Hirata, Y., Marumo, F., and Sasaki, S. (1993). Cloning and expression of apical membrane water channel of rat kidney collecting tubule. *Nature* **361**, 549–552.

Fushimi, K., Sasaki, S., and Marumo, F. (1997). Phosphorylation of serine 256 is required for cAMP-dependent regulatory exocytosis of the aquaporin-2 water channel. *J. Biol. Chem.* **272**, 14800–14804.

Futter, C. E., Gibson, A., Allchin, E. H., Maxwell, S., Ruddock, L. J., Odorizzi, G., Domingo, D., Trowbridge, I. S., and Hopkins, C. R. (1998). In polarized MDCK cells basolateral vesicles arise from clathrin-gamma-adaptin-coated domains on endosomal tubules. *J. Cell. Biol.* **141**, 611–623.

Gardner, A. M., Aviel, S., and Argon, Y. (1993). Rapid degradation of an unassembled immunoglobulin light chain is mediated by a serine protease and occurs in a pre-Golgi compartment. *J. Biol. Chem.* **268**, 25940–25947.

Garippa, R. J., Judge, T. W., James, D. E., and McGraw, T. E. (1994). The amino terminus of GLUT4 functions as an internalization motif but not an intracellular retention signal when substituted for the transferrin receptor cytoplasmic domain. *J. Cell. Biol.* **124**, 705–715.

Goji, K., Kuwahara, M., Gu, Y., Matsuo, M., Marumo, F., and Sasaki, S. (1998). Novel mutations in aquaporin-2 gene in female siblings with nephrogenic diabetes insipidus: Evidence of disrupted water channel function. *J. Clin. Endocrinol. Metab.* **83**, 3205–3209.

Griffin, C. S., and Shiels, A. (1992). Localisation of the gene for the major intrinsic protein of eye-lens-fibre cell membranes to mouse chromosome 10 by *in situ* hybridisation. *Cytogenet. Cell Genet.* **59**, 300–302.

Griffiths, G., Pfeiffer, S., Simons, K., and Matlin, K. (1985). Exit of newly synthesized membrane proteins from the trans cisterna of the Golgi complex to the plasma membrane. *J. Cell Biol.* **101**, 949–964.

Gustafson, C. E., Katsura, T., McKee, M., Bouley, R., Casanova, J. E., and Brown, D. (2000). Recycling of AQP2 occurs through a temperature- and bafilomycin-sensitive trans-Golgi-associated compartment in LLC-PK1 cells. *Am. J. Physiol.* **278**, F317–F326.

Han, Z., Wax, M. B., and Patil, R. V. (1998). Regulation of aquaporin-4 water channels by phorbol ester-dependent protein phosphorylation. *J. Biol. Chem.* **273**, 6001–6004.

Harris, H. W., Jr., Wade, J. B., and Handler, J. S. (1986). Fluorescent markers to study membrane retrieval in antidiuretic hormone-treated toad urinary bladder. *Am. J. Physiol.* **251**, C274–C284.

Harris, H. W., Jr., Handler, J. S., and Blumenthal, R. (1990). Apical membrane vesicles of ADH-stimulated toad bladder are highly water permeable. *Am. J. Physiol.* **258,** F237–F243.

Hartwig, J. H., Ausiello, D. A., and Brown, D. (1987). Vasopressin-induced changes in the three-dimensional structure of toad bladder apical surface. *Am. J. Physiol.* **253,** C707–C720.

Hayashi, M., Sasaki, S., Tsuganezawa, H., Monkawa, T., Kitajima, W., Konishi, K., Fushimi, K., Marumo, F., and Saruta, T. (1994). Expression and distribution of aquaporin of collecting duct are regulated by vasopressin V_2 receptor in rat kidney. *J. Clin. Invest.* **94,** 1778–1783.

Hays, R. M. (1983). Alteration of luminal membrane structure by antidiuretic hormone. *Am. J. Physiol.* **245,** C289–C296.

Hays, R. M., Condeelis, J., Gao, Y., Simon, H., Ding, G., and Franki, N. (1993). The effect of vaso-pressin on the cytoskeleton of the epithelial cell. *Pediatr. Nephrol.* **7,** 672–679.

Hays, R. M., Franki, N., Simon, H., and Gao, Y. (1994). Antidiuretic hormone and exocytosis: Lessons from neurosecretion. *Am. J. Physiol.* **267,** C1507–C1524.

Hobbs, H. H., Russell, D. W., Brown, M. S., and Goldstein, J. L. (1990). The LDL receptor locus in familial hypercholesterolemia: Mutational analysis of a membrane protein. *Annu. Rev. Genet.* **24,** 133–170.

Hochberg, Z., van Lieburg, A. F., Even, L., Brenner, B., Lanir, N., van Oost, B. A., and Knoers, N. V. A. M. (1997). Autosomal recessive nephrogenic diabetes insipidus caused by an aqua-porin-2 mutation. *J. Clin. Endocrinol. Metab.* **82,** 686–689.

Hochstrasser, M. (1996). Protein degradation or regulation: Ub the judge. *Cell* **84,** 813–815.

Hogg, R. J., and Husmann, D. (1993). The role of family history in predicting response to desmopressin in nocturnal enuresis. *J. Urol.* **150,** 444–445.

Ishibashi, K., Sasaki, S., Fushimi, K., Yamamoto, T., Kuwahara, M., and Marumo, F. (1997). Immuno-localization and effect of dehydration on AQP3, a basolateral water channel of kidney collecting ducts. *Am. J. Physiol.* **41,** F235–F241.

Ito, M., and Jameson, J. L. (1997). Molecular basis of autosomal dominant neurohypophyseal diabetes insipidus—Cellular toxicity caused by the accumulation of mutant vasopressin precursors within the endoplasmic reticulum. *J. Clin. Invest.* **99,** 1897–1905.

Jensen, T. J., Loo, M. A., Pind, S., Williams, D. B., Goldberg, A. L., and Riordan, J. R. (1995). Multiple proteolytic systems, including the proteasome, contribute to CFTR processing. *Cell* **83,** 129–135.

Johansson, I., Karlsson, M., Shukla, V. K., Chrispeels, M. J., Larsson, C., and Kjellbom, P. (1998). Water transport activity of the plasma membrane aquaporin pm28a is regulated by phosphorylation. *Plant Cell* **10,** 451–459.

Jung, J. S., Preston, G. M., Smith, B. L., Guggino, W. B., and Agre, P. (1994). Molecular structure of the water channel through aquaporin CHIP. The hourglass model. *J. Biol. Chem.* **269,** 14648–14654.

Kachadorian, W. A., Ellis, S. J., and Muller, J. (1979). Possible roles for microtubules and microfila-ments in ADH action on toad urinary bladder. *Am. J. Physiol.* **236,** F14–F20.

Kadowaki, T., Kadowaki, H., and Taylor, S. I. (1990). A nonsense mutation causing decreased levels of insulin receptor. *Proc. Natl. Acad. Sci. USA* **87,** 658–662.

Kalin, N., Claass, A., Sommer, M., Puchelle, E., and Tummler, B. (1999). DeltaF508 CFTR protein expression in tissues from patients with cystic fibrosis [see comments]. *J. Clin. Invest.* **103,** 1379–1389.

Kamsteeg, E. J., Wormhoudt, T. A., Rijss, J. P. L., van Os, C. H., and Deen, P. M. T. (1999b). An im-paired routing of wild-type aquaporin-2 after tetramerization with an aquaporin-2 mutant explains dominant nephrogenic diabetes insipidus. *EMBO J.* **18,** 2394–2400.

Kamsteeg, E. J., Heijnen, I., van Os, C. H., and Deen, P. M. T. (2000). The subcellular localization of an aquaporin-2 tetramer depends on the stoichiometry of phosphorylated and nonphosphorylated monomers. *J. Cell Biol.* **151,** 919–930.

Kanno, K., Sasaki, S., Hirata, Y., Ishikawa, S., Fushimi, K., Nakanishi, S., Bichet, D. G., and Marumo, F. (1995). Urinary excretion of aquaporin-2 in patients with diabetes insipidus. *N. Engl. J. Med.* **332,** 1540–1545.

Katsura, T., Verbavatz, J. M., Farinas, J., Ma, T., Ausiello, D. A., Verkman, A. S., and Brown, D. (1995). Constitutive and regulated membrane expression of aquaporin 1 and aquaporin 2 water channels in stably transfected LLC-PK1 epithelial cells. *Proc. Natl. Acad. Sci. USA* **92,** 7212–7216.

Katsura, T., Ausiello, D. A., and Brown, D. (1996). Direct demonstration of aquaporin-2 water channel recycling in stably transfected LLC-PK1 epithelial cells. *Am. J. Physiol.* **39,** F548–F553.

Katsura, T., Gustafson, C. E., Ausiello, D. A., and Brown, D. (1997). Protein kinase A phosphorylation is involved in regulated exocytosis of aquaporin-2 in transfected LLC-PK1 cells. *Am. J. Physiol.* **41,** F816–F822.

Kellenberger, S., Gautschi, I., Rossier, B. C., and Schild, L. (1998). Mutations causing Liddle syndrome reduce sodium-dependent downregulation of the epithelial sodium channel in the *Xenopus* oocyte expression system. *J. Clin. Invest.* **101,** 2741–2750.

Kim, P. S., and Arvan, P. (1998). Endocrinopathies in the family of endoplasmic reticulum (ER) storage diseases: Disorders of protein trafficking and the role of ER molecular chaperones. *Endocr. Rev.* **19,** 173–202.

King, L. S., and Agre, P. (1996). Pathophysiology of the aquaporin water channels. *Annu. Rev. Physiol.* **58,** 619–648.

Klussmann, E., Maric, K., Wiesner, B., Beyermann, M., and Rosenthal, W. (1999). Protein kinase A anchoring proteins are required for vasopressin-mediated translocation of aquaporin-2 into cell membranes of renal principal cells. *J. Biol. Chem.* **274,** 4934–4938.

Konig, N., Zampighi, G. A., and Butler, P. J. G. (1997). Characterisation of the major intrinsic protein (MIP) from bovine lens fibre membranes by electron microscopy and hydrodynamics. *J. Mol. Biol.* **265,** 590–602.

Kushmerick, C., Rice, S. J., Baldo, G. J., Haspel, H. C., and Mathias, R. T. (1995). Ion, water and neutral solute transport in Xenopus oocytes expressing frog lens MIP. *Exp. Eye Res.* **61,** 351–362.

Kuwahara, M., Fushimi, K., Terada, Y., Bai, L., Marumo, F., and Sasaki, S. (1995). cAMP-dependent phosphorylation stimulates water permeability of aquaporin-collecting duct water channel protein expressed in *Xenopus* oocytes. *J. Biol. Chem.* **270,** 10384–10387.

Kuznetsov, G., and Nigam, S. K. (1998). Folding of secretory and membrane proteins. *N. Engl. J. Med.* **339,** 1688–1695.

Lacey, P. A., Robinson, J., Collins, M. L., Bailey, D. G., Evans, C. C., Moulds, J. J., and Daniels, G. L. (1987). Studies on the blood of a Co (a-b-) proposita and her family. *Transfusion* **27,** 268–271.

Lande, M. B., Jo, I., Zeidel, M. L., Somers, M., and Harris, H. W. (1996). Phosphorylation of aquaporin-2 does not alter the membrane water permeability of rat papillary water channel-containing vesicles. *J. Biol. Chem.* **271,** 5552–5557.

Langley, J. M., Balfe, J. W., Selander, T., Ray, P. N., and Clarke, J. T. (1991). Autosomal recessive inheritance of vasopressin-resistant diabetes insipidus. *Am. J. Med. Genet.* **38,** 90–94.

Lehrman, M. A., Goldstein, J. L., Brown, M. S., Russell, D. W., and Schneider, W. J. (1985). Internalization-defective LDL receptors produced by genes with nonsense and frameshift mutations that truncate the cytoplasmic domain. *Cell* **41,** 735–743.

Lloyd, M. L., and Olsen, W. A. (1987). A study of the molecular pathology of sucrase-isomaltase deficiency. A defect in the intracellular processing of the enzyme. *N. Engl. J. Med.* **316,** 438–442.

Lorenzsonn, V., Lloyd, M., and Olsen, W. A. (1993). Immunocytochemical heterogeneity of lactase-phlorizin hydrolase in adult lactase deficiency. *Gastroenterol.* **105,** 51–59.

Lueck, A., Brown, D., and Kwiatkowski, D. J. (1998). The actin-binding proteins adseverin and gelsolin are both highly expressed but differentially localized in kidney and intestine. *J. Cell Sci.* **111,** 3633–3643.

Ma, T., Hasegawa, H., Skach, W. R., Frigeri, A., and Verkman, A. S. (1994). Expression, functional analysis, and *in situ* hybridization of a cloned rat kidney collecting duct water channel. *Am. J. Physiol.* **266,** C189–C197.

Ma, T. H., Yang, B. X., Gillespie, A., Carlson, E. J., Epstein, C. J., and Verkman, A. S. (1998). Severely impaired urinary concentrating ability in transgenic mice lacking aquaporin-1 water channels. *J. Biol. Chem.* **273,** 4296–4299.

Mandon, B., Chou, C. L., Nielsen, S., and Knepper, M. A. (1996). Syntaxin-4 is localized to the apical plasma membrane of rat renal collecting duct cells: Possible role in aquaporin-2 trafficking. *J. Clin. Invest.* **98,** 906–913.

Mandon, B., Nielsen, S., Kishore, B. K., and Knepper, M. A. (1997). Expression of syntaxins in rat kidney. *Am. J. Physiol.* **42,** F718–F730.

Maric, K., Oksche, A., and Rosenthal, W. (1998). Aquaporin-2 expression in primary cultured rat inner medullary collecting duct cells. *Am. J. Physiol.* **275,** F796–F801.

Marinelli, R. A., Pham, L., Agre, P., and LaRusso, N. F. (1997). Secretin promotes osmotic water transport in rat cholangiocytes by increasing aquaporin-1 water channels in plasma membrane—Evidence for a secretin-induced vesicular translocation of aquaporin-1. *J. Biol. Chem.* **272,** 12984–12988.

Marples, D., Knepper, M. A., Christensen, E. I., and Nielsen, S. (1995). Redistribution of aquaporin-2 water channels induced by vasopressin in rat kidney inner medullary collecting duct. *Am. J. Physiol.* **38,** C655–C664.

Marples, D., Smith, J., and Nielsen, S. (1997). Myosin-I is associated with AQP-2 water channel bearing vesicles in rat kidney and may be involved in the antidiuretic response to vasopressin [abstract]. *J. Am. Soc. Nephrol.* **8,** 62A.

Marples, D., Schroer, T. A., Ahrens, N., Taylor, A., Knepper, M. A., and Nielsen, S. (1998). Dynein and dynactin colocalize with aqp2 water channels in intracellular vesicles from kidney collecting duct. *Am. J. Physiol.* **43,** F384–F394.

Marr, N., Kamsteeg, E. J., van Os, C. H., and Deen, P. M. T. Determination of the functionality of Aquaporin-2 missense mutants in recessive NDI. *Pflugers Arch,* in press.

Martin, S., Slot, J. W., and James, D. E. (1999). GLUT4 trafficking in insulin-sensitive cells. A morphological review. *Cell. Biochem. Biophys.* **30,** 89–113.

Masur, S. K., Holtzman, E., Schwartz, I. L., and Walter, R. (1971). Correlation between pinocytosis and hydroosmosis induced by neurohypophyseal hormones and mediated by adenosine $3',5'$-cyclic monophosphate. *J. Cell Biol.* **49,** 582–594.

Mathai, J. C., and Agre, P. (1999). Hourglass pore-forming domains restrict aquaporin-1 tetramer assembly. *Biochem.* **38,** 923–928.

Mathai, J. C., Mori, S., Smith, B. L., Preston, G. M., Mohandas, N., Collins, M., Vanzijl, P. C. M., Zeidel, M. L., and Agre, P. (1996). Functional analysis of aquaporin-1 deficient red cells—The Colton-null phenotype. *J. Biol. Chem.* **271,** 1309–1313.

Maurel, C., Kado, R. T., Guern, J., and Chrispeels, M. J. (1995). Phosphorylation regulates the water channel activity of the seed-specific aquaporin alpha-TIP. *EMBO J.* **14,** 3028–3035.

Mitton, K. P., Kamiya, T., Tumminia, S. J., and Russell, P. (1996). Cysteine protease activated by expression of HIV-1 protease in transgenic mice—MIP26 (aquaporin-0) cleavage and cataract formation *in vivo* and *ex vivo*. *J. Biol. Chem.* **271,** 31803–31806.

Muggleton-Harris, A. L., Festing, M. F., and Hall, M. (1987). A gene location for the inheritance of the cataract Fraser (CatFr) mouse congenital cataract. *Genet. Res.* **49,** 235–238.

Mulders, S. M., Preston, G. M., Deen, P. M. T., Guggino, W. B., van Os, C. H., and Agre, P. (1995). Water channel properties of major intrinsic protein of lens. *J. Biol. Chem.* **270,** 9010–9016.

Mulders, S. M., Knoers, N. V. A. M., van Lieburg, A. F., Monnens, L. A. H., Leumann, E., Wuhl, E., Schober, E., Rijss, J. P. L., van Os, C. H., and Deen, P. M. T. (1997). New mutations in the AQP2 gene in nephrogenic diabetes insipidus resulting in functional but misrouted water channels. *J. Am. Soc. Nephrol.* **8,** 242–248.

Mulders, S. M., Bichet, D. G., Rijss, J. P. L., Kamsteeg, E. J., Arthus, M. F., Lonergan, M., Fujiwara, M., Morgan, K., Leijendekker, R., van der Sluijs, P., van Os, C. H., and Deen, P. M. T. (1998). An aquaporin-2 water channel mutant which causes autosomal dominant nephrogenic diabetes insipidus is retained in the Golgi complex. *J. Clin. Invest.* **102,** 57–66.

Nelson, R. D., Stricklett, P., Gustafson, C., Stevens, A., Ausiello, D., Brown, D., and Kohan, D. E. (1998). Expression of an AQP2 Cre recombinase transgene in kidney and male reproductive system of transgenic mice. *Am. J. Physiol.* **275,** C216–C226.

Nielsen, S., Smith, B. L., Christensen, E. I., and Agre, P. (1993a). Distribution of the aquaporin CHIP in secretory and resorptive epithelia and capillary endothelia. *Proc. Natl. Acad. Sci. USA* **90,** 7275–7279.

Nielsen, S., Smith, B. L., Christensen, E. I., Knepper, M. A., and Agre, P. (1993b). CHIP28 water channels are localized in constitutively water-permeable segments of the nephron. *J. Cell Biol.* **120,** 371–383.

Nielsen, S., Digiovanni, S. R., Christensen, E. I., Knepper, M. A., and Harris, H. W. (1993c). Cellular and subcellular immunolocalization of vasopressin-regulated water channel in rat kidney. *Proc. Natl. Acad. Sci. USA* **90,** 11663–11667.

Nielsen, S., Chou, C. L., Marples, D., Christensen, E. I., Kishore, B. K., and Knepper, M. A. (1995a). Vasopressin increases water permeability of kidney collecting duct by inducing translocation of aquaporin-CD water channels to plasma membrane. *Proc. Natl. Acad. Sci. USA* **92,** 1013–1017.

Nielsen, S., Marples, D., Birn, H., Mohtashami, M., Dalby, N. O., Trimble, W., and Knepper, M. A. (1995b). Expression of VAMP2-like protein in kidney collecting duct intracellular vesicles— Colocalization with aquaporin-2 water channels. *J. Clin. Invest.* **96,** 1834–1844.

Nielsen, S., Nagelhus, E. A., Amirymoghaddam, M., Bourque, C., Agre, P., and Ottersen, O. P. (1997). Specialized membrane domains for water transport in glial cells: High-resolution immunogold cytochemistry of aquaporin-4 in rat brain. *J. Neurosc.* **17,** 171–180.

Nielsen, S., Kwon, T. H., Christensen, B. M., Promeneur, D., Frokiaer, J., and Marples, D. (1999). Physiology and pathophysiology of renal aquaporins. *J. Am. Soc. Nephrol.* **10,** 647–663.

Oksche, A., Schulein, R., Rutz, C., Liebenhoff, U., Dickson, J., Muller, H., Birnbaumer, M., and Rosenthal, W. (1996a). Vasopressin V$_2$ receptor mutants that cause X-linked nephrogenic diabetes insipidus: Analysis of expression, processing, and function. *Mol. Pharmacol.* **50,** 820–828.

Oksche, A., Moller, A., Dickson, J., Rosendahl, W., Rascher, W., Bichet, D. G., and Rosenthal, W. (1996b). Two novel mutations in the aquaporin-2 and the vasopressin V$_2$ receptor genes in patients with congenital nephrogenic diabetes insipidus. *Hum. Genet.* **98,** 587–589.

Orci, L., Humbert, F., Brown, D., and Perrelet, A. (1981). Membrane ultrastructure in urinary tubules. *Int. Rev. Cytol.* **73,** 183–242.

Pearl, M., and Taylor, A. (1985). Role of the cytoskeleton in the control of transcellular water flow by vasopressin in amphibian urinary bladder. *Biol. Cell* **55,** 163–172.

Peracchia, C., and Girsch, S. J. (1989). Calmodulin site at the C-terminus of the putative lens gap junction protein MIP26. *Lens Eye Toxic. Res.* **6,** 613–621.

Phillips, M. E., and Taylor, A. (1989). Effect of nocodazole on the water permeability response to vasopressin in rabbit collecting tubules perfused in vitro. *J. Physiol. (Lond.)* **411,** 529–544.

Pind, S., Riordan, J. R., and Williams, D. B. (1994). Participation of the endoplasmic reticulum chaperone calnexin (p88, IP90) in the biogenesis of the cystic fibrosis transmembrane conductance regulator. *J. Biol. Chem.* **269,** 12784–12788.

Preston, G. M., and Agre, P. (1991). Isolation of the cDNA for erythrocyte integral membrane protein of 28 kilodaltons: Member of an ancient channel family. *Proc. Natl. Acad. Sci. USA* **88,** 11110– 11114.

Preston, G. M., Carroll, T. P., Guggino, W. B., and Agre, P. (1992). Appearance of water channels in *Xenopus* oocytes expressing red cell CHIP28 protein. *Science* **256,** 385–387.

Preston, G. M., Jung, J. S., Guggino, W. B., and Agre, P. (1993). The mercury-sensitive residue at cysteine 189 in the CHIP28 water channel. *J. Biol. Chem.* **268,** 17–20.

Preston, G. M., Smith, B. L., Zeidel, M. L., Moulds, J. J., and Agre, P. (1994). Mutations in aquaporin-1 in phenotypically normal humans without functional CHIP water channels. *Science* **265,** 1585–1587.

Purdue, P. E., Allsop, J., Isaya, G., Rosenberg, L. E., and Danpure, C. J. (1991). Mistargeting of peroxisomal L-alanine:glyoxylate aminotransferase to mitochondria in primary hyperoxaluria patients depends upon activation of a cryptic mitochondrial targeting sequence by a point mutation. *Proc. Natl. Acad. Sci. USA* **88,** 10900–10904.

Qu, D., Teckman, J. H., Omura, S., and Perlmutter, D. H. (1996). Degradation of a mutant secretory protein, alpha$_1$-antitrypsin Z, in the endoplasmic reticulum requires proteasome activity. *J. Biol. Chem.* **271,** 22791–22795.

Rapaport, D., Neupert, W., and Lill, R. (1997). Mitochondrial protein import—Tom40 plays a major role in targeting and translocation of preproteins by forming a specific binding site for the presequence. *J. Biol. Chem.* **272,** 18725–18731.

Reizer, J., Reizer, A., and Saier, M. H. J. (1993). The MIP family of integral membrane channel proteins: Sequence comparisons, evolutionary relationships, reconstructed pathway of evolution, and proposed functional differentiation of the two repeated halves of the proteins. *Crit. Rev. Biochem. Mol. Biol.* **28,** 235–257.

Repaske, D. R., Summar, M. L., Krishnamani, M. R., Gultekin, E. K., Arriazu, M. C., Roubicek, M. E., Blanco, M., Isaac, G. B., and Phillips, J. A. (1996). Recurrent mutations in the vasopressin-neurophysin II gene cause autosomal dominant neurohypophyseal diabetes insipidus. *J. Clin. Endocrinol. Metab.* **81,** 2328–2334.

Rogers, S. L., and Gelfand, V. I. (1998). Myosin cooperates with microtubule motors during organelle transport in melanophores [comment]. *Curr. Biol.* **8,** 161–164.

Rosenthal, W., Antaramian, A., Gilbert, S., and Birnbaumer, M. (1993). Nephrogenic diabetes insipidus. A V_2 vasopressin receptor unable to stimulate adenylyl cyclase. *J. Biol. Chem.* **268,** 13030–13033.

Rubenstein, R. C., Egan, M. E., and Zeitlin, P. L. (1997). *In vitro* pharmacologic restoration of CFTR-mediated chloride transport with sodium 4-phenylbutyrate in cystic fibrosis epithelial cells containing delta F508-CFTR. *J. Clin. Invest.* **100,** 2457–2465.

Rutishauser, J., Boni-Schnetzler, M., Boni, J., Wichmann, W., Huisman, T., Vallotton, M. B., and Froesch, E. R. (1996). A novel point mutation in the translation initiation codon of the pre-pro-vasopressin-neurophysin II gene: Cosegregation with morphological abnormalities and clinical symptoms in autosomal dominant neurohypophyseal diabetes insipidus. *J. Clin. Endocrinol. Metab.* **81,** 192–198.

Sabolic, I., Valenti, G., Verbavatz, J. M., Van Hoek, A. N., Verkman, A. S., Ausiello, D. A., and Brown, D. (1992). Localization of the CHIP28 water channel in rat kidney. *Am. J. Physiol.* **263,** C1225–C1233.

Sabolic, I., Katsura, T., Verbavatz, J. M., and Brown, D. (1995). The AQP2 water channel: Effect of vasopressin treatment, microtubule disruption, and distribution in neonatal rats. *J. Membr. Biol.* **143,** 165–175.

Sakuragawa, M., Kuwabara, T., Kinoshita, J. H., and Fukui, H. N. (1975). Swelling of the lens fibers. *Exp. Eye Res.* **21,** 381–394.

Sasaki, S., Fushimi, K., Saito, H., Saito, F., Uchida, S., Ishibashi, K., Kuwahara, M., Ikeuchi, T., Inui, K., Nakajima, K. *et al.* (1994). Cloning, characterization, and chromosomal mapping of human aquaporin of collecting duct. *J. Clin. Invest.* **93,** 1250–1256.

Schild, L., Lu, Y., Gautschi, I., Schneeberger, E., Lifton, R. P., and Rossier, B. C. (1996). Identification of a PY motif in the epithelial Na channel subunits as a target sequence for mutations causing channel activation found in Liddle syndrome. *EMBO J.* **15,** 2381–2387.

Schnermann, J., Chou, C. L., Ma, T., Traynor, T., Knepper, M. A., and Verkman, A. S. (1998). Defective proximal tubular fluid reabsorption in transgenic aquaporin-1 null mice. *Proc. Natl. Acad. Sci. USA* **95,** 9660–9664.

Schrier, R. W., Fassett, R. G., Ohara, M., and Martin, P. Y. (1998). Pathophysiology of renal fluid retention. *Kidney Int. Suppl.* **67,** S127–S132.

Semenza, G., and Auricchio, S. (1989). *In* "The Metabolic Basis of Inherited Disease" (C. R. Scriver, A. L. Beaudet, W. S. Sly, and D. Alle, Eds.), pp. 2975–2997. McGraw-Hill, New York.

Shi, L. B., and Verkman, A. S. (1989). Very high water permeability in vasopressin-induced endocytic vesicles from toad urinary bladder. *J. Gen. Physiol.* **94,** 1101–1115.

Shiels, A., and Bassnett, S. (1996). Mutations in the founder of the MIP gene family underlie cataract development in the mouse. *Nat. Genet.* **12,** 212–215.

Shiels, A., and Griffin, C. S. (1993). Aberrant expression of the gene for lens major intrinsic protein in the CAT mouse. *Curr. Eye Res.* **12,** 913–921.

Shimkets, R. A., Warnock, D. G., Bositis, C. M., Nelson-Williams, C., Hansson, J. H., Schambelan, M., Gill, J. R. J., Ulick, S., Milora, R. V., and Findling, J. W. (1994). Liddle's syndrome: Heritable human hypertension caused by mutations in the beta subunit of the epithelial sodium channel. *Cell* **79,** 407–414.

Shinbo, I., Fushimi, K., Kasahara, M., Yamauchi, K., Sasaki, S., and Marumo, F. (1999). Functional analysis of aquaporin-2 mutants associated with nephrogenic diabetes insipidus by yeast expression. *Am. J. Physiol.* **277,** F734–F741.

Smith, B. L., Preston, G. M., Spring, F. A., Anstee, D. J., and Agre, P. (1994). Human red cell aquaporin CHIP. I. Molecular characterization of ABH and Colton blood group antigens. *J. Clin. Invest.* **94,** 1043–1049.

Smith, C., and Neher, E. (1997). Multiple forms of endocytosis in bovine adrenal chromaffin cells. *J. Cell Biol.* **139,** 885–894.

Sollner, T., Whiteheart, S. W., Brunner, M., Erdjument Bromage, H., Geromanos, S., Tempst, P., and Rothman, J. E. (1993). SNAP receptors implicated in vesicle targeting and fusion [see comments]. *Nature* **362,** 318–324.

Staub, O., Dho, S., Henry, P. C., Correa, J., Ishikawa, T., Mcglade, J., and Rotin, D. (1996). WW domains of Nedd4 bind to the proline-rich PY motifs in the epithelial Na$^+$ channel deleted in Liddle's syndrome. *EMBO J.* **15,** 2371–2380.

Stevens, A. L., Breton, S., Gustafson, C. E., Bouley, R., Nelson, R. D., Kohan, D. E., and Brown, D. (2000). Aquaporin 2 is a vasopressin-independent, constitutive apical membrane protein in rat vas deferens. *Am. J. Physiol.* **278,** C791–C802.

Strange, K., Willingham, M. C., Handler, J. S., and Harris, H. W., Jr. (1988). Apical membrane endocytosis via coated pits is stimulated by removal of antidiuretic hormone from isolated, perfused rabbit cortical collecting tubule. *J. Membr. Biol.* **103,** 17–28.

Tamarappoo, B. K., and Verkman, A. S. (1998). Defective aquaporin-2 trafficking in nephrogenic diabetes insipidus and correction by chemical chaperones. *J. Clin. Invest.* **101,** 2257–2267.

Toriano, R., Ford, P., Rivarola, V., Tamarappoo, B. K., Verkman, A. S., and Parisi, M. (1998). Reconstitution of a regulated transepithelial water pathway in cells transfected with AQP2 and an AQP1/AQP2 hybrid containing the AQP2-c terminus. *J. Membr. Biol.* **161,** 141–149.

Tsukaguchi, H., Matsubara, H., Taketani, S., Mori, Y., Seido, T., and Inada, M. (1995). Binding-, intracellular transport-, and biosynthesis-defective mutants of vasopressin type 2 receptor in patients with X-linked nephrogenic diabetes insipidus. *J. Clin. Invest.* **96,** 2043–2050.

Turtzo, L. C., Lee, M. D., Lu, M. Q., Smith, B. L., Copeland, N. G., Gilbert, D. J., Jenkins, N. A., and Agre, P. (1997). Cloning and chromosomal localization of mouse Aquaporin 4: Exclusion of a candidate mutant phenotype, ataxia. *Genomics* **41,** 267–270.

Valenti, G., Frigeri, A., Ronco, P. M., Dettorre, C., and Svelto, M. (1996). Expression and functional analysis of water channels in a stably AQP2-transfected human collecting duct cell line. *J. Biol. Chem.* **271,** 24365–24370.

Valenti, G., Procino, G., Liebenhoff, U., Frigeri, A., Benedetti, P. A., Ahnert-Hilger, G., Nurnberg, B., Svelto, M., and Rosenthal, W. (1998). A heterotrimeric G protein of the Gi family is required for cAMP-triggered trafficking of aquaporin 2 in kidney epithelial cells. *J. Biol. Chem.* **273,** 22627–22634.

Van Hoek, A. N., Hom, M. L., Luthjens, L. H., de Jong, M. D., Dempster, J. A., and van Os, C. H. (1991). Functional unit of 30 kDa for proximal tubule water channels as revealed by radiation inactivation. *J. Biol. Chem.* **266,** 16633–16635.

Van Hoek, A. N., Yang, B., Kirmiz, S., and Brown, D. (1998a). Freeze-fracture analysis of plasma membranes of CHO cells stably expressing aquaporins 1-5. *J. Membr. Biol.* **165,** 243–254.

Van Hoek, A. N., Gustafson, C. E., and Brown, D. (1998b). Label-fracture of aquaporin-2 expressing LLC-PK1 cells: Transient appearance of vasopressin-sensitive aquaporins into existing IMP clusters [abstract]. *J. Am. Soc. Nephrol.* **9,** 27A.

van Lieburg, A. F., Verdijk, M. A. J., Knoers, N. V. A. M., van Essen, A. J., Proesmans, W., Mallmann, R., Monnens, L. A. H., van Oost, B. A., van Os, C. H., and Deen, P. M. T. (1994). Patients with autosomal nephrogenic diabetes insipidus homozygous for mutations in the aquaporin 2 water-channel gene. *Am. J. Hum. Genet.* **55,** 648–652.

van Os, C. H., Deen, P. M. T., and Dempster, J. A. (1994). Aquaporins: Water selective channels in biological membranes. Molecular structure and tissue distribution. *Biochim. Biophys. Acta* **1197,** 291–309.

Vargas-Poussou, R., Forestier, L., Dautzenberg, M. D., Niaudet, P., Dechaux, M., and Antignac, C. (1998). Mutations in the vasopressin V_2 receptor and aquaporin-2 genes in twelve families with congenital nephrogenic diabetes insipidus. *Adv. Exp. Med. Biol.* **449,** 387–390.

Verbavatz, J. M., Brown, D., Sabolic, I., Valenti, G., Ausiello, D. A., Van Hoek, A. N., Ma, T., and Verkman, A. S. (1993). Tetrameric assembly of CHIP28 water channels in liposomes and cell membranes: A freeze-fracture study. *J. Cell Biol.* **123,** 605–618.

Verbavatz, J. M., Van Hoek, A. N., Ma, T., Sabolic, I., Valenti, G., Ellisman, M. H., Ausiello, D. A., Verkman, A. S., and Brown, D. (1994). A 28 kDa sarcolemmal antigen in kidney principal cell basolateral membranes: relationship to orthogonal arrays and MIP26. *J. Cell Sci.* **107,** 1083–1094.

Verhey, K. J., and Birnbaum, M. J. (1994). A Leu-Leu sequence is essential for COOH-terminal targeting signal of GLUT4 glucose transporter in fibroblasts. *J. Biol. Chem.* **269,** 2353–2356.

Verkman, A. S., Lencer, W. I., Brown, D., and Ausiello, D. A. (1988). Endosomes from kidney collecting tubule cells contain the vasopressin-sensitive water channel. *Nature* **333,** 268–269.

Wade, J. B., Stetson, D. L., and Lewis, S. A. (1981). ADH action: Evidence for a membrane shuttle mechanism. *Ann. N.Y. Acad. Sci.* **372,** 106–117.

Ward, C. L., Omura, S., and Kopito, R. R. (1995). Degradation of CFTR by the ubiquitin-proteasome pathway. *Cell* **83,** 121–127.

Waters, M. G., Serafini, T., and Rothman, J. E. (1991). "Coatomer": A cytosolic protein complex containing subunits of non-clathrin-coated Golgi transport vesicles. *Nature* **349,** 248–251.

Yamamoto, T., Sasaki, S., Fushimi, K., Kawasaki, K., Yaoita, E., Oota, K., Hirata, Y., Marumo, F., and Kihara, I. (1995). Localization and expression of a collecting duct water channel, aquaporin, in hydrated and dehydrated rats. *Exp. Nephrol.* **3,** 193–201.

Yang, B. X., Brown, D., and Verkman, A. S. (1996). The mercurial insensitive water channel (AQP-4) forms orthogonal arrays in stably transfected Chinese hamster ovary cells. *J. Biol. Chem.* **271,** 4577–4580.

Zeitlin, P. L. (1999). Novel pharmacologic therapies for cystic fibrosis. *J. Clin. Invest.* **103,** 447–452.

Zhang, R. B., Logee, K. A., and Verkman, A. S. (1990). Expression of mRNA coding for kidney and red cell water channels in Xenopus oocytes. *J. Biol. Chem.* **265,** 15375–15378.

CHAPTER 7

Aquaporins of Plants: Structure, Function, Regulation, and Role in Plant Water Relations

Maarten J. Chrispeels,[*] **Raphael Morillon,**[*] **Christophe Maurel,**[†]
Patricia Gerbeau,[†] **Per Kjellbom,**[+] **and Ingela Johansson**[+]

[*]Division of Biology, University of California San Diego, La Jolla, California 92093-0116;
[†]Biochimie et Physiologie Moléculaire des Plantes, ENSAM/INRA/CNRS/UMII, Montpellier, France; [+]Department of Plant Biochemistry, University of Lund, Lund, Sweden

I. The Transpiration Stream
   A. Plants Absorb and Transport Prodigious Amounts of Water
   B. The Transpiration Stream Is Regulated by the Opening and Closing of Stomatal Pores in the Leaves
   C. The Structure of Plant Cells and Tissues Permits Different Pathways of Water Transport
   D. Water Movement Is Determined by Hydrostatic and Osmotic Pressure Gradients
   E. Roots Are Highly Specialized for the Uptake of Water from the Soil and Their Hydraulic Conductance Is Affected by Environmental Factors
II. Water Movement in and Between Living Tissues
   A. Cell Enlargement, Accompanied by the Formation of the Central Vacuole, Requires a Continuous Influx of Water
   B. The Vascular System of Plants Consists of Two Separate But Interconnected Systems, the Xylem and the Phloem
   C. In the Xylem, Water under Tension Moves through Hollow Vessels, and Adjacent Living Cells Are Necessary to Repair Embolisms
   D. The Organic Solutes in the Phloem Move under Pressure Generated by the Source Organ
III. Molecular Characteristics and Transport Properties of Plant Aquaporins
   A. The Aquaporins of Plants Are MIP Homologs and Resemble Those Found in Other Organisms in Their Amino Acid Sequence and Structural Features
   B. Plants Have Numerous MIP Homologs That Fall into Three Subfamilies
   C. Swelling Experiments with *Xenopus* Oocytes Demonstrated the Existence of Plant Aquaporins

Plants are sessile organisms, and wherever they find themselves, they have to adapt
to the challenges that their environment has in store for them, whether heat, cold,
drought, flood, salt, lack of nutrients, intense sunlight, excessive shade, ozone, toxic
chemicals, diseases, pests, and predators. The sessile nature of plants dominates
how they manage their two most vital processes: the assimilation of carbon dioxide
and the utilization (transpiration) of water. Terrestrial plants growing in their natural
environment are rarely free from water deficit for a period of more than a few
days and they have evolved a variety of mechanisms to endure and even thrive
under such stress (Hsiao *et al.,* 1976). Both cell-to-cell movement of water and
long-distance water transport have been the subject of numerous studies, precisely
because water is such an important factor in plant growth and in agriculture. The
recent discovery of water channel proteins or aquaporins in the membranes of plant
cells has not changed the basic paradigms of water transport. However, it has thrown

new light on the mechanisms that plants may employ to regulate water transport and conservation and on the adaptive mechanisms that plants may have evolved to manage their water economy. The goal of this chapter is to try to integrate the recent findings about aquaporins with the wealth of information that plant physiologists and biophysicists have gathered over the years about plant–water relations.

I. THE TRANSPIRATION STREAM

A. Plants Absorb and Transport Prodigious Amounts of Water

For every kilogram of organic matter that a plant makes, its root system absorbs 500 kg of water that is lost again through transpiration or evaporation from the leaves. This flow of water across the plant can be very intense, since a leaf in sunlight can transpire its weight of water in about 1 h. To prevent complete desiccation, plants evolved highly efficient barriers against water loss that allowed them to colonize terrestrial environments. The aerial parts of plants are covered with a very thin wax layer, the cuticle, which is essentially impermeable to gasses and water. Gas exchanges with the atmosphere are restricted to hundreds of stomatal pores that can be observed on the leaves and stems of plants. These pores must be open to permit carbon dioxide from the atmosphere to enter the leaves and diffuse into the photosynthetic cells of the mesophyll for subsequent assimilation into carbohydrates. However, when the stomates are open, water vapor molecules escape because there is a water vapor gradient, and the leaves are at risk of drying out. Water that is lost from the leaves must be replaced and replacement water comes ultimately from the soil where it is taken up by the roots. Thus, a constant transpiration stream passes through the plant in the daytime when the stomates are open. When there is an active transpiration stream, gradients of negative hydrostatic pressure (tensions) within the plant vascular tissues pull the water up and represent the basic driving force for water transport and uptake from the soil and through the plant.

In other situations, osmotic gradients drive water transport in plant tissues. For example, at night, when the stomates of all but the desert plants are closed, the transpiration stream stops, but water continues to be taken up by the roots, though at a much-reduced rate, because they continue to accumulate nutrient ions from the soil. These solutes lower the water potential and create the driving force for further water absorption. The entire root functions as an osmometer and the uptake of water creates in the stele, or central cylinder of the root, a positive hydrostatic pressure called root pressure. This nighttime uptake of water continues until the plant is fully hydrated causing the cells to reach full turgor. At least two other examples of internal water flows in the plant are caused by osmotic gradients: (1) Water flows into the expanding (growing) cells, and this constant influx of water is needed to sustain growth; (2) plants have a tissue, called the phloem, that is specialized for the movement of organic solutes from sites of production such as the leaves to sites of utilization such as the roots, flowers, and fruits. When organic solutes are loaded into this part of the vascular system by pumps and transporters, the increase in osmotic pressure causes water to accompany the transport of solutes across the membranes of the transfer cells, thereby creating an aqueous solute stream.

B. The Transpiration Stream Is Regulated by the Opening and Closing of Stomatal Pores in the Leaves

During its passage through the plant, water passes through tissues composed of living cells in the roots and leaves, but most of the journey of the water molecules is through a system of hollow connected capillary tubes in the xylem of the roots, stems, and petioles of the plant. The hollow tubes have thick walls of cellulose and lignin, and these walls are all that remains of the cells that synthesized them and formed the tubes initially. In flowering plants, the individual cells are called vessel elements, and the pipes, representing vertical rows of cells, are called vessels.

The rate at which water travels up a stem depends on the transpirational demand placed on the xylem by evaporation from the leaves, and of course on the cross section of the xylem pipes, which measure 10–100 μm in diameter (and up to 300 μm, in some tree species). Poiseuille's law indicates, however, that only the smallest vessels in the protoxylem can represent a consistent hydraulic resistance for long-distance water transport (Steudle and Peterson, 1998).

Because the epidermis is covered with a cuticle, transpiration is restricted to small cavities located underneath the stomata and called the stomatal chambers. Water escapes in the form of water vapor by diffusion of water molecules from the cell wall surfaces of the mesophyll cells that are exposed to the intercellular spaces. Cell walls are about 75% water by mass and resemble a very dense aqueous gel (Cosgrove, 1997), and the force that moves water through the plant is generated at these air–water interfaces within the leaf. The difference in potential between the water in intercellular spaces and that in the vapor in the stomatal chamber defines a gradient that determines the rate of evaporation when the stomates are fully open. Stomates are small pores that are bordered by two specialized epidermal cells called guard cells that have unevenly thickened cell walls (Fig. 1). Changes in turgor pressure within the guard cells cause the two adjacent walls to come together or pull farther apart, thereby changing the aperture of the stomate. The total loss of water through the stomates is referred to as the stomatal conductance. Stomatal conductance is subject to elaborate regulation via mechanisms that change the turgor of the guard cells. Guard cell turgor is regulated by environmental signals such as (blue) light, which causes stomates to open, and internal signals such as the hormone abscisic acid (ABA), which causes stomates to close.

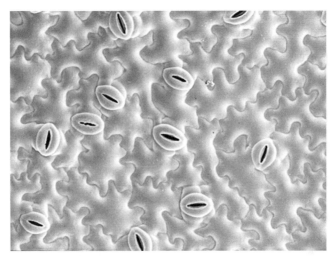

FIGURE 1   Cryo-scanning electron microscopy of the surface of a squash leaf. Micrograph shows epidermal cells and numerous open stomates bordered by guard cells. [From Berndt, M.-L., McCully, M. E., and Canny, M. J. (1999). Is xylem embolism and refilling involved in rapid wilting and recovery of plants following root cooling and rewarming? *Plant Biology* **1**, 506–515.]

Calculations of the total plant resistance to the transport of water showed that the resistance at low water flux, when there is little or no transpiration stream, is about 30 times greater than at high water flux. This means that the hydraulic architecture of a plant is such that it allows plants to function at different transpiration regimes, the resistance to water being adjusted to the demand. As explained in the next section, water transport across living tissues, in particular in the leaves, determines these properties to a large extent.

C. The Structure of Plant Cells and Tissues Permits Different Pathways of Water Transport

Living plant cells resemble those of other eukaryotes in many ways: The protoplasm, consisting of a nucleus surrounded by a dense cytoplasm, is enclosed by a plasma membrane, and individual plasma membranes of adjacent cells are separated by an extracellular matrix. However, plant cells also have unique features that have profound consequences on their water balance and for water transport in tissues.

First, unlike other eukaryotic cells, the cytoplasms of adjacent cells are connected by cytoplasmic bridges or plasmodesmata that traverse the extracellular matrix, or cell wall. These plasmodesmata probably permit the passage of water and ions, but what proportion of the total water flows from one cell to another, actually passing through plasmodesmata, is not known. Second, a large portion (often up to 85%) of the volume of most living plant cells is taken up by one or more vacuoles whose content is separated from the cytoplasm by a vacuolar membrane or tonoplast. This means that water exchange between the cell and the exterior must be mediated through two membranes in series, the plasma membrane and the tonoplast. Third, the extracellular matrix of plant cells is much thicker than that of animal cells and forms a cell wall. These walls are mechanically highly resistant. They restrict the osmotic expansion of the cell and because of the tension generated, they provide a very solid armature for the cell and tissues. The cell walls of most living cells are made out of hydrophilic macromolecules and are quite freely permeable to water. This cell wall compartment is also referred to as the apoplast, to distinguish it from the symplast, the cytoplasmic continuum that ties all the protoplasts together in one huge network because of the presence of plasmodesmata. Generally speaking, cell walls provide a hydrophilic environment. An exception is found in the conducting tubes of the xylem whose lignified walls are impermeable to water.

When water flows through a tissue of living cells it may flow through the apoplast, never entering the cells (apoplastic route), or it may flow from cell to cell by the cellular path. This cellular path has two components: the transmembrane path, with water entering and leaving the cell through the plasma membrane, and the symplastic path (Fig. 2). Water that travels along the symplastic path through the plasmodesmata does not need to pass through the plasma membranes. Transport through the vacuole may also be involved, but water flow may bypass the vacuole through cytoplasmic strands and, as we will see later, the tonoplast of most plant cells seems anyway to be very permeable to water. The relative importance of the apoplastic, transmembrane, and symplastic pathways has not been clearly established, but water flow across a living tissue probably involves a combination of these three pathways. It is likely that an active transpiration stream creates hydrostatic gradients that favor greater utilization of the apoplastic path, whereas in the absence of transpiration, osmotic gradients will drive water along the cellular path.

D. Water Movement Is Determined by Hydrostatic and Osmotic Pressure Gradients

What determines the rate at which water moves through a membrane or through a cell? With respect to their permeability to water, membranes have two important intrinsic properties: P_f, the osmotic water permeability and P_d, the diffusional water permeability. P_d is a measure of the free diffusion of water across the membrane in the absence of any imposed gradient, and thus in the absence of net exchange of water across the membrane. P_f, on the other hand, is the permeability to water when there is either an osmotic or a hydrostatic gradient (or both) and refers to the directional flow of water across the membrane that is observed under these

Pathways of water transport

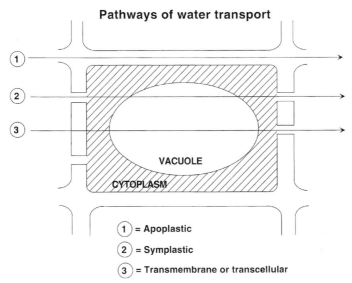

① = Apoplastic

② = Symplastic

③ = Transmembrane or transcellular

FIGURE 2 Pathways of water transport through plant cells. Water can move through the cell walls or apoplast (path 1) or through the cells. When moving through the cells, it can go through the plasmodesmata, with or without going through the vacuole (path 2), or through the plasma membrane (path 3).

conditions. For a lipid bilayer, P_d is more or less equal to P_f, but for many biological membranes P_f is greater than P_d. The notion of P_f is related to that of hydraulic conductivity (L_p) because these two variables are proportional to each other according the following equation: $P_f = L_p RT / V_w$, where R is the universal gas constant, T is the absolute temperature, and V_w is the partial molar volume of water. The hydraulic conductivity is determined in an experiment in which a membrane or a cell is given an osmotic challenge and one measures the net transmembrane volume flow J_v. To understand how this is done, we need to return to the forces that cause water to move.

Water movement across a membrane is determined by the difference in water potential (ψ) between the two sides of the membrane. Water potential is a concept developed by plant physiologists and corresponds to the general concept of chemical potential applied to water molecules. In living plant cells, the water potential gradient mostly depends on two components, the hydrostatic and osmotic pressure gradients, respectively called ΔP and $\Delta \pi$.

Hydrostatic pressure is caused by the resistance of the cell wall to cell expansion, which would be driven by an osmotic influx of water into the cell. Because of the high mechanical resistance of plant cell walls, turgor pressures up to 10 bars can be recorded. However, the wall is somewhat elastic, and limited but significant volume changes of a few percent in relative cell volume can accompany any change in turgor. The internal osmotic pressure is caused by the presence in the cell of proteins but more importantly of organic solutes and minerals. The osmotic gradient $\Delta\pi = RT\Delta C$, where R and T have their usual values, creates the osmotic driving force $\Delta\Psi_{osm} = \sigma\Delta\pi$ where σ is the reflection coefficient. The reflection coefficient is a measure of the selectivity of the membrane for the solute. If the membrane is much less permeable (several orders of magnitude) to the solute than to water, then σ is maximal and equal to 1. If the solute permeates the membrane as rapidly as water, then σ will be zero. The motive force that moves water through a membrane equals $\Delta p - \sigma\Delta\pi$. When cells are given an osmotic challenge, the volume flow of water, J_v, is given by the following equation: $J_v = L_p \cdot A \cdot (\Delta P - \sigma\Delta\pi)$ where A is the cell surface area. We saw earlier that water also moves through the xylem vessels. These dead cells do not have membranes and water movement is strictly related to Δp, the hydrostatic pressure (or tension). The hydraulic conductivity of certain organs such as the root is in the range of 10^{-7} m s^{-1} MPa^{-1}, similar to what can be observed for a single cell. In contrast, the conductivity of the xylem pipes, as calculated from Poiseuille's law, is much higher than through a living cell and it can be calculated that much greater pressure or water potential gradients are needed to move water across living cells than through the hollow tubes of the xylem. Using some reasonable assumptions, it can be calculated that the gradient to move water through a 100-μm-long cell with two membranes is 10^{10} times greater than to move water to a hollow 40-μm xylem vessel (Taiz and Zeiger, 1998).

E. Roots Are Highly Specialized for the Uptake of Water from the Soil and Their Hydraulic Conductance Is Affected by Environmental Factors

We saw earlier that water that is lost from leaves by transpiration is replaced by water that is taken up by the roots from the soil. The efficient uptake of water is made possible by the large surface area that the entire root system exposes to the soil solution and soil particles. To reach the xylem vessels at the center of the root, water must travel through the epidermis, cortex, pericycle, endodermis, and vascular parenchyma (Fig. 3). Water can take an apoplastic route or use a cellular path. Once water reaches the endodermis, the apoplastic pathway is blocked by the Casparian strip, a band of suberin, a hydrophobic wax, in the radial walls of the cells of the endodermis. Because of this blockage, all water must take the cellular path before it arrives at the endodermis or once it reaches this cell layer. The cells in the root where water crosses cell membranes and goes from the apoplast to the symplast are not yet known.

The true purpose of the Casparian strip is to prevent the backflow of mineral nutrients and other solutes. Minerals are taken up by channel proteins and transporters of root epidermal and cortex cells, and the concentration of minerals in and around

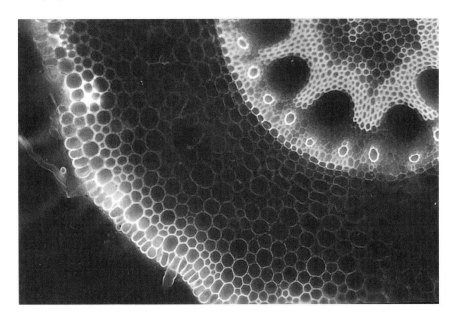

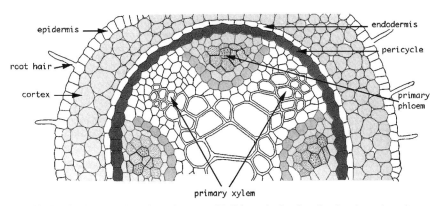

FIGURE 3 Cross-sections through a root. (A) Schematic drawing showing the various tissues of a root, as labeled. The cortex is drawn with fewer cell layers than are normally present in a root. (B) Fluorescence micrograph of a nodal root of maize, 7 cm from the root tip. The intensity is a reflection of the natural fluorescence of the cell walls. [From Fig. 2A in McCully, M. E., and Canny, M. J. (1985). Localization of translocated 14C in roots and root exudates of field-grown maize. *Physiologia Plantarum* **65**, 380–392.]

the cells of the stele is much greater than in the soil solution. Once the minerals are loaded in the xylem elements (which have no membranes) the nutrients might leak out through the apoplastic path. However, the Casparian strip prevents this. As a result, the entire multicellular pathway from root hair to vessel element behaves as if it were a single semipermeable membrane, and it is possible to calculate its hydraulic conductivity. Roots generally have a high hydraulic conductance compared to leaves, possibly because the membranes of the root cells contain abundant aquaporins (see later). In nontranspiring conditions, the uptake of mineral nutrients from the soil and the presence of organic solutes causes the root to have a high osmotic pressure and this permits the rapid uptake of water and the buildup of root pressure on the inside of the endodermis. Thus, when a young plant is cut, xylem sap may continue to ooze out of the stump for quite some time because of this root pressure. The pressure can be measured by sealing a small manometer over the stump. The pressure is typically 0.1 MPa, but pressures as high as 0.5 MPa have been measured.

Water deficit caused by withholding irrigation water or supplementing the growth medium with salt or an osmoticum causes an almost immediate reduction in growth and cell enlargement. The plants respond in a variety of ways, including a change in the hydraulic conductivity of the membranes. Maize seedlings grown in a nutrient solution supplemented with 100 mM NaCl have a root L_p that is about half that of plants grown without the salt. The effect of salt was much greater at the cellular level, with salt causing a 3- to 6-fold decrease in cellular L_p (Azaizeh and Steudle, 1991; Azaizeh *et al.*, 1992). The authors of this study interpret these results to show that in the salt-grown plants, water flow is mostly around the cells (through the apoplast) under hydrostatic conditions. Tyerman *et al.* (1989), on the other hand, found that salt did not affect the root cell L_p of tobacco, a plant that grows reasonably well at high salinity. This raises the question that salt may affect the abundance of water channel proteins differently in different species.

There is considerable evidence that anoxia reduces the ability of the roots to take up water from the soil. Birner and Steudle (1993) used a root pressure probe to show that anoxia depresses the hydraulic conductivity of maize roots substantially, causing L_p to reach a low level after several hours of anoxia. This low value could be ascribed to an effect of anoxia on the L_p of the membranes.

II. WATER MOVEMENT IN AND BETWEEN LIVING TISSUES

A. Cell Enlargement, Accompanied by the Formation of the Central Vacuole, Requires a Continuous Influx of Water

A unique feature of plant growth is that cell division, which normally occurs in meristems, is followed by an enormous enlargement—up to 1000-fold in some cases–of the daughter cells. Because of the continuous activity of plant meristems, newly formed cells escape the cell division zone and their enlargement is accompanied and followed by cell differentiation. Meristematic cells have numerous small ($1–3$ μM in diameter) vacuoles, and these cells generate new vacuoles *de novo,* so that after cell division each daughter cell has a more or less equal complement of vacuoles. When a meristematic cell escapes the apical meristem, it first enlarges isodiametrically, and somewhat later it begins to elongate. There is a significant increase in cytoplasmic constituents, but most of the change in cell size can be accounted for by an increase in the volume of the vacuole. It is assumed that the influx of water into the expanding vacuoles depends on an osmotic gradient between the vacuole and the cytoplasm. Cell growth requires coordinated increases of cytoplasmic and vacuolar volumes, membrane synthesis, and recycling. Finally, most mature plant cells have a large central vacuole surrounded by a thin layer of cytoplasm that is appressed against the plasma membrane and the cell wall. The large vacuole with its dilute aqueous solution (most vacuoles contain very little protein or other macromolecules) is therefore also the potential water reservoir of the cell. If the cytoplasm has to readjust its osmotic potential, water can flow from the vacuole to the cytoplasm across the tonoplast.

Direct measurements of water potential along the radius of rapidly growing (elongating) seedlings showed that water potential was highest near the xylem vessels and lowest near the epidermis (Nonami and Boyer, 1993). The same water potential gradient was not observed in tissues that had ceased elongation. This gradient will drive a radial water transport into the elongating cells to sustain their enlargement.

It has been disputed whether L_p, the hydraulic conductivity of the membranes, is ever limiting for cell enlargement, or whether the loosening or relaxation of the cell wall matrix is the only critical factor in this process. Because cell enlargement is such an important component of growth, it is a highly regulated process. It has been demonstrated that proteins in the cell wall, such as expansins and endo-β-glucanases, respond to intracellular signals and alter the cross-linking of the cell wall matrix to permit cell enlargement. This whole process is accompanied by the rapid synthesis of components of the plasma membrane and the tonoplast to sustain the increase in cellular volume. When plants grow in dry soil and experience water deficit, the growth of the shoot is greatly inhibited, but the roots continue to grow almost normally. In such roots, elongation is confined to a narrower zone near the tip (Liang *et al.,* 1997). Under normal and droughted conditions, the maintenance of turgor, the influx of water, and the modification of the wall to permit elongation must all be carefully coordinated.

B. The Vascular System of Plants Consists of Two Separate But Interconnected Systems, the Xylem and the Phloem

Plants contain two entirely separate systems to transport water, mineral nutrients, and organic molecules. The xylem, discussed above, transports primarily water and minerals taken up from the soil and delivers these to the leaves; the phloem carries organic molecules throughout the plant (shoot to root, or root to shoot). Each system has specialized cell types that are best adapted to its function. The xylem and phloem always lie adjacent to each other and they appear to be functionally connected.

The xylem of flowering plants has two major cell types: vessel elements that are fused end-to-end into vessels and parenchyma cells that are interspersed with the vessels. Vessel elements are both long and wide. When the cross walls separating individual cells are broken down as part of the differentiation process, the vessel elements are connected and become vessels. The lateral cell walls are very thick and encrusted with lignin, a hydrophobic polymer. Adjacent vessels are connected by pits, very small pores that form where the wall has not been thickened and does not contain lignin. Such pores can act as one-way valves for water if gas fills the adjacent vessel. Because of the high surface tension of water, gas in an adjacent vessel cannot pass through the pore and push the water away, but water can enter the gas-filled vessel and slowly flow over the inside wall surface pushing away the gas.

The phloem also consists of tubes, called sieve tubes, in which the original sieve tube elements remain separated by a perforated sieve plate. Unlike the xylem vessels, these cells retain some elements of the cytoplasm, including plasma membranes, but they lack nuclei. Densely cytoplasmic companion cells with numerous mitochondria lie next to every sieve tube element. Companion cells are thought to provide energy for the transport of organic nutrients in the sieve tubes. At both ends of the sieve tube lie transfer cells that function in the loading and unloading of organic molecules and water into the sieve tube. Transfer cells have a highly convoluted plasma membrane and cell wall, the function of which seems to be to maximize the surface area for entry or exit of solutes. The phloem transports the products of carbon, nitrogen, and sulfur assimilation from the leaves, considered to be source organs, to sink organs such as roots, flowers, and fruits. Breakdown of macromolecules in storage organs such as roots, cotyledons, or bark also produces organic solutes that are transported to developing stems or leaves via the phloem.

C. In the Xylem, Water under Tension Moves through Hollow Vessels, and Adjacent Living Cells Are Necessary to Repair Embolisms

We have seen already that the driving force that moves water up a tall tree is not pressure from the root, although modest root pressures can be measured, but the tension that is created in the leaves by the constant evaporation of water through the leaf stomata. This tension is easily propagated through the xylem vessels because these "cells" lack membranes and cytoplasm. It is precisely because water has a certain tensile strength as a result of intermolecular hydrogen bonds, that the tension that is created in the leaves can be propagated all the way down to the roots. This explanation for the movement of water through the xylem vessels is known as the cohesion-tension theory.

Because xylem vessels have thick cell walls, they are particularly well adapted to resist collapsing under tension or pressure. Water under tension is physically unstable, and when water is under tension, dissolved gasses will escape into the vapor phase and form a gas bubble. Gasses have no tensile strength and the bubble causes the water column to break and the vessel to cavitate. This in turn causes water transport through that particular vessel to stop. Recent evidence shows that cavitation occurs frequently and has a diurnal rhythm. Canny (1997) developed a method to determine the number of vessels in a stem that are filled with water or filled with gas. The method relies on quick-freezing the stem to liquid nitrogen temperature and then examining a section through the stem by scanning electron microscopy. Their results show that embolisms in sunflower petioles start increasing soon after sunrise and that by 9:00 a.m. as much as 40% of the vessels may have embolisms. Subsequently, the embolisms are repaired as water enters the cells through the pits, and by nightfall most of the cavitated vessels have been restored. The mechanism that permits embolism repair remains to be investigated (Wei et al., 1999), but it is possible that the compensating pressure in the living cells that surround the xylem vessels (Canny, 1995) or osmotic exudation by these cells, as facilitated by aquaporins in their membranes (Holbrook and Zwieniecki, 1999), plays a major role in repairing the embolisms.

D. The Organic Solutes in the Phloem Move under Pressure Generated by the Source Organ

The principal function of the transfer cells that are found at both ends of the phloem is to load and unload sugars and amino acids into and from the sieve tubes. The loading of solutes into the sieve tubes in the source organs creates an osmotic pressure difference across the plasma membrane that causes water to enter the sieve tubes at the site of loading. This water comes from the xylem. The unloading of solutes at the other end causes a drop in osmotic pressure relative to the surrounding parenchyma cells and water exits the phloem at the site of solute unloading. This water enters the xylem. The influx of water at one end and its efflux at the other end creates a pressure difference in the phloem and a passive flow that transports substances from source to sink. Such passive transport can be observed by measuring dye movement in the phloem (Oparka et al., 1994).

The idea of an internal circulation of water between phloem and xylem was first proposed by the German plant physiologist Ernst Muench in 1927 (Fig. 4).

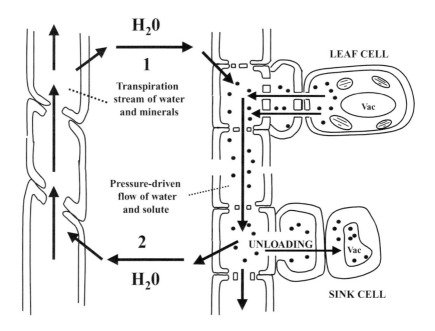

XYLEM VESSEL   PHLOEM SIEVE TUBE

FIGURE 4   Diagram showing the circulation of water between the xylem and the phloem. At the source, solutes (black dots) are actively loaded into the phloem sieve tubes via the companion cells. Water enters the phloem cells osmotically (path 1) and the high pressure causes the bulk flow of the sugar solution in the phloem (downward in the figure). When solutes are unloaded in the sink cells, water leaves the phloem (path 2) as the osmotic pressure decreases. This cycling of water between xylem and phloem is accompanied by an upward movement of water in the transpiration stream.

Köckenberger *et al.* (1997) developed a noninvasive and nondestructive method to measure the volume of this flow in castor bean seedlings. The method relies on a flow-sensitive NMR microimaging technique. In these seedlings, which have little transpiration, but an intense solute transport from the source organs (the cotyledons) to the sinks (the growing root and shoot), the volume flow in the phloem was found to be about half of the volume flow in the xylem. The difference between the two volume flows could be accounted for by evaporation. The average velocity in the xylem (1.7 m·h^{-1}) was slightly lower than in the phloem (2.0 m·h^{-1}). These results demonstrated for the first time that water circulates between the phloem and xylem even when there is no evaporation or transpiration. Because the phloem sieve tubes and the xylem vessels are separated by living parenchyma cells with cell membranes, the resistance to flow between the two systems can be modulated.

III. MOLECULAR CHARACTERISTICS AND TRANSPORT PROPERTIES OF PLANT AQUAPORINS

A. The Aquaporins of Plants Are MIP Homologs and Resemble Those Found in Other Organisms in Their Amino Acid Sequence and Structural Features

The first plant MIP homolog, nodulin 26 (Nod26), was identified in 1987, three years after the identification of MIP and well before the molecular characterization of CHIP28/AQP1 (Fortin *et al.*, 1987). Early work on Nod26 and later on plant tonoplast intrinsic proteins (α-TIP) (Johnson *et al.*, 1990) helped in recognizing that MIP homologs define a large superfamily with members in prokaryotes and eukaryotes, at a time when the function of these proteins remained unclear.

All plant MIPs can be unambiguously classified in one large family, together with their animal and microbial counterparts, on the basis of typical sequence features (Park and Saier, 1996). Although the overall degree of homology between these proteins can be rather low (<30%), their hydropathy profile invariably shows six hydrophobic domains, interpreted as putative membrane-spanning segments. These domains, numbered 1–6, are connected by five extramembrane loops, A–E. Because of an early intragenic gene duplication, MIPs are formed by a structural motif repeated in tandem. This motif contains three putative membrane-spanning domains, and because of this odd number, the protein is formed by two homologous halves oriented in the opposite direction within the membrane. MIPs also exhibit a typical sequence signature with some residues being highly conserved. The most striking motif is a perfectly conserved Asn-Pro-Ala-Val-Thr (NPAVT) sequence in the second connecting loop (B). A similar NPA motif is found in the fifth (E) loop. These NPA motifs are the most distinctive features of MIPs and only a few proteins, such as yeast Fps1 and plant AtNLM1, have a substitution within this sequence (Park and Saier, 1996; Weig *et al.*, 1997).

Overall comparison of MIP sequences shows that the highest sequence conservation is found within the membrane core of the protein. The hydrophilic N- and C-terminal tails are much more divergent. In all plant MIPs identified so far, these tails are fairly short, up to 50 residues for the N-terminal tail in the PIP1 subfamily, and the proteins exhibit a fairly similar molecular weight of 25–31 kDa. In contrast, some other MIPs, such as yeast Fps1 or the *big brain* gene product from *Drosophila* have large N- and C-terminal extensions and an overall molecular weight greater than 70 kDa (Park and Saier, 1996).

There is relatively little information about the topology of plant MIPs. Work by Miao *et al.* (1992) on Nod26 demonstrated the functionality of a glycosylation site at position 150 in the C loop, placing this loop on the extracytoplasmic side of the membrane. Phosphorylation of plant MIPs in the N-terminal tail (at residues Ser 7 and Ser 23 of bean α-TIP), in the second connecting loop (B) (at residues Ser 99 and Ser 115, of α-TIP and spinach PM28a, respectively), and in the C-terminal domain (at residues Ser 274 and Ser 262 of PM28a and Nod26, respectively) has been reported (Johansson *et al.,* 1998; Maurel *et al.,* 1995; Weaver and Roberts, 1992). This indicates that these parts of the protein face the cytosolic side of the membrane. Proteolytic cleavage of aquaporin homologs in right-side-out plasma membrane vesicles of red beet also established that loops C and E face the extracytoplasmic (apoplastic) space (Barone *et al.,* 1997, 1998). Altogether, these data conform to the topological model determined in great detail for mammalian aquaporin-1 (AQP1) (see Borgnia *et al.,* 1999, for a review). It is thus agreed that all MIPs have six transmembrane domains connected by five loops, with the N and C termini bathing in the cytosol. Whether the protein is targeted to the plasma membrane, the tonoplast, or peribacteroid membrane, the extracytoplasmic side of the protein will be located in the apoplastic, vacuolar, or peribacteroid space, respectively. More specifically, a molecular model has been proposed for AQP1 in which the two NPA motifs plunge deep into the membrane to form an aqueous pore (Fig. 5). This so-called hourglass model has been confirmed and refined by electron cryocrystallography of the protein (see Heymann *et al.,* 1998, for a review). Electron cryocrystallography of bean α-TIP shows that this aquaporin also forms tetramers and that the subunit has a similar structure to that postulated for mammalian aquaporins (Daniels *et al.,* 1999).

The propensity of plant MIPs to aggregate in dimers or even larger complexes, after they have been solubilized in detergents such as SDS, has been reported by several groups (Barone *et al.,* 1998; Daniels *et al.,* 1994; Johnson *et al.,* 1989; Miao *et al.,* 1992). In many cases, these interactions are favored by heating and are resistant to SDS. It was recently shown, in the case of a vacuolar MIP from spinach leaves and a plasma membrane MIP of red beet, that these aggregates can be dissociated by a treatment with high urea concentrations or sulfhydryl modification reagents, respectively (Barone *et al.,* 1998; Karlsson *et al.,* 2000). It is not clear whether this phenomenon reveals an artefactual biochemical property or relevant aspects of plant MIP biogenesis and assembly. The specificity of this reaction is suggested by the observation that heteromers of close MIPs that are coexpressed in a same red beet plasma membrane preparation could not be observed (Barone *et al.,* 1998). A good correlation between the oligomeric structure of animal and bacterial MIPs after solubilization in mild detergents and their transport selectivity has also recently been reported (Lagrée *et al.,* 1998, 1999). It was concluded that solute facilitators (aquaglyceroporins) would exhibit a monomeric structure whereas water-selective aquaporins function as tetramers.

Some animal AQPs can exhibit a very high level of organization in their native membrane and form orthogonal arrays due to the ordered clustering of tetrameric

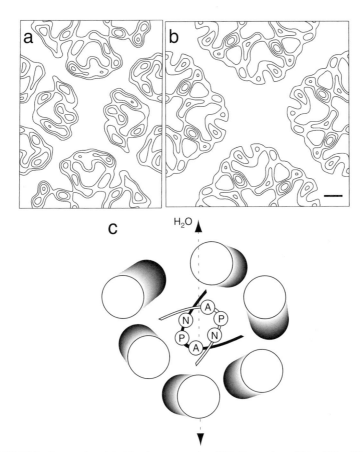

FIGURE 5   Structural model of the plant aquaporin α-TIP. Comparison of (a) α-TIP and (b) AQP1 projection structures at 7.7-Å resolution. AQP1 projection structure derived from the 3D dataset of electron diffraction results of Cheng *et al.*, 1997. α-TIP projection structure derived from the dataset of Daniels *et al.*, 1999. One unit cell is shown for each. Bar-10 Å. The similarity between the two projection structures is remarkable, which is not unexpected. Higher resolution datasets permit the construction of a model as shown in (c). (See also Fig. 12.)

units (Yang *et al.*, 1996). These arrays were initially visualized by freeze–fracture electron microscopy in the vasopressin-regulated epithelia of amphibian urinary bladder (Chevalier *et al.*, 1974) and provided one of the first hints at the molecular existence of water channels in animal membranes. Such structure could have been expected in some plant membranes, which exhibit similarly high water permeability and MIP expression levels. However, inspection of plant membranes by freeze–fracture electron microscopy never revealed any of these motifs.

B. Plants Have Numerous MIP Homologs That Fall into Three Subfamilies

Since the early discovery of MIP, an increasing number of homologs have been identified in various organisms (Park and Saier, 1996). Because of their abundance and tight regulation by endogenous and environmental stimuli, numerous plant MIP genes have simply been uncovered as genes that are specifically responsive to factors such as light, drought, or salinity (Kaldenhoff *et al.,* 1993; Yamaguchi-Shinozaki *et al.,* 1992). Sequencing programs of expressed sequence tags (ESTs) in model plant species also proved to be efficient for the discovery of novel MIP sequences (Weig *et al.,* 1997). These sequences have been tentatively grouped into contigs, but the low accuracy of certain sequences and the fact that some sequences do not clearly overlap limit the resolution of this method. The present count of MIP genes in *Arabidopsis* is around 35. With the genomic sequence of this plant being completed in 2000, the final count should be known very soon with maybe a few more MIP genes to be expected. Other plant species may have an even greater MIP diversity and their number in maize has been estimated to be around 35–40 (Chaumont, Jung, and Chrispeels, unpublished results, 2000).

From the numerous sequences available, it appears that all known plant MIPs define three independent branches in the MIP family tree (Park and Saier, 1996). One subfamily comprises the so-called tonoplast intrinsic proteins (TIPs) because their founding members have been identified in the vacuolar membrane (Höfte *et al.,* 1992; Johnson *et al.,* 1990). The second subfamily contains plasma membrane intrinsic proteins (PIPs), some of them being expressed at the plasma membrane (Daniels *et al.,* 1994; Kammerloher *et al.,* 1994). The third subfamily is formed by close homologs of Nod26 and has members in legume and nonlegume species (Fortin *et al.,* 1987; Weig *et al.,* 1997). This family is referred to as the Nod-like MIPs or NLMs (Fig. 6). Thus each subfamily is defined by typical sequence motifs and can be distinguished by slight variations in overall gene organization (Schäffner, 1998). For instance, TIPs have shorter C, D, and E loops than PIPs, and the N terminus of a subclass of PIPs (PIP1) is longer than any of the others. Interestingly, the percentage of sequence similarity between members of each subfamily is greater than that between MIPs of the same species but of a different subfamily. It has been difficult, however, to unambiguously distinguish orthologs in different plant species, because our knowledge of MIP sequences remains incomplete and because diversification of the MIP family likely occurred during recent evolution.

It is more likely, in contrast, that the emergence of the three plant MIP subfamilies has occurred very early during evolution since most plant species examined so far have genes in each of the subfamilies. Because these subfamilies are plant specific,

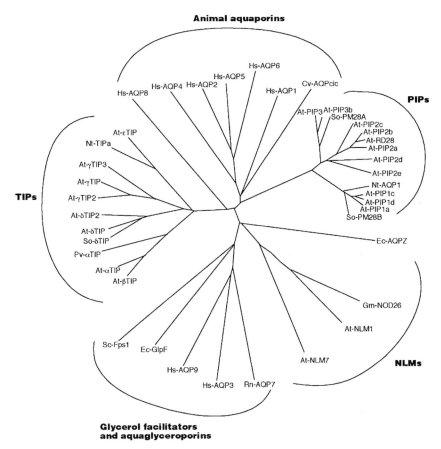

FIGURE 6 Phylogenetic analysis of MIPs of different species. Most MIP family members are either aquaporins or their specificity is unknown. The plant tonoplast intrinsic proteins (TIPs) fall into four different subfamilies, whereas the plant plasma membrane intrinsic proteins (PIPs) fall into two subfamilies. The glycerol facilitators of *E. coli* (Ec-GlpF) and yeast (Sc-Fps1) transport glycerol. The mammalian aquaglyceroporins AQP3, AQP7, and AQP9 transport glycerol as well as water. The plant MIP Nod26, belonging to the NLM (Nod26-like MIP) subfamily, as well as the PIP and TIP homologs Nt-AQP1 and Nt-TIPa, have been reported to be permeable to small uncharged solutes in addition to water. At, *Arabidopsis thaliana*; Cv, *Cicadella viridis*; Ec, *Escherichia coli*; Gm, *Glycine max*; Hs, *Homo sapiens*; Nt, *Nicotiana tabacum*; Pv, *Phaseolus vulgaris*; Rn, *Rattus norvegicus*; Sc, *Saccharomyces cerevisiae*; So, *Spinacia oleracea*. (Adapted from Johansson *et al.*, 2000.)

they may have evolved after plant and animals diverged from a common ancestor. An exception to this rule came with the discovery of a novel animal aquaporin, named AQP8, which displays the most identity with plant γ-TIP (Koyama *et al.*, 1997). The evolutionary significance of this finding is not yet clear.

The nomenclature of plant aquaporins has not yet been standardized. Most authors in the field conform to the initial distinction between TIPs and PIPs (Schäffner, 1998). The denomination of the third family is less accepted but, in *Arabidopsis,* the name NLM for nodulin-26-like MIP has been agreed to by several authors (Kjellbom *et al.,* 1999). More confusion arises when it comes to defining subclasses. The distinction between PIP1 and PIP2 has been largely used but members belonging to the PIP2 subfamily, such as RD28 and PIP3, will have to be renamed (Kjellbom *et al.,* 1999; Schäffner, 1998). New letters will have to used to describe the newly identified members of each subclass. In contrast to PIPs, TIP subtypes have been defined by greek letters (Daniels *et al.,* 1996; Höfte *et al.,* 1992; Kjellbom *et al.,* 1999); the identification of paralogs renders this denomination quite confusing and letters rather than figures should be used at the moment to distinguish them. These issues will be clarified, in *Arabidopsis* at least, after the full genomic sequence becomes available. All family members will have to be renamed according to a consistent rule. However, because of the diversity of plant MIPs, the correspondence between species homologs will surely be more complex than in animals, and it will be difficult to avoid using specific names for each species.

C. Swelling Experiments with Xenopus Oocytes Demonstrated the Existence of Plant Aquaporins

Initial efforts to elucidate the function of plant MIPs remained unsuccessful due to repeated failures to assign ion or solute transport properties to these proteins using vacuoles from transgenic plants (Maurel and Chrispeels, unpublished, 1992). The link between MIPs and water transport was first established by Peter Agre and colleagues in their landmark study on the functional characterization of CHIP28/AQP1 in *Xenopus* oocytes (Preston *et al.,* 1992). Previous research by Fischbarg *et al.* (1989) and Zhang and Verkman (1991) had established that these cells provide a nice system for the functional expression of water channels. First, their size makes these cells easy to handle and inject individually with exogenous mRNA. Exogenous mRNA is then efficiently translated, usually leading to high expression levels of the encoded protein. Secondly, the low cell surface-to-volume ratio, which follows from large cell size, and the lack of endogenous water channels gives native oocytes transferred to a hypotonic medium very slow swelling kinetics that can easily be followed on individual oocytes by video microscopy. In the presence of a 150-mOsm gradient, native oocytes will increase their size by less than 1% min. The striking observation is that these kinetics can be strongly modified after expression of exogenous aquaporins. After expression of *Arabidopsis* γ-TIP, for instance, oocytes increased their volume by >40% in less than 2 min and then ruptured (Maurel *et al.,* 1993) (Fig. 7). As documented

Swelling assay in *Xenopus* oocytes

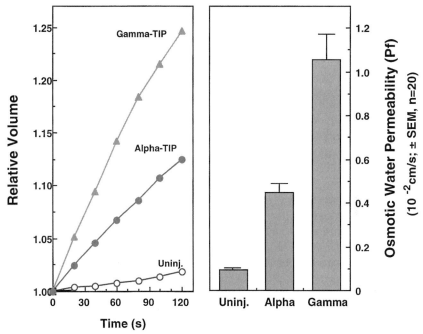

FIGURE 7 Functional expression of plant tonoplast aquaporins in *Xenopus* oocytes. Oocytes were injected with complementary RNAs specific for *Arabidopsis* γ-TIP (Maurel *et al.*, 1993) or bean α-TIP (Maurel *et al.*, 1995). Uninjected oocytes were taken as controls. *Left panel*: After 3D incubation in isosmotic conditions, oocytes were exposed from $t = 0$ to a hypotonic solution (for experimental details, see Maurel *et al.*, 1995). The enhanced swelling rate of oocytes injected with γ-TIP or α-TIP, as compared to controls, reflects the appearance of a path for facilitated diffusion across the oocyte membrane. *Right panel*: Osmotic water permeability values can be deduced from the swelling rate of individual oocytes. Because the plasma membrane represents the default pathway for membrane proteins in animal cells, it is assumed that at least some plant tonoplast proteins are expressed at the oocyte surface. The swelling assay exclusively reflects the activity of aquaporins expressed at the cell surface. (Modified from Maurel *et al.*, 1995.)

for mammalian aquaporins, there can be considerable differences in the activity of individual MIPs, and some plant MIPs have been found to have little or no activity in the oocyte assay (Yamada *et al.*, 1995; Chaumont *et al.*, 2000. Whether this means that they are truly inactive or cannot be activated by the *Xenopus* oocyte is unknown. Nevertheless, studies on plant aquaporins and ion channels using *Xenopus* oocytes have proved that these cells generally provide a very useful system for the functional analysis of plant membrane channels, whether they are present in the plant vacuolar or plasma membrane.

D. Molecular Mechanisms of Water Permeation and Its Inhibition by Mercury Ions

Mercury derivatives are the only known wide-spectrum inhibitors of water channels in biological membranes. The inhibitory effect of these compounds was used initially to provide functional evidence for water channels in animal and plant membranes (Macey, 1984; Wayne and Tazawa, 1990) and more recently to probe aquaporin structure and function. In the case of AQP1, for instance, it was demonstrated that mercury ions specifically oxidize a Cys residue at position 189, in the E-loop close to the second NPA motif (Preston *et al.*, 1993; Zhang *et al.*, 1993). It was proposed that steric hindrance of the mercury somehow occludes the pore. In support for this, similar inhibitory effects can be obtained by substituting Cys 189 residue by large size residues (Preston *et al.*, 1993; Zhang *et al.*, 1993). With the exception of AQP3, which carries a mercury sensitive site in the N-terminal tail (Kuwahara *et al.*, 1997), the blocking effects of mercury on most mammalian aquaporins investigated are also exerted through Cys residues located in the E loop or in the other NPA containing loop B (Bai *et al.*, 1996; Shi and Verkman, 1996). These observations support the hourglass model of AQP1 and indicate a critical role for these two loops in defining the aqueous pore.

Similar to their animal counterparts, most plant aquaporins can be inhibited by mercury. Mercury-resistant aquaporins have been described, however, in the *Arabidopsis* and tobacco plasma membrane (Biela *et al.*, 1999; Daniels *et al.*, 1994). Site-directed mutagenesis of Cys residues in *Arabidopsis* δ-TIP and γ-TIP indicated that the site of mercury action is located at positions 116 and 118, respectively, in the third transmembrane segment of the protein (Daniels *et al.*, 1996). In agreement with this, studies by Barone *et al.* (1997, 1998) on aquaporin homologs of red beet plasma membrane suggest that mercury ions bind to Cys residues of transmembrane domains 2 and 3. These authors also showed that mercury binding induces a conformational change in the protein that results in an altered sensitivity of the N tail and second extracytoplasmic (C) loop to trypsin digestion (Barone *et al.*, 1998). Interestingly, the site(s) of mercury action on plant aquaporin differ from those identified in their animal counterparts. Because animal and plant aquaporins likely share similar structures, these findings indicate that many regions of the protein participate in the pore structure and in its function. The importance of the B and E loop for plant aquaporin function was demonstrated by confering mercury sensitivity to the mercury-resistant aquaporin RD28 after introduction of Cys residues in these domains (Daniels *et al.*, 1994).

In conclusion, mercury inhibition has confirmed the existence of a narrow pore in aquaporins but does not provide any direct support for the single file model for water transport. Mutation studies also indicate that Cys residues in the pore region are not crucial per se in aquaporin function. Atomic resolution of aquaporin structure will surely provide critical information as to the molecular mechanisms of water transport and further enhance the discovery of novel aquaporin inhibitors. The existence of mercury-resistant aquaporins, in addition to the numerous cellular targets of mercury, tell us to be cautious with its use in physiological studies of mercury as a general blocker of water channels.

E. Some Plasma Membrane MIPs Transport Neutral Solutes, Others Do Not Transport Water in Oocytes, and the Specificity Remains to Be Discovered

Oocyte expression also provides a convenient system to assay the transport specificity of aquaporins. Electrophysiological measurements by the two-electrode voltage clamp technique can be easily performed because of the large size of these cells. A thorough analysis of oocytes expressing γ-TIP did not reveal any novel membrane current, neither in isotonic conditions nor in hypotonic conditions (Maurel *et al.*, 1993). These results showed that the pore of this aquaporin excludes all ions, even in conditions where a strong solvent (water) drag may operate. Lack of an ion transport activity has also been reported in a few other aquaporins, such as bean α-TIP or tobacco NtAQP1 (Biela *et al.*, 1999; Maurel *et al.*, 1995). In particular, a high resistance of these channels to proton diffusion is consistent with the maintenance of large pH gradients across the vacuolar and plasma membrane of plant cells. An ion channel activity has been reported for Nod26 but, similar to bovine MIP, this activity could only be detected after reconstitution of the protein in artificial membranes (Weaver *et al.*, 1994). Expression of Nod26 in *Xenopus* oocytes revealed a permeability to water and small neutral solutes such as glycerol, but not to charged molecules (Dean *et al.*, 1999; Rivers *et al.*, 1997). The significance of ion transport by MIP homologs in artificial membranes has thus been questioned because abnormal aggregation or subunit assembly of these highly hydrophobic proteins may lead to the formation of artefactual ion-permeable pores. In contrast, the recent discovery that mammalian AQP6 exhibit an anion channel activity after expression in *Xenopus* oocytes (Yasui *et al.*, 1999) will surely spark renewed interest in the electrophysiological analysis of aquaporins for those plant aquaporins in particular that display no or a poor water transport activity (Chaumont *et al.*, 2000).

Plant aquaporins have homologs in bacteria, yeast, and mammals that transport small neutral solutes like glycerol and urea. Some of them, such as GlpF, the glycerol facilitator of the inner membrane of *Esherichia coli* even function as exclusive solute transporters (Maurel *et al.*, 1994). Such activity can be easily assayed through the uptake of radiolabeled solutes in *Xenopus* oocytes. Whereas those plant aquaporins that were first characterized proved to be selective for water, solute transporting aquaporins have have been recently identified in the plasma membrane (NtAQP1), the tonoplast (NtTIPa), and the peribacteroid membrane (Nod26) (Biela *et al.*, 1999; Dean *et al.*, 1999; Gerbeau *et al.*, 1999) of plant cells. These proteins exhibit a slightly distinct selectivity profile, NtAQP1 and Nod26 being permeable to glycerol while NtTIPa is mostly permeable to urea. These findings are interesting in at least two respects. First, they demonstrate the existence of solute transporting proteins in novel aquaporin subfamilies that belong to the larger branch of orthodox aquaporins (Agre *et al.*, 1998). This branch was initially assumed to contain water-selective channels exclusively because all solute-permeable aquaporins identified in animals and microorganisms define a distinct family branch of so-called aquaglyceroporins (Agre *et al.*, 1998). The distinction between these two subfamilies has allowed certain authors to identify five key residues that would critically determine the transport selectivity of MIPs (Froger *et al.*, 1998). This idea has been supported by the finding that mutations in two of these residues present in the insect aquaporin CIC into the residues found in GlpF transmute the water-selective channel CIC into a solute-permeable channel (Lagrée *et al.*, 1999). The identification of solute-permeable plant aquaporin suggests however that these residues are not necessary for solute transport. More specifically, the identification of pairs of homologs that exhibit different selectivity will help in further refining the molecular determinants of water selectivity.

Secondly, although glycerol and urea transport have as far as we know only a marginal significance in plant physiology, the discovery of plant aquaporins that are permeable to these solutes raises critical questions about the dual activity of aquaporins in water and solute transport (Biela *et al.*, 1999; Dean *et al.*, 1999; Gerbeau *et al.*, 1999). Physiologically relevant solutes that are transported by aquaporins remain to be discovered, and compatible solutes of small size that normally accumulate in drought-stressed plants may be good candidates. The transport of gaseous substances by aquaporins has also been described, and increased permeability of plant membranes to CO_2 or NH_3 may be crucial for leaf and symbiotic root nodule physiology, respectively (Tyerman *et al.*, 1999).

IV. SUBCELLULAR LOCATION OF PLANT AQUAPORINS

A. MIPs Are Abundant Proteins

A typical feature of plant aquaporins is that their expression is highly regulated, in response to developmental, hormonal, and environmental stimuli. The first plant MIP, Nod26, was identified as a protein specifically expressed during the establishment of the soybean-*Rhizobium* symbiosis (Fortin *et al.,* 1987). The plasma membrane aquaporin gene RD28 and the tobacco TIP homolog TobRB7 were characterized as dehydration-induced or root-specific genes, respectively (Conkling *et al.,* 1990; Yamaguchi-Shinozaki *et al.,* 1992). The identification of these genes, by means of screening procedures of moderate sensitivity, was also possible because these genes were expressed at high levels.

MIPs generally represent major protein constituents of plant membranes and this has been a crucial feature for their identification by biochemical and immunological methods. Interestingly, these are the same charactersitics that led to identifying MIP and CHIP 28 in the membranes of lens fiber and erythrocytes, respectively (for a review, see van Os *et al.,* 1994). For instance, bean α-TIP was chosen as a model to study tonoplast protein biogenesis in seeds because it is the most abundant intrinsic polypeptide of the protein storage vacuoles (PSVs) (Johnson *et al.,* 1989). Five *Arabidopsis* PIPs were identified by an immunoscreening procedure because antibodies directed against purified plasma membrane mostly recognized these proteins (Kammerloher *et al.,* 1994). In spinach leaves, aquaporin PM28a can represent up to 15% of plasma membrane proteins (Johansson *et al.,* 1996).

Despite a molecular structure reminiscent of membrane channels, this abundance initially suggested a role for plant MIPs in membrane stabilization, in particular when plant cells undergo stringent dehydration or temperature cycles (Johnson *et al.,* 1989). Although these ideas cannot be completely discarded, the finding that most MIPs function as water channel can now easily rationalize their abundance. To be significant in terms of volume exchange, the flows of water observed in plant cells must correspond to billions and billions of transported water molecules. Although the intrinsic water permeability of aquaporins is high (up to 2×10^{-13} cm^3 s^{-1}; Yang and Verkman, 1997), high expression levels of these proteins are needed to mediate such a flow. In these respects, water channels differ from ion channels, which can generate an electrical signal with a few thousand transported ions.

B. Aquaporins Are Targeted to the Plasma Membrane, the Vacuolar Membrane, and the Endomembranes

The idea that the two major subfamilies of plant MIPs coincide with a well-defined cell localization followed from early studies showing that TIPs localize in the tonoplast (Höfte *et al.,* 1992), whereas PIPs are mostly found in the plasma membrane (Daniels *et al.,* 1994; Kammerloher *et al.,* 1994). The cell localization of TIPs has been consistently determined in several systems, using membrane fractionation and immunocytochemistry. For instance, bean α-TIP is specifically targeted to the membrane of PSVs in seeds (Johnson *et al.,* 1989). When expressed in the leaves of transgenic tobacco, the protein was localized in the corresponding membrane, that is, the tonoplast of central vacuoles (Höfte *et al.,* 1991).

Careful immunolocalization studies of TIPs have confirmed earlier observations that plant cells contain different types of vacuoles. Paris *et al.* (1996) showed that, in pea root tips, antibodies raised against bean seed α-TIP recognized a vacuolar membrane compartment that contains lectins and could be identified as PSVs. Whether these antibodies recognize the same protein in the root tips as in the seeds is not known. Antibodies specific for beet TIP-Ma27 recognized, in the same pea root cells, acidic lytic vacuoles (LVs), which are also specifically labeled by antibodies to the protease aleurain. More recently, antibodies against δ-TIP, another TIP homolog, helped to define in potato tuber and epidermal cells a novel type of vacuole called Δ-vacuoles (Jauh *et al.,* 1998). The latter vacuoles specifically contain abundant vacuolar storage proteins that also have enzymatic or other acitivities such as protease inhibitors, acid phosphatases, or lipoxygenases.

Motor cells in the pulvinus of *Mimosa pudica* also contain two distinct types of vacuoles, the tannin vacuole and the aqueous vacuole. A higher density of γ-TIP-like aquaporins was detected in the latter vacuoles, consistent with the role played by these vacuoles in cell volume adjustment during leaf movement, (Fleurat-Lessard *et al.,* 1997). However, during leaf movement the critical change in water permeability should occur in the plasma membrane and may result from the sudden opening of PIP aquaporins. Altogether, these data show that TIP isoforms participate in the differentiation of vacuolar subtypes, with specialized functions in storage, degradation, or osmoregulation. Conversely, these findings provide a partial explanation for the diversity of TIP proteins in plants. What remains to be determined is to what extent these proteins differ in their transport and regulatory properties and how these fit into the specialization of their respective vacuolar subtypes. These findings also raise important questions about the cellular mechanisms that allow the differential targeting of homologous proteins to distinct subcellular compartments. Recent work by Jiang and Rogers (1998) was interpreted to show the existence of two distinct targeting pathways for vacuolar membrane proteins. One is direct from the ER to the PSVs while the other one leads to LVs via the Golgi. Using chimeric marker proteins, they concluded that information contained within both membrane-spanning domains and cytoplasmic tails determines the ultimate location of the protein.

The localization of PIPs in the plasma membrane initially relied on membrane purification by aqueous two-phase partitioning (Daniels *et al.,* 1994; Kammerloher *et al.,* 1994). The idea that PIPs are exclusively in this membrane was recently challenged by Barkla *et al.* (1999). These authors identified several PIP isoforms, MIP-A, MIP-B, and MIP-C, in *Mesembryanthemunm crystallinum* and raised specific antibodies against the proteins. When the antibodies were used to probe membrane fractions separated by sucrose gradients, it came as a surprise that each antibody revealed a specific profile. Proteins immunoreactive to either of the three antibodies were detected in a tonoplast-enriched fraction and peaked in distinct fractions of the sucrose gradient. Although the corresponding membrane compartments were not identified precisely, these data suggest that different PIP isoforms can reside in distinct subcellular membranes and that the plasma membrane does not represent the major cell localization of some PIPs. Further work is needed to rule out the possibility of cross-reactivity of these antibodies with TIPs.

Immunocytochemical studies have also indicated the presence of TIP-like proteins in the plasma membrane of pea cotyledon cells (Robinson *et al.*, 1996a), but given the abundance of TIP and PIP isoforms and the use of antibodies across species, it is not possible to draw firm conclusions from such experiments. At present, the multiple localization of aquaporins in various compartments of the cell secretory pathway is not clear. It could reflect the routing of these proteins during biogenesis and degradation. Alternatively, it could represent the existence for these proteins of a shuttle mechanism, according to the paradigm of vasopressin-dependent regulation of AQP2 in the renal collecting duct (Sasaki *et al.*, 1998). In such a model, stimulation of the plant cell by hormonal, developmental, or environmental stimuli would lead to mobilization of aquaporins from a storage compartment to their target membrane (Barkla *et al.*, 1999).

A puzzling question also concerns the subfamily defined by Nod26. Consistent data showed this protein to be exclusively localized in the peribacteroid membrane of symbiotic root nodules (Miao *et al.*, 1992) and to be absent from the corresponding plasma membrane. The peribacteroid membrane delimits a specialized intracellular compartment and controls the exchanges between the bacteroid and the host plant cell. Thus, the cell localization of Nod26 homologs found in plant species without symbiotic root nodules remains an open question (Weig *et al.*, 1997).

V. WHAT DO THE WATER AND SOLUTE TRANSPORT PROPERTIES OF MEMBRANES TELL US ABOUT THE AQUAPORINS IN THOSE MEMBRANES?

A. How Are the Water and Solute Transport Properties of Membranes Measured?

Measurement of water transport in biological systems raises difficult biophysical problems, and this question has been a matter of intense investigation for more than a century! Early measurements of the permeability of plant cell membranes to water, such as those conducted by Höfler, as early as 1918, relied on the osmotic shrinkage and swelling of plant cells. Although these types of measurements have been widely used, the presence in plant tissues of barriers such as cell walls that limit solute diffusion renders their interpretation difficult. To avoid these problems,

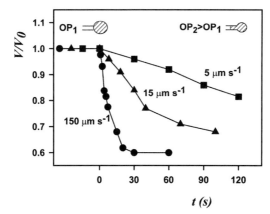

FIGURE 8 Variation in the speed of relative volume change of protoplasts in a single preparation of protoplasts obtained from *Arabidopsis* leaves. The volume of individual protoplasts is measured first in an isotonic solution. The medium is changed to be hypertonic and the variation in volume with time is measured. Some protoplasts shrink extremely rapidly—in one or two seconds—others shrink slowly over the course of 2 or more minutes. Measurements performed as described by Ramahaleo *et al.,* 1999 (Unpublished data of R. Morillon and M. Chrispeels, 1999).

Ramahaleo *et al.* (1999) recently adapted the technique to isolated protoplasts or vacuoles. The objects are handled by a micropipette and a gentle suction is applied to generate a slight membrane tension and to keep the object with a well-defined geometric shape. Changes in protoplast or vacuole volume upon transfer to a hyperosmotic medium can be followed by videomicroscopy, and water permeability values are derived from the rate of volume adjustment (Ramahaleo *et al.,* 1999). The results show a 100-fold variation in the osmotic water permeability of protoplasts derived from the leaves of a single population of plants (see also Fig. 8).

Attempts have been made to label water molecules using radiotracers or NMR techniques (for a review, see Maurel, 1997). An intrinsic drawback of these techniques is that they allow the determination of the diffusional water permeability (P_d) rather than the osmotic water permeability (P_f), whereas the latter is the most significant biologically. Important artifacts caused by unstirred layers have also led to erroneous conclusions with these techniques.

The invention of the cell pressure probe provided an original approach to investigate the water transport properties of plant cells (for reviews, see Steudle, 1993; Tomos and Leigh, 1999). As we have seen before, the water balance of plant cells is largely determined by the mechanical properties of their cell walls, which allow the buildup of a turgor or hydrostatic pressure within the cell. The cell pressure probe allows one to measure and perturbate this pressure. Although this technique can be challenging when applied to small cells, its principle is quite simple. A micropipette filled with oil, a largely uncompressible liquid, is inserted within the cell and connected to a pressure transducer. Observation of the meniscus at the interface of the oil and the cell sap or cell solution allows one to follow accurately any volume flow between the interior of the cell and the probe. The internal volume of the probe can also be finely adjusted by a motor-driven rod. This allows one to counterbalance the hydrostatic pressure present in the cell and this pressure can be measured, once the cell volume has been stabilized at its initial volume. By rapidly moving the meniscus, a sudden change in pressure can also be imposed. The change in pressure recorded for a given change in volume (or meniscus displacement) allows the determination of the elasticity of the cell wall. The kinetics of cell pressure relaxation, observed at constant probe volume (meniscus immobilized), are generally exponential and the half-time of this relaxation provides another water-relation parameter of the cell (Steudle, 1989, 1993). These parameters, together with the cell surface-to-volume ratio, allow the determination of the hydraulic conductivity (L_p) of the cell, in other terms its intrinsic permeability to water. Because in most cases the probe is inserted in the vacuole, this L_p value integrates the resistance in series of the vacuolar and plasma membrane. In addition, the resistance of the latter is determined by the respective contributions of the lipid membrane (with or without aquaporins) and the plasmodesmatal channels.

The efficiency with which a solute can be transported across a given membrane barrier is primarily indicated by its permeability coefficient. The reflection coefficient more specifically defines the osmotic efficiency of this solute and can provide a set of complementary indications as to the mechanisms of solute transport. The value of this coefficient is usually comprised between 1 (full osmotic effects) and 0 (no osmotic effects) (Steudle, 1989). Its value depends on the respective permeabilities of the membrane to the solute and to water. Moreover it takes into account possible frictional interactions between solute and water during transport. This component can become critical when the two types of molecules have to permeate across the same narrow pore (Henzler and Steudle, 1995; Schütz and Tyerman, 1997). The cell pressure probe can be useful to characterize the transport characteristics of highly permeable solutes in plant cells (Steudle, 1993). For these measurements one uses the biphasic cell, pressure relaxations induced by osmosis in the presence of various solutes. The first phase reflects the osmotic movement of water in or out of the cell, whereas the second phase reflects the solute equilibration. Both solute permeability and reflection coefficients can be accurately determined from these type of measurements.

The water and solute transport properties of membrane vesicles purified from plant cells can also be studied by stopped-flow spectrophotometry. This technique allows a membrane vesicle suspension to be very efficiently mixed (within a few milliseconds) with a solution of a distinct osmotic potential or solute concentration (Verkman, 1995). This triggers a solute and/or osmotic water flow across the vesicle membrane. The accompanying change in vesicle volume can then be detected by alteration in the light scattering properties of the vesicle suspension. Alternatively, a vesicle-entrapped fluorophore, with concentration-dependent quenching properties, can be used as a vesicle volume reporter. The kinetics of vesicle volume adjustment, which usually occur within seconds, together with the size of the vesicles can be used to calculate the intrinsic permeability properties of the membrane. The application of stopped-flow techniques to plant membranes has been very useful to provide biophysical evidence for active water and solute channels in plant membranes (Gerbeau *et al.*, 1999; Maurel *et al.*, 1997b; Niemietz and Tyerman, 1997; Rivers *et al.*, 1997). Although the purification of membranes can possibly alter some of their properties, this technique can also give access to the differential permeability properties of subcellular membranes (Gerbeau *et al.*, 1999; Maurel *et al.*, 1997b; Niemietz and Tyerman, 1997).

B. Discovery of Water Channels Validates Earlier Work Showing That Plants Regulate Hydraulic Conductivity at the Cellular Level

During the last 20 years, pressure probe and other types of water transport assays have been used to investigate various aspects of plant–water relations and have been applied to many plant species and cell types (Maurel, 1997; Steudle, 1989). What have we learned from these types of experiments? First, they give a very complex picture with a large variability in water permeability values and with different cell types exhibiting very different water permeabilities. For instance, cortex cells of elongating pea epicotyls can be 400 times more permeable to water than the correponding epidermal cells (Cosgrove and Steudle, 1981). Large variations between plant species were also found (Steudle, 1989). Such variation was very confounding and continuing efforts were made to refine the methodology and understand the technical drawbacks such as the difficulty of estimating the size of the cell that is being investigated. Now, the discovery of channels provides a molecular explanation for the very high water permeability of certain membranes.

Second, water transport measurements by the various techniques described above have clearly shown that plant cells are able to vary the water transport properties of their membranes. Environmental factors can trigger such changes and, for instance, an increase in soil salinity reduces by 4- to 6-fold the water permeability of cortex cells in maize roots (Azaizeh *et al.*, 1992). Recent studies by Ramahaleo *et al.* (1999) clearly show that the osmotic water permeability (P_f) of root protoplasts also significantly varies during root development. Initially, these observations were discussed in terms of physiological integration, but few attempts have been made to understand the molecular and cellular basis of such regulation. It seems quite likely that regulation of aquaporin abundance and activity are involved in these adjustments.

C. By Which Criteria Can We Detect the Presence of Active Water Channels?

The existence of a very high membrane permeability to water ($P_f > 100\ \mu\text{m}\cdot\text{s}^{-1}$) only hints at the activity of water channels, and biophysicists have defined criteria by which they unambiguously identify a pore-mediated water transport (Finkelstein, 1987). The permeation of water across a lipid membrane necessitates a series of processes that apply to individual water molecules: solubilization in the lipid phase, diffusion through the lipid bilayer, and resolubilization in the aqueous extramembrane space. In contrast, when water permeates a transmembrane water channel, several water molecules are present at the same time in the pore. This is reflected by an apparent cooperativity when an osmotic or a hydrostatic driving force is exerted on the membrane. Water transport (permeability) becomes then much more efficient as compared to the free diffusion of water molecules along the pore observed in the absence of a driving force. Thus, in such membranes, the osmotic water permeability (P_f) is several-fold greater than the diffusional water permeability (P_d), whereas the P_f and P_d of a lipid membrane are equal.

This criterion has been historically very important for demonstrating the existence of water channels, long before their molecular structure was elucidated (Finkelstein, 1987; Koefoed-Johnsen and Ussing, 1953). However, it has led to erroneous conclusions because of artefacts in these measurements due to the presence of unstirred layers (Dainty, 1963), and the interpretation of a P_f/P_d ratio, as an estimate of the number of water molecules in the pore (Finkelstein, 1987) should be taken with caution. Nevertheless, comparative measurements of P_f and P_d in a few plant membrane systems have been recently reported, as evidence for active water channels in these membranes (Henzler and Steudle, 1995; Niemietz and Tyerman, 1997; Rivers *et al.*, 1997). Henzler and Steudle (1995) followed a novel procedure for determining P_d in Chara internodal cells. Using a pressure probe, these authors were able to show that heavy water (D_2O) shows osmotic effects, because it permeates with a slightly smaller efficiency than normal water (H_2O). The biphasic pressure relaxation observed in these conditions was interpreted to yield P_d as the cell permeability to this somewhat special solute. The P_f of the Chara membrane was also determined by means of the cell pressure probe using a classical hydrostatic or osmotic procedure. Another approach used with wheat root tonoplast and soybean peribacteroid membranes was to determine both the P_f and P_d of purified membrane vesicles by stopped-flow spectrophotometry (Niemietz and Tyerman, 1997; Rivers *et al.*, 1997). The diffusion of heavy water into the vesicles was followed by specific changes in the fluorescence of a vesicle-entrapped fluorophore (ANTS) and was used to calculate P_d. These studies, which yielded P_f/P_d ratios between 7, were all indicative of active water channels.

Because it occurs by a solubility-diffusion mechanism, the transport of water or solute across a lipid membrane necessitates that each molecule cross a high energy barrier. In other words, this process displays a high activation energy, which results in a high dependence of transport on temperature. In contrast, facilitated transport across an aqueous pore requires a lesser activation energy and is thus less dependent on temperature. This dependence can thus be very useful in analyzing the nature of the path by which water or solutes are transported. It was used to provide evidence for water channels (Maurel, 1997; Niemietz and Tyerman, 1997; Wayne and Tazawa, 1990) as well as for solute channels (Gerbeau *et al.,* 1999; Hertel and Steudle, 1997) in purified membranes and in entire cells.

Because most molecular water channels can be blocked by mercurials, these compounds have been useful in providing evidence for such channels in plant membranes (Maurel *et al.,* 1997b; Niemietz and Tyerman, 1997; Wayne and Tazawa, 1990; Zhang and Tyerman, 1999). However, the effects of mercury derivatives can be very unspecific and these compounds can affect numerous targets, especially when the drug is applied at the cell level. A strong depolarization of the cell membrane potential, in response to mercury concentrations as low as 5 μM and concomitant to inhibition of the cell hydraulic conductivity, has been described in wheat root and in *Chara* internodal cells (Schütz and Tyerman, 1997; Zhang and Tyerman, 1999). In the latter cells, higher mercury concentrations ($>100 \mu M$) significantly inhibited cell respiration and cytoplasmic streaming and cell death was observed after 1 h (Schütz and Tyerman, 1997). Ramahaleo *et al.* (1999) showed that protoplasts die rapidly in 50 μM $HgCl_2$. On the other hand, some AQPs are known to be resistant to mercury derivatives (Biela *et al.,* 1999; Daniels *et al.,* 1994). These chemicals are therefore of dubious value in probing the contribution of water channels in membrane permeability in plants.

Recently, Steudle and Henzler revisited earlier studies on the transport of organic solutes in *Chara* cells (Henzler and Steudle, 1995; Steudle and Henzler, 1995). Using the blocking effects of mercury, together with cell pressure probe measurements, they could deduce the contribution of water channels to the transport of organic molecules such as acetone, formamide, or alcohols of different molecular size. Because most of these molecules are highly lipophilic, the contribution of water channels to the overall membrane permeability was minor. These authors could show, however, that the reflection coefficient associated with the water channel path was much below unity. This was interpreted as evidence for strong interactions of solutes with water within the pores of aqueous transmembrane channels. This type of study, however, does not provide hints about the molecular identity of these solute and water channels. In this respect, there is good correspondence between the solute transport activity of aquaporin Nod26 and NtTIPa in *Xenopus* oocytes and the solute permeability of their membranes in which these channels are found: the tobacco cell tonoplast and the peribacteroid membrane of symbiotic soybean root nodules, respectively (Dean *et al.,* 1999; Gerbeau *et al.,* 1999; Rivers *et al.,* 1997). This correspondence indicated that aquaporins can contribute to a major extent to the high permeability of some plant membranes to small neutral solutes.

D. Antisense Expression of Aquaporins Provides Direct Evidence That They Are Involved in Transmembrane Water Flow in Planta

All studies mentioned earlier indicate the presence of active water channels in plant membranes but do not provide a direct connection to the molecular identity of these channels. Reverse genetic approaches can be helpful to address this question. Kaldenhoff *et al.* (1998) recently described *Arabidopsis* plants expressing an antisense *pip1c* transgene. The plants had a reduced expression of this and closely related genes of the *pip1* subfamily. Osmotic swelling of leaf protoplasts revealed a reduced P_f in protoplasts isolated from antisense lines as compared to control protoplasts, similar to the reduction obtained after mercury treatment (1 mM HgCl$_2$) of the same protoplasts. These experiments provide the first evidence that aquaporins significantly contribute to water transport across plant membranes.

VI. THE MULTIPLE ROLES OF AQUAPORINS: HOW TO LINK WATER TRANSPORT PROPERTIES TO FUNCTIONS AT THE CELLULAR AND TISSUE LEVEL

A. Isolated Vesicles Derived from the PM and the TP Exhibit Distinct Water Transport Properties, the TP Vesicles Being by Far the Most Permeable

Because of its large size and dilute content, the vacuole usually represents the major container of water within the plant cell. This morphological observation should not lead to the naive idea that the vacuole can spare some excess water, retain it tightly within the cell, and release it at will when the demand of the cell becomes critical. The water permeability of cell membranes, even devoid of water channels, is such that equilibration between cytosol and vacuole will occur within minutes. The water potential gradient between the two compartments ultimately determines water exchanges between them, and cytosol and vacuole must be at water potential equilibrium most of the time. It has also been demonstrated that the properties of the cell wall rather than the size of the vacuole determine water storage capacity of plant cells (Steudle, 1989). The more elastic the cell wall, the more water can be exchanged between the cell and its exterior without a dramatic change in turgor. This elasticity may have some relevance during water deficit when cell types with a critical function for plant survival have to be supplied with water from neighboring tissues.

The idea that water transport across the vacuolar membrane has a marginal significance for plant cell water relations began to be questioned after the discovery of aquaporins. Although the activity of these proteins in their native membranes remained to be determined, the striking abundance of aquaporin homologs in the tonoplast (TP) suggested that at least in certain conditions the transport of water across this membrane must be highly efficient. It also became apparent that very little was known about the respective water permeabilities of the plasma membrane (PM) and the TP. Because of the presence of specific aquaporins in the two compartments, both of them should be highly regulated, in a distinct manner.

These ideas have been confirmed by direct measurements of water transport in isolated PM and TP vesicles. Maurel *et al.* (1997b) purified plasma membrane and tonoplast vesicles from tobacco suspension cells by free-flow electrophoresis. Niemietz and Tyerman (1997) used aqueous two-phase partitioning to purify PM vesicles from wheat roots. The fraction depleted in PM vesicles contains intracellular membranes and was taken as a TP-enriched fraction. In both studies, stopped-flow measurements revealed a low water permeability for purified PM vesicles, associated with a high E_a, suggesting that water is predominantly transported across the lipid bilayer in these vesicles. In contrast, TP vesicles showed a much higher permeability, with a reduced E_a. In particular, a 100-fold difference in water permeability was observed between PM and TP vesicles isolated from the same tobacco cells (Maurel *et al.,* 1997b) (Fig. 9). Mercury inhibition of water transport in TP vesicles from tobacco cell suspension and wheat root, and a high P_f/P_d in the latter vesicles, provided other evidence for the activity of water channels in the plant tonoplast. Water transport measurements in vacuoles isolated from onion, red beet, petunia, and rape, with P_f values greater than 200 $\mu m \cdot s^{-1}$ confirmed the notion that water transport across the TP of most plant species is highly efficient, due to the activity of mercury-sensitive water channels (Morillon and Lassalles, 1999). The interpretation of this property, however, is not straightforward.

At the tissue level, it remains unclear whether plant cell vacuoles represent a significant resistance to water flow. First, the contribution of the transcellular path to the overall flow may not be critical in most tissues. More generally, cell walls but also plasmodesmata, and within the cell, cytoplasmic strands, provide extensive bypasses to vacuoles. Because of these, the water permeability of vacuoles will hardly determine the overall flow rate of water across tissues. In a complex path of joining cells, the resistance of vacuolar membranes still remains significant in determining the drop in water potential on both sides of the membrane. This follows from the basic relationship between flow on the one hand and hydraulic conductance and driving force on the other hand. In optimal growing conditions, enormous amounts of water flow across living tissues, in particular roots and leaves. This flow is driven by the transpiration regime of the plants. It possibly fluctuates in response to environmental changes and concomitant modulation of stomatal aperture. In these conditions, it can be critical that the cytosol and vacuole of any plant cell remain in perfect osmotic equilibrium at all times. In case of an intense water flow this will only be the case if the vacuolar membrane poses no significant hydraulic resistance.

The significance of a high water permeability of the TP is also puzzling at the single cell level. It is generally assumed that the plasma membrane must represent

Stopped-flow light scattering

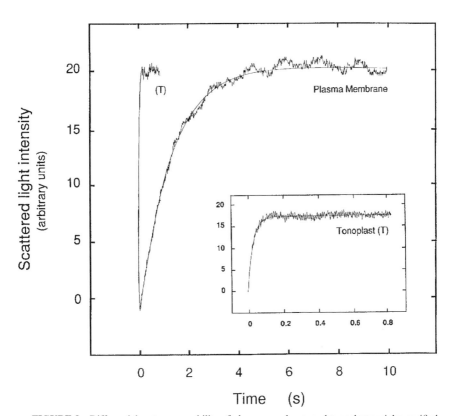

FIGURE 9 Differential water permeability of plasma membrane and tonoplast vesicles purified from tobacco suspension cells, as revealed by stopped-flow light scattering measurements. Membrane vesicles enriched in plasma membrane or tonoplast (T) were purified from tobacco suspension cells by free-flow electrophoresis (for experimental details, see Maurel *et al.*, 1997). Purified vesicles were exposed, in a stopped-flow apparatus, to a sudden increase in external osmolarity at $t = 0$. The increase in light scattering reflects the osmotic shrinking of the vesicles. Note that the two membrane preparations exhibit strikingly different time courses of shrinking, since volume adjustment was completed in less than 0.2 s for tonoplast vesicles (inset) and took more than 5 s in plasma membrane vesicles. The following osmotic water permeability values (P_f) were deduced from the rate of osmotic shrinking and the size of the vesicles: plasma membrane, $P_f = 6.2\ \mu\mathrm{m} \cdot \mathrm{s}^{-1}$; tonoplast: $P_f = 682\ \mu\mathrm{m} \cdot \mathrm{s}^{-1}$. (Modified from Maurel *et al.*, 1997b.)

the most significant membrane barrier to regulate water exchanges between the cell and its bathing medium. Simple calculations show that if the water permeability of the PM remains low, an additional resistance at the tonoplast will marginally determine the rate of cell equilibration. However, additional constraints for cell osmoregulation are imposed by the compartmentation of plant cells in cytosol and vacuole. The cytosol is the most critical compartment for most cell metabolic functions and requires very fine regulation of its volume and osmotic potential. Because it also has to mediate water exchange between the vacuole and the cell exterior, the cytosol is subject to possible volume fluctuations if water mobilization into and from the vacuole becomes limiting. A swelling or shrinking can then be expected in case of a sudden water influx or efflux, respectively. This behavior can be nicely shown by simulating water transport in a model plant cell (Tyerman *et al.,* 1999). Although these processes are only transient, they can represent very rapid and drastic changes well beyond the constraints that can be felt at the whole cell or vacuole level. Along these ideas, it appears that a nonlimiting water transport at the TP can efficiently damp these volume fluctuations. It remains to be established whether such fast kinetic perturbations of water potential are physiologically significant. The sudden rehydration of droughted root tissues, the propagation of cavitation shocks in xylem parenchyma cells, or the mechanical deformation of rigid cells may represent critical situations in this respect. In contrast to what is observed in plant cells, vacuolar membranes and PM-targeted secretory vesicles purified from yeast cells both showed a reduced, lipid-mediated, water permeability (Coury *et al.,* 1999). It is not clear whether this difference reflects different regulation properties or distinct physiological functions for the vacuolar membrane of yeast and plants.

B. Is the PM the Limiting Barrier for Water Exchange?

It appears as a paradox that no water channel activity could be detected in plasma membrane vesicles from either tobacco suspensions or wheat roots (Maurel *et al.,* 1997b; Niemietz and Tyerman, 1997), whereas plasma membrane aquaporin genes are so abundantly expressed in plants. Strong expression of at least three PIP genes in tobacco suspension cells indeed suggests a tight regulation of water transport at the PM of these cells (Maurel *et al.,* unpublished data, 1999). The idea that some functional properties of PM aquaporins may be lost during PM vesicle isolation was recently substantiated by Zhang and Tyerman (1999). These authors characterized the hydraulic conductivity of wheat root cortex cells by means of a pressure probe. The same root material had been previously used for measuring the water permeability of purified PM vesicles (Niemietz and Tyerman, 1997). Surprisingly, the hydraulic conductivity of the intact cells was found to be higher than what could have been expected from the P_f values measured in purified PM vesicles. It was calculated that the P_f values of PM vesicles would have to be increased up to 10-fold and mercury inhibition of water transport at the PM would had to be hypothesized to account for the observed whole-cell behavior. The mechanisms that lead to water channel inactivation during PM vesicle isolation are unknown and may correspond to an alteration in protein phosphorylation levels (Johansson *et al.*, 1998) or dissociation from cellular elements such as the cytoskeleton (Wayne and Tazawa, 1988).

The idea that PM aquaporins can be upregulated in certain conditions runs counter to the notion that the TP always remains more permeable to water than the PM. Simulation studies showed, however, that satisfactory cytosol osmoregulation can be achieved as long as the TP has a 5- to 10-fold higher P_f than the PM (Fig. 10). In tobacco cells, for instance, the 100-fold difference observed in purified vesicles would allow a large amplitude of regulation at the PM. A simple model would be to propose that the TP constantly exhibit a high water permeability, whereas water transport at the PM can show marked regulation and determine the overall exchange of water between the cell and its exterior.

C. Mercury Inhibition of Water Transport across Plant Tissues Suggests a Role for Aquaporins in Regulating Transcellular Water Flow

Evidence for active water channels at the tissue level, by contrast to the membrane and cell levels, can hardly be established based on biophysical criteria. Mercury and its derivatives are the only chemicals available to block water channels but, as discussed above, their use is subject to serious drawbacks. Notwithstanding these, and following the molecular identification of the first plant aquaporins, several research groups have gone on with exploring the effects of mercury on water transport in whole tissues or plants.

A single study dealt with elongating tissues (Hejnowicz and Sievers, 1996) whereas all others focused on water channels in roots (Amodeo *et al.*, 1999; Carvajal *et al.*, 1996, 1999; Maggio and Joly, 1995; Quintero *et al.*, 1999; Tazawa *et al.*, 1997). This is consistent with the central role that the root hydraulics is thought to play in the water relation of plants. Also, a variety of methods have been developed to measure water transport in this organ. Excised roots can be adjusted in a pressure bomb and the pressure-to-flow relationship allows one to determine a hydrostatic component of the root hydraulic conductivity (L_p) (Maggio and Joly, 1995). Alternatively, the spontaneous exudation flow of excised roots, together with the osmotic pressure of the sap, can be used to determine their osmotic L_p (Carvajal *et al.*, 1996, 1999). Tazawa *et al.* (1997) recently described a new trans-root osmosis method using barley plants. They placed 5-day-old excised roots in a device initially developed to measure the L_p of giant internodal *Chara* cells by transcellular osmosis. A similar approach was taken by Amodeo

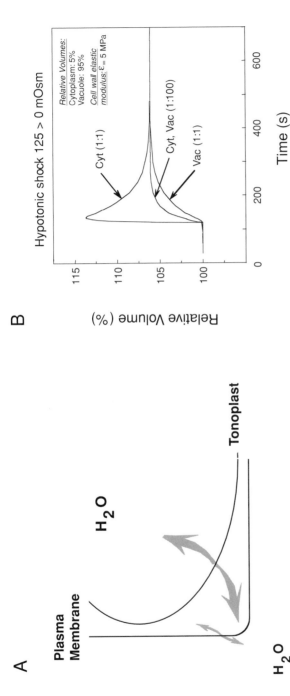

FIGURE 10 Simulations of water transport in a model plant cell. (A) Schematic view of a plant cell showing that a high water permeability of the tonoplast allows efficient mobilization of water out of and into the vacuole. Thus, a water permeability, higher in the tonoplast than the plasma membrane, suggests a critical role for the vacuole in the osmoregulation of the cytosol. (B) This idea is supported by water transport kinetics in a model plant cell. The osmotic water permeability (P_f) of the plasma membrane was set to a constant value of 6 $\mu m \cdot s^{-1}$ (see Maurel et al., 1997), whereas two P_f values were considered for the tonoplast: $P_f = 6$ $\mu m \cdot s^{-1}$ (the PM:TP ratio is then 1:1) or $P_f = 600$ $\mu m \cdot s^{-1}$ (the PM:TP ratio is then 1:100). The latter cell configuration is derived from P_f values experimentally determined in membrane vesicles purified from tobacco suspension cells (Maurel et al., 1997). For further details on the parameter values and equations used in the simulation, see Fig. 3 in Tyerman et al. (1999). When subjected to a hypotonic shock (125 to 0 mOsm), the model cell shows a limited expansion, of about 5% for $t > 400$ s, because of the mechanical resistance of the cell wall. However, in the 1:1 configuration the cytoplasm shows a dramatic increase in volume (by up to 15%) before water finally enters the vacuole. This behavior is not observed in the 1:100 configuration because transfer of water from the cytosol to the vacuole is not limiting. Note that in this model, and in contrast to what is observed for the cytoplasm, the kinetics of vacuolar volume adjustment are fairly similar, regardless of the water permeability of the tonoplast.

et al. (1999) who placed sugar beet slices between two chambers that were then submitted to differential osmotic or hydrostatic pressures. Water flow across the slice reflects transport along the axial or radial pathways of the sugar beet root, respectively, depending on the plane of cutting the root. Also, Lu and Neumann (1999) recently described a novel osmotic jump method to measure the root-to-leaf L_p of intact rice seedlings, by following the effects on the leaf elongation rate of changing water potentials in the root solution.

Numerous artefacts and experimental difficulties can be encountered with these techniques, and they have been discussed by several authors (Tazawa *et al.*, 1997; Zimmermann and Steudle, 1998). Measurement of root L_p under a hydrostatic pressure gradient is certainly the safest approach but corresponds to conditions where the contribution of the apoplastic path is possibly predominant (Steudle and Peterson, 1998). Measurement of root water transport in response to a osmotic driving force is more dubious but will likely reveal a higher contribution of the membrane (water channel) path (Amodeo *et al.*, 1999).

Because of these experimental difficulties, the effects of mercury must rely on several crucial controls. First, it should be ensured that the blocking effects occur within minutes and are restricted to proteinaceous cell structures. Because of this, they must be reversed by reducing agents. Such reversibility, however, was not observed in some studies (Tazawa *et al.*, 1997). A second important control is to show that solute transport across the root barrier was not modified. General metabolic effects can profoundly alter root behavior in general, and its osmotic pumping activity in particular. In this respect, several authors checked that K^+ transport in the root sap was not altered after mercury treatment (Carvajal *et al.*, 1999; Maggio and Joly, 1995).

All of these precautions being taken, Maggio and Joly (1995) using excised tomato roots were the first to show that water-channel-mediated transport may contribute about 60% to the root L_p (Fig. 11). Carvajal *et al.* (1996) used a lower mercury concentration (50 μM HgCl$_2$ instead of 500 μM) and had similar results in wheat roots. They found that the mercury-sensitive path may correspond to the variations in conductivity observed between roots of plants cultured in different conditions. Under nutrient (N or P) deprivation conditions, the L_p of wheat roots was low and mercury did not affect this residual water conductance. In contrast, in the presence of a complete supply of nutrients, the root L_p was enhanced up to 3- to 6-fold. Mercury reduced the root L_p to the residual value observed in nutrient-starved plants. Similar results were obtained in paprika roots after a saline treatment (Carvajal *et al.*, 1999). The sensitivity of root exudation to mercury disappeared after the treatment, concomitant with a reduction in L_p. Lu and Neumann (1999) also reported that the hydraulic conductance of rice seedlings is reduced after a moderate water stress. But, at variance with the previous work, mercury inhibition of water transport could be detected in water-stressed plants but not in fast-growing controls. Other endogenous or external stimuli, such as ABA, free

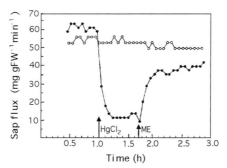

FIGURE 11   Pressure-induced water flux through a single HgCl₂-treated root system. Tomato plants were detopped and the root system bathed in 0.5 m*M* HgCl₂ (solid circles) by injecting the inhibitor into the solution at the time of the vertical arrow. The sap flow from the cut stump was measured continuously in the control (open circles) and the treated plant. Injection of mercaptoethanol (ME) largely reversed the effect of mercuric chloride. This experiment allows one to calculate J_v, which was inhibited by 71.5% by mercuric chloride. (Adapted from Maggio and Joly, 1995.)

calcium in the soil, night and day cycles, and anoxia, are known to affect root L_p and it was proposed that these effects are mediated by the regulation of root water channels (Henzler *et al.*, 1999; Quintero *et al.*, 1999; Steudle and Peterson, 1998).

Despite some experimental uncertainties, inhibitory effects of mercury have now been consistently reported by several groups (Amodeo *et al.*, 1999; Carvajal *et al.*, 1996, 1999; Lu and Neumann, 1999; Maggio and Joly, 1995; Quintero *et al.*, 1999; Tazawa *et al.*, 1997) and these reports clearly establish that water transport across plant roots comprises a mercury-sensitive component. This property has usually been interpreted as functional evidence for water channels in plant tissues. Although this interpretation seems reasonable, other experimental approaches will be needed to substantiate this conclusion. Finally, rather than providing direct evidence for aquaporin water transport, the blocking effects of mercury show at the very least that cellular functions and metabolism are needed to achieve optimal transport of water across plant tissues. This substantiates the importance of the cell-to-cell path and possibly that of aquaporins.

D. Plants with Down-Regulated Aquaporins Have Unusual Phenotypes

Reverse genetics also offers an attractive approach to address the role of aquaporins at the whole-plant level. Because of the high number of closely related aquaporin genes, antisense approaches were the first to be undertaken. Kaldenhoff *et al.* (1998) showed that a single antisense *pip1c*-derived construct was able to target several genes of the *pip1* subfamily in *Arabidopsis,* but these effects were incomplete and could not be quantified in detail. Nevertheless, specific effects of the transgene were observed, such as a reduced P_f in leaf protoplasts (see above). The transgenic lines also displayed interesting developmental alterations. While their aerial parts were similar to those of controls, the root fresh weight of anti-sense plants was increased up to 4-fold. The interpretation of this phenotype is not straightforward. It has been proposed that enhanced root growth in the antisense lines reflects compensation by the plants for a reduced root L_p. This explanation is attractive because it is simple but it will need further experimental support. In particular, the root L_p of these antisense plants needs to be measured, and it needs to be established that the change in L_p compared to control plants corresponds to the observed L_p dramatic increase in root surface area. Indeed, the root-to-shoot ratio of plants is highly dependent on a large number of factors, including water availability, mineral nutrition, and hormonal production at the root but also at the shoot level. Nevertheless, the most important outcome of these studies was to show that aquaporins can profoundly influence plant growth and development.

Other strategies are now being used. Several groups are searching for single aquaporin gene knockout mutants. These mutants were generated after insertion in the plant genome of a transposon or a *Agrobacterium tumefaciens* transferred DNA. Because these mutants were generated by random insertion of the transgene, the mutant of interest has to be screened out of a large collection of mutant lines using a PCR strategy (Bouchez and Höfte, 1998). If phenotypes can be resolved in these mutant plants, this approach will be helpful in deciphering the specific function of single aquaporin genes. This will surely help us in understanding the meaning of the high variety of aquaporin genes in plants.

VII. REGULATION OF AQUAPORIN EXPRESSION AND WATER TRANSPORT ACTIVITY

Currently at least 30 MIP family members have been identified in *Arabidopsis* (Johansson *et al.,* 2000; Kjellbom *et al.,* 1999; Weig *et al.,* 1997). A few MIPs have been shown to be specific for water, and it is likely that the majority of those not tested for transport specificity will turn out to be aquaporins as well. Many MIPs are expressed in developmentally regulated and cell-type-specific manners. Some are induced by drought whereas others are down-regulated by drought, and the expression of yet others is influenced by plant hormones and light. Some MIPs are localized to one or a few specific cell types, for example, guard cells (Sarda *et al.,* 1997) and meristematic cells of root tips (Chaumont *et al.,* 1998), whereas others are constitutively expressed in many cell types of many organs. In the latter case (i.e., constitutively expressed aquaporins), individual aquaporins may constitute as much as 20% of total plasma membrane or vacuolar membrane protein (Johansson *et al.,* 1996, 1998; Karlsson *et al.,* 2000). Within certain tissues, aquaporin expression is usually correlated with cells known to be associated with large fluxes of water, for example, guard cells and expanding/elongating cells. Furthermore, aquaporin expression has also been shown in cells where water is expected to enter or exit in order for water to flow across tissues. Such examples are the endodermal cell layer and the xylem parenchyma cells. Water needs to enter, and eventually exit from, the endodermal cells and to exit from the xylem parenchyma cells in order to flow from the soil to the xylem vessels of the root. However, because all cells need to maintain a cytosolic osmoregulation in order to be able to perform normal metabolic processes, it is likely that all cells express one or several aquaporins.

Knowledge of the expression patterns of different aquaporins is essential for discerning the physiological role of aquaporins at the whole plant level. The expression patterns will point to cells in which rapid transmembrane water transport is important. Expression patterns of members of different aquaporin subgroups will tell us whether they are present in the same cells or if they are expressed in different cell types. Together with information regarding the post-translational regulation of individual members of the different subgroups this will reveal whether aquaporins are not only buffering osmotic fluctuations of the cytosol, but also necessary for the bulk flow of water in plants.

A. Distribution of Aquaporins in the Plant

Organ-specific expression has been demonstrated for α-TIP of bean cotyledons (Johnson *et al.,* 1989) and TobRB7 of tobacco roots (Conkling *et al.,* 1990; Yamamoto *et al.,* 1991). However, most MIPs are generally expressed in more than one organ/tissue, although, within specific tissues, expression is often limited to, or excluded from, certain cell types.

A number of aquaporins are expressed in elongating or expanding cells. The spinach δ-TIP homolog So-δTIP (Karlsson *et al.,* 2000), the radish γ- and δ-TIP homologs γ- and δ-VM23 (Higuchi *et al.,* 1998), as well as the cauliflower γ-TIP homologs BobTIP26-1 and -2 (Barrieu *et al.,* 1999) are all expressed in expanding cells in different tissues. Cell enlargement is associated with an extensive influx of water into the cell made possible by the high osmotic potential of the vacuole. To sustain normal metabolic processes in the cytosol during cell expansion, a tight osmoregulation of the cytosol must be exerted. Aquaporins of vacuolar and plasma membranes are likely to be involved in buffering osmotic fluctuations of the cytosol associated with cell expansion in developing tissues.

Many aquaporins are also expressed in cells where transcellular water transport is expected to occur. This is the case for endodermal cells where So-δTIP, a spinach δ-TIP homolog, is expressed (Karlsson *et al.,* 2000) and for cells around vascular bundles where At-δTIP and At-PIP1b are expressed (Daniels *et al.,* 1996;

Kaldenhoff *et al.*, 1995). In particular, So-δTIP and MIPA, an ice plant PIP1 homolog, are expressed in xylem parenchyma cells (Karlsson *et al.*, 2000; Yamada *et al.*, 1995) and cells surrounding the phloem express Zm-TIP1, which is a maize γ-TIP homolog (Barrieu *et al.*, 1998). Aquaporins expressed in xylem parenchyma cells are likely to be involved in maintaining the transpiration stream by feeding water into the xylem vessels, thereby increasing the hydraulic conductivity of roots (Holbrook and Zwieniecki, 1999; Tyree *et al.*, 1999).

There is a high sequence similarity between δ-TIPs of different subgroups regardless of species. However, their expression patterns are totally different in different species. For example, *Arabidopsis* δ-TIP is expressed only in shoots (Daniels *et al.*, 1996), sunflower δ-TIPs exclusively in guard cells in leaves (Sarda *et al.*, 1997), and radish δ-TIP in tap roots, petioles, and cells of leaf veins but not in mesophyll cells or in root cells (Higuchi *et al.*, 1998). Thus, it appears as if expression patterns, and therefore probably also the physiological role, of different TIPs cannot be extrapolated from one species to another solely based on sequence similarities. It is the C-terminal regions of TIPs that are most dissimilar to PIPs and NLMs and synthetic peptides corresponding to these regions have therefore been used to raise TIP-specific antibodies. Because these C-terminal regions are only weakly conserved between δ-TIPs of different species such antisera will most likely not recognize corresponding TIPs in other species, for example, δ-TIP orthologs (Jauh *et al.*, 1999). However, the apparent lack of homology might also reflect that the true orthologs have not yet been identified.

B. Transcriptional Regulation of MIP Genes

A few aquaporin homologs have been shown to be encoded by drought-induced genes. This is the case for clone 7a, a pea PIP1 homolog (Guerrero *et al.*, 1990), for TRAMP, a tomato PIP1 homolog (Fray *et al.*, 1994), as well as for the cauliflower γ-TIP homologs BobTIP26-1 and -2 (Barrieu *et al.*, 1999). Also RD28, an *Arabidopsis* PIP2 homolog (Yamaguchi-Shinozaki *et al.*, 1992), is induced by desiccation, although it is also relatively abundant in nonstressed plants (Daniels *et al.*, 1994). There seem to be two separate signal transduction pathways for gene induction when plants experience drought stress. One pathway is independent of abscissic acid (ABA) and the other involves ABA (Shinozaki and Yamaguchi-Shinozaki, 1997). The genes encoding RD28 and TRAMP were shown to be ABA independent, whereas the gene encoding clone 7a was activated by exogenous ABA. Drought-responsive promoter elements (DREs; TACCGACAT) and ABA-responsive promoter elements (ABREs; T/CACGTGGC) have been identified and one would expect to find these elements in the genes induced by the two separate drought-induced signaling pathways, although additional regulatory *cis*-acting DNA elements might have gone unnoticed so far.

As we have seen, some of the *Arabidopsis* MIP homologs are abundantly expressed in nonstressed plants, some are expressed at low levels, and others are induced upon drought stress. Some MIP homologs have only been identified as genomic clones, and currently, 8 out of 34 identified *Arabidopsis* MIP genes have no corresponding ESTs (expressed sequence tags), which might suggest that their messages are not very abundant. Thus, the expression pattern for many MIPs might be quite specific. This specificity could either be associated with a spatially restricted expression or it could be associated with an expression induced by specific stimuli such as developmental cues, hormones, light, microbial infections, as well as different stresses. However, when evaluating expression data available in the literature it is crucial to consider that changes in mRNA levels are not always reflected at the protein level. Parameters such as mRNA stability and protein turnover have to be considered. Immunolocalization is a more accurate basis for statements concerning actual protein levels because message and protein levels may not correlate at all, as has been shown for several constitutively expressed aquaporins (e.g., Johansson *et al.*, 1996).

Spatially restricted expression can be exemplified by SunTIP7 and SunTIP20, two highly similar sunflower tonoplast aquaporins, which are expressed in guard cells (Sarda *et al.*, 1997). In addition, the SunTIP7 transcript shows diurnal fluctuations, with a peak at noon, and low levels at dawn, at dusk, and during the night when stomata are closed. This transcript also increases during drought stress. Thus, an increase in the SunTIP7 message seems to correlate well with the increased water efflux from guard cells necessary for stomatal movement and closure. The mRNA of the *Lotus* (*L. japonicus*) homolog corresponding to the aquaporin At-PIP1a of *Arabidopsis* plasma membranes also varies in abundance diurnally (Henzler *et al.*, 1999). The message is expressed in root cells and the fluctuations of the hydraulic conductivity of the roots closely resemble the diurnal variations in the abundance of the message. The decrease in aquaporin message levels occurred 2 to 4 h before the decrease in hydraulic conductivity (Clarkson *et al.*, 2000).

The plant hormones ABA and gibberellins have been shown to influence the expression of several aquaporin genes. The aquaporin At-γTIP (Maurel *et al.*, 1993) of the vacuolar membrane of wild-type *Arabidopsis* plants is expressed in cells undergoing elongation (Ludevid *et al.*, 1993). In a gibberellin-deficient mutant of *Arabidopsis,* exogenous gibberellins increase the expression of At-γTIP (Phillips and Huttly, 1994). Treatment of the mutant with exogenous gibberellins also induces stem elongation by increasing the rate of cell division and elongation.

The expression of two aquaporins, γ-VM23 and δ-VM23, of the vacuolar membrane of radish cells is light regulated (Higuchi *et al.*, 1998). The transcripts can be detected in expanding leaves and in hypocotyles growing in the dark. White light up-regulates the messages in cells of cotyledons and roots and down-regulates the messages in hypocotyls, which also stop elongating upon the switch from dark to light. Thus, the expression seems to be correlated with cells in expanding tissues, that is, with cell elongation. The *Arabidopsis pip1b* gene is activated by light as well, although it is also activated by plant hormones (Kaldenhoff *et al.*, 1993, 1996).

One vacuolar MIP homolog, Mt-AQP1, of the legume *Medicago truncatula* is induced upon infection by the mycorrhiza fungus *Glomus mosseae* (Krajinski *et al.*, 2000; Roussel *et al.*, 1997). The gene transcript was induced in arbuscular mycorrhiza and when the cRNA was transiently expressed in *Xenopus* oocytes, Mt-AQP1 was shown to be an aquaporin. Similarly, infection of tobacco roots by root-knot nematodes induces a root-specific aquaporin, TobRB7, in the vacuolar membrane (Opperman *et al.*, 1994). Interestingly, in the promoter of *tobrb7*, separate sequences are responsible for the constitutive root expression under normal conditions and for the induced expression upon nematode infection. By influencing the gene expression in root cells of the host plant, nematodes induce the formation of feeding sites in the roots. Much in the same way, *Bradyrhizobium japonicum* triggers soybean host cells to form root nodules in which nitrogen is fixed symbiotically. The plant-derived peribacteroid membrane encloses the bacteroids and separates them from the host cell cytosol. One specific MIP homolog, Nod26, is expressed in the peribacteroid membrane and has been shown to transport water as well as small uncharged solutes (Dean *et al.*, 1999; Miao *et al.*, 1992; Rivers *et al.*, 1997).

C. Post-Translational Regulation of Aquaporin Activity by Phosphorylation

Regulation of aquaporins has not only been observed at the transcriptional level but also at the post-translational level. MIP homologs have been shown to undergo phosphorylation and glycosylation as well as proteolytic processing (Higuchi *et al.*, 1998; Inoue *et al.*, 1995; Johansson *et al.*, 1998; Johnson and Chrispeels, 1992; Miao *et al.*, 1992; Weaver *et al.*, 1991). An aquaporin model with potential phosphorylation sites is shown in Fig. 12.

Soybean Nod26, kidney bean α-TIP, and spinach PM28A have been shown to be phosphorylated *in vivo* (Johansson *et al.*, 1998; Johnson and Chrispeels, 1992; Miao *et al.*, 1992). α-TIP is phosphorylated at serine-7 in the N-terminal tail, whereas both Nod26 and PM28A are phosphorylated in the C-terminal tail at serine-262 and serine-274, respectively. Using the *Xenopus* oocyte expression system, it has been shown that phosphorylation of PM28A and α-TIP at these sites causes an increase in water transport activity (Johansson *et al.*, 1998; Maurel *et al.*, 1995). Whether phosphorylation of serine-262 in the C terminus of Nod26 has any effect on water transport activity is not known. A C-terminal consensus phosphorylation site, at positions corresponding to serine-274 of PM28A and serine-262 of Nod26, is present in the PIP2 and NLM subfamily members, but

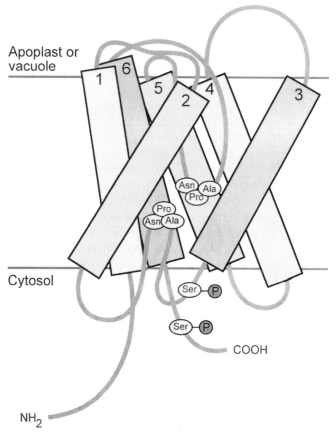

FIGURE 12   Putative structure of plant MIPs. For plasma membrane intrinsic proteins (PIPs), for tonoplast intrinsic proteins (TIPs), and probably also for the Nod26-like MIPs (NLMs), both the N- and C-terminal regions are located at the cytosolic side of the membrane. The other side is either exposed to the apoplast (PIPs) or to the vacuolar compartment (TIPs). The corresponding side of Nod26 of the peribacteroid membrane is exposed to the intermembrane space of nodules. The cellular location of the remaining NLM homologs is not known. The NPA boxes are fully conserved in the PIP and TIP homologs, while the NLMs show a considerable variation especially in the second NPA box. Also indicated are the putative phosphorylation sites regulating the water transport actvity. The phosphorylation site in the C-terminal region is conserved in all PIP2 and NLM homologs, while the phosphorylation site in the first cytosolic loop is conserved in all PIP homologs and also in TIPs, although most TIPs have a threonine instead of a serine at this position. (Adapted from Heymann *et al.*, 1998, and Kjellbom *et al.*, 1999.)

not in the PIP1 and TIP subfamily members (Johansson *et al.,* 2000). Within the PIP2 and NLM subfamilies, activity may be regulated by phosphorylation of the C-terminal serine residues and this mode of regulation may be specific for these two subfamilies and absent from the PIP1 and TIP homologs. For the NLMs, the amino acid environment around the serines corresponding to the phosphorylated serine in Nod26 is partly different from the corresponding sites in PIP2 homologs, for example, PM28A. The serines in the C-terminal regions of the NLMs are followed by two hydrophobic amino acid residues and one basic residue, instead of one hydrophobic and one basic residue as in the PIP2 subgroup. However, two NLMs, NLM 6 and 7 (accession numbers CAB39791 and AFF14664), have the same sequence of nonpolar and basic amino acids downstream of the serine, as compared to the PIP2 homologs. Mammalian AQP2 is also phosphorylated in the C terminus, although this phosphorylation seems not to be involved in regulating the water transport activity of the protein, rather it seems to regulate targeting of periplasmic vesicles to the plasma membrane (Lande *et al.,* 1996). The *in vivo* phosphorylated serine-7 of kidney bean α-TIP is not conserved in the putative orthologues At-αTIP of *Arabidopsis* (Höfte *et al.,* 1992) and MP23 and MP28 of pumpkin (Inoue *et al.,* 1995). Furthermore, At-αTIP has been reported not to be phosphorylated *in vivo* (Maurel *et al.,* 1997a), and it remains to be established whether these TIPs are true orthologs.

   In vivo, a high apoplastic water potential triggers a phosphorylation of serine-274 in the C-terminal region of PM28A (Johansson *et al.,* 1996, 1998). At low apoplastic water potential, such as during drought stress, serine-274 is less phosphorylated. Oocyte expression studies with mutant forms of PM28A suggest that dephosphorylation of serine-274 lowers the water transport activity of PM28A. At drought stress this may provide the plant with additional time to adjust to the low water potential (e.g., by transcriptionally activating genes involved in osmolyte synthesis pathways).

   In addition to the highly conserved serine-274 in the C-terminal region of PIP2 and NLM subfamily members, a highly conserved serine is also present in the first cytosolic loop. This serine is located in a consensus phosphorylation sequence Arg/X-Lys-X-Ser-X-X-Arg/Lys, which is present at this position in all PIP1 and PIP2 homologs. A similar motif, Arg-X-Ser-X-X-Arg, is present at corresponding positions in most TIPs, including α-TIP of kidney bean. The corresponding sequence in γ-TIPs is Thr-X-X-Arg. Results from oocyte expression studies of kidney bean α-TIP and PM28A of spinach indicate that phosphorylation at this conserved serine results in an increase in water channel activity (Johansson *et al.,* 1998; Maurel *et al.,* 1995). So far, phosphorylation of this residue has not been demonstrated *in vivo,* although, the oocyte results strongly indicate that phosphorylation/dephosphorylation at this site regulates aquaporin water transport activity.

   In purified spinach plasma membrane vesicles, PM28A is phosphorylated on serine-274. The protein kinase responsible is plasma membrane associated and

Ca^{2+} dependent (Johansson *et al.,* 1996). Also kidney bean α-TIP is, in a Ca^{2+}-dependent manner, phosphorylated by a protein kinase associated with the vacuolar membrane (Johnson and Chrispeels, 1992). Furthermore, Nod26 is phosphorylated in a Ca^{2+}-dependent manner by a protein kinase associated with the peribacteroid membrane in root nodules (Weaver *et al.,* 1991). As in mammalian cells, Ca^{2+} is a known intracellular messenger in plant cells. Calcium-dependent protein kinases (also called calmodulin-like domain protein kinases, CDPKs) have been identified in higher plants (Roberts 1993; Roberts and Harmon, 1992) and seem to replace mammalian protein kinase C and Ca^{2+}/calmodulin-dependent protein kinases. CDPKs depend on micromolar or submicromolar concentrations of Ca^{2+} and bind Ca^{2+} directly without the involvement of effector molecules such as calmodulin. Other types of Ca^{2+}-regulated protein kinases, different from CDPKs, may exist in plants, although no Ca^{2+}/calmodulin-stimulated protein kinase activities have so far been identified in plants (Satterlee and Sussman, 1998).

In the C-terminal, regions Nod26 and PM28A are phosphorylated on serine residues in the sequence Lys-X-X-Ser-X-X-Lys and Lys-X-X-X-Ser-X-Arg, respectively. These are motifs recognized by several CDPKs (Olah *et al.,* 1989; Polya *et al.,* 1989; Roberts and Harmon, 1992). Thus, the substrate specificity seems to differ for different CDPKs. The consensus phosphorylation sites of CDPKs and mammalian protein kinase C seem to overlap since PM28A can be phosphorylated *in vitro* by protein kinase C (Johansson *et al.,* 2000).

In vivo, PM28A is not only phosphorylated at serine-274 but also at serine-277 (Johansson *et al.,* 1998). Serine-277 is never phosphorylated without serine-274 being phosphorylated as well. Some protein kinases recognize consensus phosphorylation sites containing phosphorylated amino acids. As in the case of phosphorylation of serine-277 of PM28A, casein kinase-1 phosphorylates serines in sequences containing phosphoserine at position -3 (Pinna and Ruzzene, 1996). Both serine-274 and serine-277 are conserved among plant MIPs belonging to the PIP2 homologs.

D. A Model for the Involvement of Aquaporins in Osmosensing

In order for plant cells to respond to water deficit the cells must be able to monitor the change in osmotic potential. Osmosensors, probably located at the cell surface, initiate a signal transduction pathway leading to adaptive responses. In yeast, two-component systems are involved in osmosensing (Chang and Stewart, 1998). A eukaryotic two-component system consists of one plasma membrane protein and a histidine kinase, the extracellular domain of which senses the change in osmotic potential thereby triggering the autophosphorylation of a histidine residue located in the cytosolic domain of the protein. The phosphoryl group is then transferred to an aspartate residue in the same domain. In yeast, the stress signal is then further transmitted via two intermediate phosphoproteins (YPD1 and SSK1) to a MAPK (mitogen-activated protein kinase) cascade, ultimately affecting the transcriptional regulation of stress-responsive genes. A plant homolog, ATHK1, one of two known osmosensory proteins in yeast, SLN1, has been identified in *Arabidopsis* (Urao *et al.*, 1999). The function of ATHK1 as an osmosensor was demonstrated by functional complementation of a yeast mutant. It is not known by which mechanism ATHK1 senses changes in osmolarity.

In well-watered plants the apoplastic water potential is high and the cytosolic Ca^{2+} concentration is obviously sufficiently high to activate CDPKs responsible for phosphorylating aquaporins, for example, PM28A (Johansson *et al.*, 1998). The signal transduction pathway could involve osmosensors similar to ATHK1, and Ca^{2+} could be released from intracellular stores, for example, the vacuole. Alternatively, Ca^{2+} could originate from the apoplast and a Ca^{2+} channel in the plasma membranes could act as osmosensor itself. Ca^{2+}-specific, stretch-activated channels have been identified in guard cells of *Vicia faba* (Cosgrove and Hedrich, 1991). Whether the aquaporins of the vacuolar membrane are opened or closed at high apoplastic water potential is not known. However, oocyte expression studies suggest that the water transport activity of α-TIP is regulated by phosphorylation/dephosphorylation similarly to how PM28A is regulated, although phosphorylation occurs at a different site. At low apoplastic water potentials, the model (Fig. 13) suggests that the Ca^{2+} concentration in the cytosol drops, thereby inactivating the plasma membrane aquaporins. In this situation, vacuolar aquaporins might allow water to flow into the cytosol in order to counteract a decreased cytosolic water potential. This would either require that the water transport activities of TIPs are activated by lower Ca^{2+} concentrations as compared to PIPs, or that the Ca^{2+} concentration at the vacuolar membrane is transiently higher as compared to the Ca^{2+} concentration at the plasma membrane, following a drop in apoplastic water potential. If ATHK1-type osmosensors are involved in the signal transduction pathway, the putative downstream MAPK cascade might at low apoplastic water potentials target the nucleus and influence the transcription of osmotic stress-responsive genes, for example, genes encoding proteins involved in osmolyte production. This would constitute a second line of defense against water deficit and more immediate post-translational regulation of aquaporins, for example, phosphorylation/dephosphorylation events, being the first line of defense allowing the plant to mobilize adaptive responses.

It is likely that the roots sense the osmotic stress first and transfer this information to the shoot in the form of an as yet unidentified signal. The synthesis of ABA in roots is increased during drought stress and ABA has been detected in xylem sap. However, stomatal closure has been reported to occur before any increased levels of ABA can be detected (Trejo and Davies, 1991). Thus, ABA is probably not the only substance functioning as a root-to-shoot signal. Contrary to ABA, the flow of

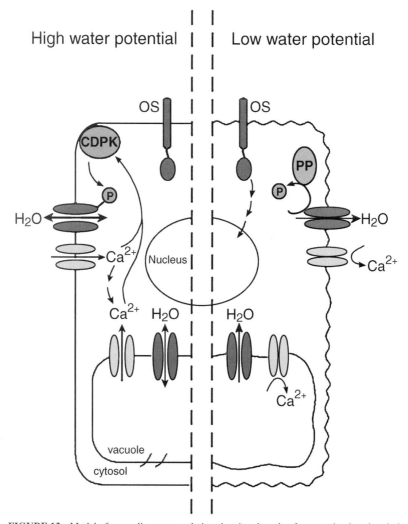

FIGURE 13 Model of cytosolic osmoregulation showing the role of aquaporin phosphorylation. When the plant cell senses a water potential decrease, the plasma membrane aquaporins become dephosphorylated, thereby lowering the water permeability of the membrane. The aquaporins of the vacuolar membrane remain open, due to phosphorylation at the consensus phosphorylation site in the first cytosolic loop, and allow water to flow into the cytosol to compensate for water lost to the apoplast. At high apoplastic water potential, a plasma membrane localized Ca^{2+}-dependent protein kinase (CDPK) is actively phosphorylating plasma membrane aquaporins, thereby increasing their water transport activity. The Ca^{2+} concentrations needed to activate the plasma membrane CDPK may arise from Ca^{2+} entering from the apoplast or from the vacuole, as indicated, but Ca^{2+} could also originate from the ER (not shown). It is not known whether the two-component osmosensor is only involved in the secondary response, the drought-stress-induced gene activation (as indicated), or if it also influences the primary stress response, the post-translational regulation of aquaporins. (Adapted from Johansson *et al.*, 1998, and Kjellbom *et al.*, 1999.)

cytokinins from roots is depressed during drought stress and may therefore act as a negative message to the shoot (Jackson, 1997). Interestingly, one of the histidine kinase homologs identified in plants is implicated in the perception of cytokinin signal transduction (Kakimoto, 1996).

Hormone-regulated translocation of intracellular vesicles to the plasma membrane has been demonstrated for vesicles containing the mammalian aquaporins AQP1 and AQP2 (Nielsen *et al.,* 1995). In plants, regulation of aquaporins by vesicle shuttling, similar to AQP1 and 2 translocation, has not yet been demonstrated. However, some PIPs and NLMs have been immunolocalized to intracellular membrane vesicles speculated to be transient structures by which translocation of aquaporins can be mediated (Barkla *et al.,* 1999; Robinson *et al.,* 1996b; U. Johanson *et al.,* unpublished, 2000).

Acknowledgments

Grants to P. K. from SJFR, NFR, the EU-Biotech program (B104-CT98-0024) and the Swedish Strategic Network for Plant Biotechnology are gratefully acknowledged.

References

Agre, P., Bonhivers, M., and Borgnia, M. J. (1998). The aquaporins, blueprints for cellular plumbing systems. *J. Biol. Chem.* **273,** 14659–14662.

Amodeo, G., Dorr, R., Vallejo, A., Sutka, M., and Parisi, M. (1999). Radial and axial water transport in the sugar beet storage root. *J. Exp. Bot.* **50,** 509–516.

Azaizeh, H., and Steudle, E. (1991). Effects of salinity on water transport of excised maize (*Zea mays* L.) roots. *Plant Physiol.* **97,** 1136–1145.

Azaizeh, H., Gunse, B., and Steudle, E. (1992). Effects of NaCl and CaCl$_2$ on water transport across root cells of maize (*Zea mays* L.) seedlings. *Plant Physiol.* **99,** 886–894.

Bai, L., Fushimi, K., Sasaki, S., and Marumo, F. (1996). Structure of aquaporin-2 vasopressin water channel. *J. Biol. Chem.* **271,** 5171–5176.

Barkla, B.J., Vera-Estrella, R., Pantoja, O., Kirch, H.-H., and Bohnert, H. J. (1999). Aquaporin localization—how valid are the TIP and PIP labels. *Trends Plant Sci.* **4,** 86–88.

Barone, L. M., Shih, C., and Wasserman, B. P. (1997). Mercury-induced conformational changes and identification of conserved surface loops in plasma membrane aquaporins from higher plants—Topology of PMIP31 from *Beta vulgaris* L. *J. Biol. Chem.,* **272,** 30672–30677.

Barone, L. M., Mu, H. H., Shih, C. J., Kashlan, K. B., and Wasserman, B. P. (1998). Distinct biochemical and topological properties of the 31-and 27-kilodalton plasma membrane intrinsic protein subgroups from red beet. *Plant Physiol.* **118,** 315–322.

Barrieu, F., Chaumont, F., and Chrispeels, M. J. (1998). High expression of the tonoplast aquaporin ZmTIP1 in epidermal and conducting tissues of maize. *Plant Physiol.* **117,** 1153–1163.

Barrieu, F., Marty-Mazars, D., Thomas, D., Chaumont, F., Charbonnier, M., and Marty, F. (1999). Desiccation and osmotic stress increase the abundance of mRNA of the tonoplast aquaporin BobTIP26-1 in cauliflower cells. *Planta* **209,** 77–86.

Biela, A., Grote, K., Otto, B., Hoth, S., Hedrich, R., and Kaldenhoff, R. (1999). The *Nicotiana tabacum* plasma membrane aquaporin NtAQP1 is mercury-insensitive and permeable for glycerol. *Plant J.* **18,** 565–570.

Birner, T. P., and Steudle, E. (1993). Effects of anaerobic conditions on water and solute relations, and on active transport in roots of maize (*Zea mays* L.). *Planta* **190,** 474–483.

Borgnia, M., Nielsen, S., Engel, A., and Agre, P. (1999). Cellular and molecular biology of the aquaporin water channels. *Annu. Rev. Biochem.,* **68,** 425–458.

Bouchez, D., and Höfte, H. (1998). Functional genomics in plants. *Plant Physiol.* **118,** 725–732.

Canny, M. J. (1995). A new theory for the ascent of sap. Cohesion supported by tissue pressure. *Ann. Botany,* **75,** 343–357.

Canny, M. J. (1997). Vessel contents during transpiration—embolisms and refilling. *Am. J. Botany.,* **85,** 1223–1230.

Carvajal, M., Cooke, D. T., and Clarkson, D. T. (1996). Responses of wheat plants to nutrient deprivation may involve the regulation of water-channel function. *Planta* **199,** 372–381.

Carvajal, M., Martinez, V., and Alcaraz, C. F. (1999). Physiological function of water channels as affected by salinity in roots of paprika pepper. *Physiol. Plant.* **105,** 95–101.

Chang, C., and Stewart, R. C. (1998). The two-component system. *Plant Physiol.* **117,** 723–731.

Cheng, A., van Hoek, A. N., Yeager, M., Verkman, A. S., and Mitra, A. K. (1997). Three-dimensional organization of a human water channel. *Nature,* **387,** 627–630.

Chaumont, F., Barrieu, F., Herman, E. M., and Chrispeels, M. J. (1998). Characterization of a maize tonoplast aquaporin expressed in zones of cell division and elongation. *Plant Physiol.* **117,** 1143–1152.

Chaumont, F., Barrieu, F., Jung, R., and Chrispeels, M. J. (2000). Plasma membrane intrinsic proteins from maize cluster in two sequence subgroups with differential aquaporin activity. *Plant Physiol.* **122,** 1025–1034.

Chevalier, J., Bourguet, J., and Hugon, J.S. (1974). Membrane-associated particles: Distribution in frog urinary bladder epithelium at rest and after oxytocin treatment. *Cell Tissue Res.* **152,** 129–140.

Clarkson, D. T., Carvajal, M., Henzler, T., Waterhouse, R. N., Smyth, A. J., Cooke, D. T., and Steudle, E. (2000). Root hydraulic conductance: Diurnal aquaporin expression and the effects of nutrient stress. *J. Exp. Bot.* **51,** 61–70.

Conkling, M. A., Cheng, C.-L., Yamamoto, Y. T., and Goodman, H. M. (1990). Isolation of transcriptionally regulated root-specific genes of tobacco. *Plant Physiol.* **93,** 1203–1211.

Cosgrove, D. J. (1997). Assembly and enlargement of the primary cell wall in plants. *Ann. Rev. Cell Develop. Biol.* **13,** 171–201.

Cosgrove, D. J., and Hedrich, R. (1991). Stretch-activated chloride, potassium, and calcium channels coexisting in plasma membranes of guard cells of *Vicia faba* L. *Planta* **186,** 143–153.

Cosgrove, D. J., and Steudle, E. (1981). Water relations of growing pea epicotyl segments. *Planta* **153,** 343–350.

Coury, L. A., Hiller, M., Mathai, J. C., Jones, E. W., Zeidel, M. L., and Brodsky, J. L. (1999). Water transport across yeast vacuolar and plasma membrane-targeted secretory vesicles occurs by passive diffusion. *J. Bacteriol.,* **181,** 4437–4440.

Dainty, J. (1963). Water relations of plant cells. *Adv. Botanical Res.* **1,** 279–326.

Daniels, M. J., Mirkov, T. E., and Chrispeels, M. J. (1994). The plasma membrane of *Arabidopsis thaliana* contains a mercury-insensitive aquaporin that is a homolog of the tonoplast water channel protein TIP. *Plant Physiol.* **106,** 1325–1333.

Daniels, M. J., Chaumont, F., Mirkov, T. E., and Chrispeels, M. J. (1996). Characterization of a new vacuolar membrane aquaporin sensitive to mercury at a unique site. *Plant Cell* **8,** 587–599.

Daniels, M. J., Chrispeels, M. J., and Yeager, M. (1999). Projection structure of a plant vacuole membrane aquaporin by electron cryo-crystallography. *J. Mol. Biol.* **294,** 1337–1349.

Dean, R. M., Rivers, R. L., Zeidel, M. L., and Roberts, D. M. (1999). Purification and functional reconstitution of soybean Nodulin 26. An aquaporin with water and glycerol transport properties. *Biochemistry* **38,** 347–353.

Finkelstein, A. (1987). Water Movement through Lipid Bilayers, Pores, and Plasma Membranes. Theory and Reality. Distinguished Lecture Series of the Society of General Physiologists. New York: John Wiley & Sons, Vol. 4, pp. 1–228.

Fischbarg, J., Kuang, K., Hirsch, J., Lecuona, S., Rogozinski, L., Silverstein, S. C., and Loike, J. (1989). Evidence that the glucose transporter serves as a water channel in J774 macrophages. *Proc. Natl. Acad. Sci. USA* **86,** 8397–8401.

Fleuratlessard, P., Frangne, N., Maeshima, M., Ratajczak, R., Bonnemain, J. L., and Martinoia, E. (1997). Increased expression of vacuolar aquaporin and H+-ATPase related to motor cell function in Mimosa pudica L. *Plant Physiol.,* **114,** 827–834.

Fortin, M. G., Morrison, N. A., and Verma, D. P. S. (1987). Nodulin-26, a peribacteroid membrane nodulin is expressed independently of the development of the peribacteroid compartment. *Nucleic Acids Res.* **15,** 813–824.

Fray, R. G., Wallace, A., Grierson, D., and Lycett, G. W. (1994). Nucleotide sequence and expression of a ripening and water stress-related cDNA from tomato with homology to the MIP class of membrane channel proteins. *Plant Mol. Biol.* **24,** 539–543.

Froger, A., Tallur, B., Thomas, D., and Delamarche, C. (1998). Prediction of functional residues in water channels and related proteins. *Prot. Sci.* **7,** 1458–1468.

Gerbeau, P., Güclü, J., Ripoche, P., and Maurel, C. (1999). Aquaporin Nt-TIPa can account for the high permeability of tobacco cell vacuolar membrane to small neutral solutes. *Plant J.* **18,** 577–587.

Guerrero, F. D., Jones, J. T., and Mullet, J. E. (1990). Turgor-responsive gene transcription and RNA levels increase rapidly when pea shoots are wilted. Sequence and expression of three inducible genes. *Plant Mol. Biol.* **15,** 11–26.

Hejnowicz, Z., and Sievers, A. (1996). Reversible closure of water channels in parenchymatic cells of sunflower hypocotyl depends on turgor status of the cells. *J. Plant Physiol.* **147,** 516–520.

Henzler, T., and Steudle, E. (1995). Reversible closing of water channels in *Chara* internodes provides evidence for a composite transport model of the plasma membrane. *J. Exp. Botany* **46,** 199–209.

Henzler, T., Waterhouse, R. N., Smyth, A. J., Carvajal, M., Cooke, D. T., Schäffner, A. R., Steudle, E., and Clarkson, D. T. (1999). Diurnal variations in hydraulic conductivity and root pressure can be correlated with the expression of putative aquaporins in the roots of *Lotus japonicus. Planta* **210,** 50–60.

Hertel, A., and Steudle, E. (1997). The function of water channels in *Chara*: The temperature dependence of water and solute flows provides evidence for composite membrane transport and for a slippage of small organic solutes across water channels. *Planta* **202,** 324–335.

Heymann, J. B., Agre, P., and Engel, A. (1998). Progress on the structure and function of aquaporin-1. *J. Struct. Biol.* **121,** 191–206.

Higuchi, T., Suga, S., Tsuchiya, T., Hisada, H., Morishima, S., Okada, Y., and Maeshima, M. (1998). Molecular cloning, water channel activity and tissue specific expression of two isoforms of radish vacuolar aquaporin. *Plant Cell Physiol.* **39,** 905–913.

Höfler, K. (1918). Permeabilitätsbestimmung nach der plasmometrischen Methode. *Ber. Dtsch. Bot. Ges.* **36,** 414–422.

Höfte, H., Faye, L., Dickinson, C., Herman, E. M., and Chrispeels, M. J. (1991). The protein-body proteins phytohemagglutinin and tonoplast intrinsic protein are targeted to vacuoles in leaves of transgenic tobacco. *Planta* **184,** 431–437.

Höfte, H., Hubbard, L., Reizer, J., Ludevid, D., Herman, E. M., and Chrispeels, M. J. (1992). Vegetative and seed-specific forms of tonoplast intrinsic protein in the vacuolar membrane of *Arabidopsis thaliana. Plant Physiol.* **99,** 561–570.

Holbrook, N. M., and Zwieniecki, M. A. (1999). Embolism repair and xylem tension: Do we need a miracle? *Plant Physiol.* **120,** 7–10.

Hsiao, T. C., Acevedo, E., Fereres, E., and Henderson, D. H. (1976). Water stress, growth and osmotic adjustment. *Phil. Trans. R. Soc. Lond.* B **273,** 479–500.

Inoue, K., Takeuchi, Y., Nishimura, M., and Hara-Nishimura, I. (1995). Characterization of two integral membrane proteins located in the protein bodies of pumpkin seeds. *Plant Mol. Biol.* **28,** 1089–1101.

Jackson, M. (1997). Hormones from roots as signals for the shoots of stressed plants. *Trends Plant Sci.* **2**, 22–28.

Jauh, G. Y., Fischer, A. M., Grimes, H. D., Ryan, C. A., and Rogers, J. C. (1998). Delta-tonoplast intrinsic protein defines unique plant vacuole functions. *Proc. Natl. Acad. Sci. USA* **95**, 12995–12999.

Jauh, G.-Y., Phillips, T. E., and Rogers, J. C. (1999). Tonoplast intrinsic protein isoforms as markers for vacuolar functions. *Plant Cell* **11**, 1867–1882.

Jiang, L., and Rogers, J. C. (1998). Integral membrane protein sorting to vacuoles in plant cells: Evidence for two pathways. *J. Cell. Biol.* **143**, 1183–1199.

Johansson, I., Larsson, C., Ek, B., and Kjellbom, P. (1996). The major integral proteins of spinach leaf plasma membranes are putative aquaporins and are phosphorylated in response to Ca^{2+} and apoplastic water potential. *Plant Cell* **8**, 1181–1191.

Johansson, I., Karlsson, M., Shukla, V. K., Chrispeels, M. J., Larsson, C., and Kjellbom, P. (1998). Water transport activity of the plasma membrane aquaporin PM28A is regulated by phosphorylation. *Plant Cell* **10**, 451–459.

Johansson, I., Karlsson, M., Johanson, U., Larsson, C., and Kjellbom, P. (2000). The role of aquaporins in cellular and whole plant water balance. *Biochim. Biophys. Acta.* **1465**, 324–342.

Johnson, K. D., and Chrispeels, M. J. (1992). Tonoplast-bound protein kinase phosphorylates tonoplast intrinsic protein. *Plant Physiol.* **100**, 1787–1795.

Johnson, K. D., Herman, E. M., and Chrispeels, M. J. (1989). An abundant, highly conserved tonoplast protein in seeds. *Plant Physiol.* **91**, 1006–1013.

Johnson, K. D., Höfte, H., and Chrispeels, M. J. (1990). An intrinsic tonoplast protein of protein storage vacuoles in seeds is structurally related to a bacterial solute transporter (GlpF). *Plant Cell* **2**, 525–532.

Kakimoto, T. (1996). CKI1, a histidine kinase homolog implicated in cytokinin signal transduction *Science* **274**, 982–985.

Kaldenhoff, R., Kölling, A., and Richter, G. (1993). A novel blue light- and abscisic acid-inducible gene of *Arabidopsis thaliana* encoding an intrinsic membrane protein. *Plant Mol. Biol.* **23**, 1187–1198.

Kaldenhoff, R., Kölling, A., Meyers, J., Karmann, U., Ruppel, G., and Richter, G. (1995). The blue light-responsive *AthH2* gene of *Arabidopsis thaliana* is primarily expressed in expanding as well as in differentiating cells and encodes a putative channel protein of the plasmalemma. *Plant J.* **7**, 87–95.

Kaldenhoff, R., Kölling, A., and Richter, G. (1996). Regulation of the *Arabidopsis thaliana* aquaporin gene AthH2 (PIP1b). *J. Photochem. Photobiol.* **36**, 351–354.

Kaldenhoff, R., Grote, K., Zhu, J.-J., and Zimmermann, U. (1998). Significance of plasmalemma aquaporins for water-transport in *Arabidopsis thaliana*. *Plant J.* **14**, 121–128.

Kammerloher, W., Fischer, U., Piechottka, G. P., and Schäffner, A. R. (1994). Water channels in the plant plasma membrane cloned by immunoselection from a mammalian expression system. *Plant J.* **6**, 187–199.

Karlsson, M., Johansson, I., Bush, M., McCann, M. C., Maurel, C., Larsson, C., and Kjellbom, P. (2000). An abundant TIP expressed in mature highly vacuolated cells. *Plant J.* **21**, 83–90.

Kjellbom, P., Larsson, C., Johansson, I., Karlsson, M., and Johanson, U. (1999). Aquaporins and water homeostasis in plants. *Trends Plant Sci.* **4**, 308–314.

Köckenberger, W., Pope, J. M., Xia, Y, Jeffrey, K. R., Komor, E., and Callaghan, P. T. (1997). A non-invasive measurement of phloem and xylem water flow in castor bean seedlings by nuclear magnetic resonance microimaging. *Planta* **201**, 53–63.

Koefoed-Johnsen, V., and Ussing, H. H. (1953). The contributions of diffusion and flow to the passage of D_2O through living membranes. Effect of neurohypophyseal hormone on isolated anuran skin. *Acta Physiol. Scand.* **28**, 60–76.

Koyama, Y., Yamamoto, T., Kondo, D., Funaki, H., Yaoita, E., Kawasaki, K., Sato, N., Hatakeyama, K., and Kihara, I. (1997). Molecular cloning of a new aquaporin from rat pancreas and liver. *J. Biol. Chem.* **272**, 30329–30333.

Krajinski, F., Biela, A., Schubert, D., Gianinazzi-Pearson, V., Kaldenhoff, R., and Franken, P. (2000). Arbuscular mycorrhiza development regulates the mRNA abundance of Mtaqp1 encoding a mercury-insensitive aquaporin of Medicago truncatula. *Planta* in press.

Kuwahara, M., Gu, Y., Ishibashi, K., Marumo, F., and Sasaki, S. (1997). Mercury-sensitive residues and pore site in AQP3 water channel. *Biochemistry* **36**, 13973–13978.

Lagrée, V., Froger, A., Deschamps, S., Pellerin, I., Delamarche, C., Bonnec, G., Gouranton, J., Thomas, D., and Hubert, J. F. (1998). Oligomerization state of water channels and glycerol facilitators. Involvement of loop E. *J. Biol. Chem.* **273**, 33949–33953.

Lagrée, V., Froger, A., Deschamps, S., Hubert, J. F., Delamarche, C., Bonnec, G., Thomas, D., Gouranton, J., and Pellerin, I. (1999). Switch from an aquaporin to a glycerol channel by two amino acids substitution. *J. Biol. Chem.* **274**, 6817–6819.

Lande, M. B., Jo, I., Zeidel, M. L., Somers, M., and Harris, Jr., H. W. (1996). Phosphorylation of aquaporin-2 does not alter the membrane water permeability of rat papillary water channel-containing vesicles. *J. Biol. Chem.* **271**, 5552–5557.

Liang, B. M., Sharp, R. E., and Baskin, T. I. (1997). Regulation or growth anisotrophyin well-watered and water-stressed maize roots. 1. Spatial distribution of longitudinal, radial, and tangential expansion rates. *Plant Physiol.* **115**, 101–111.

Lu, Z., and Neumann, P. M. (1999). Water stress inhibits hydraulic conductance and leaf growth in rice seedlings but not the transport of water via mercury-sensitive water channels in the root. *Plant. Physiol.* **120**, 143–151.

Ludevid, D., Höfte, H., Himelblau, E., and Chrispeels, M. J. (1992). The expression pattern of the tonoplast intrinsic protein γ-TIP in *Arabidopsis thaliana* is correlated with cell enlargement. *Plant Physiol.* **100**, 1633–1639.

Macey, R. I. (1984). Transport of water and urea in red blood cells. *Am. J. Physiol.* **246**, C195–C203.

Maggio, A., and Joly, R. J. (1995). Effects of mercuric chloride on the hydraulic conductivity of tomato root systems. Evidence for a channel-mediated water pathway. *Plant Physiol.* **109**, 331–335.

Maurel, C. (1997). Aquaporins and water permeability of plant membranes. *Ann. Rev. Plant Physiol. Plant Mol. Biol.* **48**, 399–429.

Maurel, C., Chrispeels, M., Lurin, C., Tacnet, F., Geelen, D., Ripoche, P., and Guern, J. (1997a). Function and regulation of seed aquaporins. *J. Exp. Bot.* **48**, 421–430.

Maurel, C., Reizer, J., Schroeder, J. I., and Chrispeels, M. J. (1993). The vacuolar membrane protein γ-TIP creates water specific channels in *Xenopus* oocytes. *EMBO J.* **12**, 2241–2247.

Maurel, C., Reizer, J., Schroeder, J. I., Chrispeels, M. J., and Saier, M. H. J. (1994). Functional characterization of the *Escherichia coli* glycerol facilitator, GlpF, in *Xenopus* oocytes. *J. Biol. Chem.* **269**, 11869–11872.

Maurel, C., Kado, R. T., Guern, J., and Chrispeels, M. J. (1995). Phosphorylation regulates the water channel activity of the seed-specific aquaporin α-TIP. *EMBO J.* **14**, 3028–3035.

Maurel, C., Tacnet, F., Güclü, J., Guern, J., and Ripoche, P. (1997b). Purified vesicles of tobacco cell vacuolar and plasma membranes exhibit dramatically different water permeability and water channel activity. *Proc. Nat. Acad. Sci. USA* **94**, 7103–7108.

Miao, G.-H., Hong, Z., and Verma, D. P. S. (1992). Topology and phosphorylation of soybean nodulin-26, an intrinsic protein of the peribacteroid membrane. *J. Cell Biol.* **118**, 481–490.

Morillon, R., and Lassalles, J.-P. (1999). Osmotic water permeability of isolated vacuoles. *Planta* **210**, 80–84.

Nonami, H., and Boyer, J. S. (1993). Direct demonstration of a growth induced water potential gradient. *Plant Physiol.* **102**, 13–19.

Nielsen, S., Chou, C.-H., Marples, D., Christensen, E. I., Kishore, B. K., and Knepper, M. A. (1995). Vasopressin increases water permeability of kidney collecting duct by inducing translocation of aquaporin-CD water channels to plasma membrane. *Proc. Natl. Acad. Sci. USA* **92,** 1013–1017.

Niemietz, C. M., and Tyerman, S. D. (1997). Characterization of water channels in wheat root membrane vesicles. *Plant Physiol.* **115,** 561–567.

Olah, Z., Bogre, L., Lehel, C., Farago, A., Seprodi, J., and Dudits, D. (1989). The phosphorylation site of Ca^{2+}-dependent protein kinase from alfalfa. *Plant. Mol. Biol.* **12,** 453–461.

Oparka, K. J., Duckett, C. M., Prior, D. A. M., and Fisher, D. B. (1994). Real-time imaging of phloem unloading in the root tip of *Arabidopsis. Plant J.* **6,** 759–766.

Opperman, C. H., Taylor, C. G., and Conkling, M. A. (1994). Root-knot nematode-directed expression of a plant root-specific gene. *Science* **263,** 221–223.

Paris, N., Stanley, C. M., Jones, R. L., and Rogers, J. C. (1996). Plant cells contain two functionally distinct vacuolar compartments. *Cell* **85,** 563–572.

Park, J. H., and Saier, M. H. (1996). Phylogenetic characterization of the MIP family of transmembrane channel proteins. *J. Membr. Biol.* **153,** 171–180.

Phillips, A. L., and Huttly, A. K. (1994). Cloning of two gibberellin-regulated cDNAs from *Arabidopsis thaliana* by subtractive hybridization: Expression of the tonoplast water channel, γ-TIP, is increased by GA₃. *Plant Mol. Biol.* **24,** 603–615.

Pinna, L. A., and Ruzzene, M. (1996). How do protein kinases recognize their substrates. *Biochim. Biophys. Acta* **1314,** 191–225.

Polya, G. M., Morrice, N., and Wettenhall, R.E.H. (1989). Substrate specificity of wheat embryo calcium-dependent protein kinase. *FEBS Lett.* **253,** 137–140.

Preston, G. M., Carroll, T. P., Guggino, W. B., and Agre, P. (1992). Appearance of water channels in *Xenopus* oocytes expressing red cell CHIP28 protein. *Science* **256,** 385–387.

Preston, G. M., Jung, J. S., Guggino, W. B., and Agre, P. (1993). The mercury-sensitive residue at cysteine 189 in the CHIP28 water channel. *J. Biol. Chem.* **268,** 17–20.

Quintero, J. M., Fournier, J. M., and Benlloch, M. (1999). Water transport in sunflower root systems: Effects of ABA, Ca^{2+} status and $HgCl_2$. *J. Exp. Bot.* **50,** 1607–1612.

Ramahaleo, T., Morillon, R., Alexandre, J., and Lassalles, J.-P. (1999). Osmotic water permeability of isolated protoplasts. Modifications during development. *Plant Physiol.* **119,** 885–896.

Rivers, R. L., Dean, R. M., Chandy, G., Hall, J. E., Roberts, D. M., and Zeidel, M. L. (1997). Functional analysis of Nodulin 26, an aquaporin in soybean root symbiosomes. *J. Biol. Chem.* **272,** 16256–16261.

Roberts, D. M. (1993). Protein kinases with calmodulin-like domains: novel targets of calcium signals in plants. *Curr. Opin. Cell Biol.* **5,** 242–246.

Roberts, D. M., and Harmon, A. C. (1992). Calcium-modulated proteins: Targets of intracellular calcium signals in higher plants. *Annu. Rev. Plant Physiol. Plant Mol. Biol.* **43,** 375–414.

Robinson, D. G., Haschke, H. P., Hinz, G., Hoh, B., Maeshima, M., and Marty, F. (1996a). Immunological detection of tonoplast polypeptides in the plasma membrane of pea cotyledons. *Planta* **198,** 95–103.

Robinson, D. G., Sieber, H., Kammerloher, W., and Schäffner, A. R. (1996b). PIP1 aquaporins are concentrated in plasmalemmasomes of *Arabidopsis thaliana* mesophyll. *Plant Physiol.* **111,** 645–649.

Roussel, H., Bruns, S., Gianinazzi-Pearson, V., Hahlbrock, K., and Franken, P. (1997). Induction of a membrane intrinsic protein-encoding mRNA in arbuscular mycorrhiza and elicitor stimulated cell suspension cultures of parsley. *Plant Sci.* **126,** 203–210.

Sarda, X., Tousch, D., Ferrare, K., Legrand, E., Dupuis, J. M., Casse-Delbart, F., and Lamaze, T. (1997). Two TIP-like genes encoding aquaporins are expressed in sunflower guard cells. *Plant J.* **12,** 1103–1111.

Sasaki, S., Ishibashi, K., and Marumo, F. (1998). Aquaporin-2 and -3: Representatives of two subgroups of the aquaporin family colocalized in the kidney collecting duct. *Annu. Rev. Physiol.* **60,** 199–220.

Satterlee, J. S., and Sussman, M. R. (1998). Unusual membrane-associated protein kinases in higher plants. *J. Membr. Biol.* **164,** 205–213.

Schäffner, A. R. (1998). Aquaporin function, structure, and expression: Are there more surprises to surface in water relations? *Planta* **204,** 131–139.

Schütz, K., and Tyerman, S. D. (1997). Water channels in *Chara corallina. J. Exp. Bot.* **48,** 1511–1518.

Shi, L.-B., and Verkman, A. S. (1996). Selected cysteine point mutations confer mercurial sensitivity to the mercurial-insensitive water channel MIWC/AQP-4. *Biochemistry* **35,** 538–544.

Shinozaki, K., and Yamaguchi-Shinozaki, K. (1997). Gene expression and signal transduction in water-stress response. *Plant Physiol.* **115,** 327–334.

Steudle, E. (1989). Water flow in plants and its coupling to other processes: An overview. *Methods Enzymol.* **174,** 183–225.

Steudle, E. (1993). Pressure probe techniques: basic principles and application to studies of water and solute relations at the cell, tissue and organ level. *In* "Water Deficits: Plant Responses from Cell to Community" (J.A.C. Smith and H. Griffiths, eds.), pp. 5–36. Bios Scientific Publishers Ltd, Oxford, UK.

Steudle, E., and Henzler, T. (1995). Water channels in plants: Do basic concepts of water transport change? *J. Exp. Botany,* **46,** 1067–1076.

Steudle, E., and Peterson, C. A. (1998). How does water get through roots? *J. Exp. Bot.* **49,** 775–788.

Taiz, L., and Zeiger, E. (1998). "Plant Physiology," 2nd ed. Chapter 4. Sinauer Associates.

Tazawa, M., Ohkuma, E., Shibasaka, M., and Nakashima, S. (1997). Mercurial-sensitive water transport in barley roots. *J. Plant Res.* **110,** 435–442.

Tomos, A. D., and Leigh, R. A. (1999). The presure probe: A versatile tool in plant cell physiology. *Annu. Rev. Plant Physiol. Plant Mol. Biol.* **50,** 447–472.

Trejo, C. L., and Davies, W. J. (1991). Drought-induced closure of *Phaseolus vulgaris* L. stomata precedes leaf water deficit and any increase in xylem ABA concentration. *J. Exp. Biol.* **42,** 1507–1515.

Tyerman, S. D., Oats, P., Gibbs, J., Dracup, M., and Greenway, H. (1989). Turgor-volume regulation and cellular water relations of *Nicotiana tabacum* roots grown in high salinities. *Austral. J. Plant Physiol.* **16,** 517–531.

Tyerman, S. D., Bohnert, H. J., Maurel, C., Steudle, E., and Smith, J. A. (1999). Plant aquaporins: Their molecular biology, biophysics and significance for plant water relations. *J. Exp. Bot.* **50,** 1055–1071.

Tyree, M. T., Salleo, S., Nardini, A., Lo Gullo, M. A., and Mosca, R. (1999). Refilling of embolized vessels in young stems of laurel. Do we need a new paradigm. *Plant Physiol.* **120,** 11–21.

Urao, T., Yakubov, B., Satoh, R., Yamaguchi-Shinozaki, K., Seki, M., Hirayama, T., and Shinozaki, K. (1999). A transmembrane hybrid-type histidine kinase in Arabidopsis functions as an osmosensor. *Plant Cell* **11,** 1743–1754.

van Os, C. H., Deen, P. M. T., and Dempster, J. A. (1994). Aquaporins: Water selective channels in biological membranes. Molecular structure and tissue distribution. *Biochimica Biophysica Acta* **1197,** 291–309.

Verkman, A. S. (1995). Optical methods to measure membrane transport processes. *J. Membr. Biol.* **148,** 99–110.

Wayne, R., and Tazawa, M. (1988). The actin cytoskeleton and polar water permeability in characean cells. *Protoplasma. Suppl.* **2,** 116–130.

Wayne, R., and Tazawa, M. (1990). Nature of the water channels in the internodal cells of *Nitellopsis. J. Membr. Biol.* **116,** 31–39.

Weaver, C. D., Crombie, B., Stacey, G., and Roberts, D. M. (1991). Calcium-dependent phosphorylation of symbiosome membrane proteins from nitrogen-fixing soybean nodules. *Plant Physiol.* **95,** 222–227.

Weaver, C. D., and Roberts, D. M. (1992). Determination of the site of phosphorylation of nodulin 26 by the calcium-dependent protein kinase from soybean nodules. *Biochemistry* **31,** 8954–8959.

Weaver, C. D., Shomer, N. H., Louis, C. F., and Roberts, D. M. (1994). Nodulin 26, a nodule-specific symbiosome membrane protein from soybean, is an ion channel. *J. Biol. Chem.* **269,** 17858–17862.

Wei, C. F., Tyree, M. T., and Steudle, E. (1999). Direct measurement of xylem pressure in leaves of intact maize plants. A test of the cohesion-tension theory taking hydraulic architecture into consideration. *Plant Physiol.* **121,** 1191–1205.

Weig, A., Deswarte, C., and Chrispeels, M. J. (1997). The major intrinsic protein family of Arabidopsis has 23 members that form three distinct groups with functional aquaporins in each group. *Plant Physiol.* **114,** 1347–1357.

Yamada, S., Katsuhara, M., Kelly, W. B., Michalowski, C. B., and Bohnert, H. J. (1995). A family of transcripts encoding water channel proteins: Tissue-specific expression in the common ice plant. *Plant Cell* **7,** 1129–1142.

Yamaguchi-Shinozaki, K., Koizumi, M., Urao, S., and Shinozaki, K. (1992). Molecular cloning and characterization of 9 cDNAs for genes that are responsive to desiccation in *Arabidopsis thaliana*: Sequence analysis of one cDNA clone that encodes a putative transmembrane channel protein. *Plant Cell Physiol.* **33,** 217–224.

Yamamoto, Y. T., Taylor, C. G., Acedo, G. N., Cheng, C.-L., and Conkling, M. A. (1991). Characterization of *cis*-acting sequences regulating root-specific gene expression in tobacco, *Plant Cell* **3,** 371–382.

Yang, B., Brown, D., and Verkman, A. S. (1996). The mercurial insensitive water channel (AQP-4) forms orthogonal arrays in stably transfected chinese hamster ovary cells. *J. Biol. Chem.* **271,** 4577–4580.

Yang, B. X., and Verkman, A. S. (1997). Water and glycerol permeabilities of aquaporins 1-5 and MIP determined quantitatively by expression of epitope-tagged constructs in *Xenopus* oocytes. *J. Biol. Chem.* **272,** 16140–16146.

Yasui, M., Hazama, A., Kwon, T.-H., Nielsen, S., Guggino, W. B., and Agre, P. (1999). Rapid gating and anion permeability of an intracellular aquaporin. *Nature* **402,** 184–187.

Zhang, R., and Verkman, A. S. (1991). Water and urea permeability properties of *Xenopus* oocytes: expression of mRNA from toad urinary bladder. *Am. J. Physiol.* **260,** C26–C34.

Zhang, R., van Hoek, A. N., Biwersi, J., and Verkman, A. S. (1993). A point mutation at cysteine-189 blocks the water permeability of rat kidney water channel CHIP28k. *Biochemistry* **32,** 2938–2941.

Zhang, W.-H., and Tyerman, S. D. (1999). Inhibition of water channels by $HgCl_2$ in intact wheat root cells. *Plant Physiol.* **120,** 849–857.

Zimmermann, H. M., and Steudle, E. (1998). Apoplastic transport across young maize roots: Effect of the exodermis. *Planta* **206,** 7–19.

CHAPTER 8

Microbial Water Channels and Glycerol Facilitators

Gerald Kayingo,* Roslyn M. Bill,[†] Guiseppe Calamita,[+] Stefan Hohmann,[†] and Bernard A. Prior*

*Department of Microbiology, University of Stellenbosch, Matieland 7602, South Africa; [†]Department of Cell and Molecular Biology/Microbiology, Göteborg University, SE-40530 Göteborg, Sweden; [+]Dipartimento di Fisiologia Generale e Ambientale, Università degli Studi di Bari, 70126 Bari, Italy

I. INTRODUCTION

In contrast to the majority of cells from multicellular organisms, microbial cells are in direct contact with a highly variable environment. Hence, bacteria, fungi, algae, and protozoa must be able to respond to a wealth of widely varying conditions. For instance, fungi such as yeasts tolerate pH values from about 3–8, and many bacteria are productive over a range of more than 30°C. In particular, microorganisms can live and proliferate at variable water activities and under different nutritional conditions. This inevitably requires the ability to adjust transport processes for the uptake and/or efflux of water, osmolytes, nutrients, and metabolic end products.

Transmembrane transport in unicellular microorganisms is mediated by different systems that are classified according to their mode of function into channels, pores, facilitators, carriers, porters, or pumps (André, 1995; Nikaido and Saier, 1992; Paulsen *et al.*, 1998a, 1998b; Saier, 1994, 1998; Saier *et al.*, 1999). Pores and channels allow free passage of solutes across the membrane whereas carriers and porters possess specific binding sites via which the solute traverses the membrane. Whereas proteins that catalyze facilitated diffusion (facilitators) do not involve energy coupling and therefore cannot operate against a substrate concentration, pumps couple metabolic energy during active transport. Most transport proteins studied so far fall into relatively few families, which are characterized by conserved motifs and/or similar topology. For instance, the major facilitator superfamily (MFS) constitutes a huge number of proteins for the uptake of sugars, amino acids, ions, and other compounds (Pao *et al.*, 1998). Facilitated transport by these proteins can be coupled as symport or antiport to a proton gradient, thereby allowing transport against a substrate concentration gradient. Another major class of transport proteins is the ATP binding cassette (ABC) transporters, which use the energy derived from ATP hydrolysis for active transport of many different substrates into or out of the cell (van Veen and Konings, 1997, 1998). All microbial genomes sequenced so far contain genes encoding many MFS and ABC transporters: the genome of the gram-negative bacterium *Escherichia coli* encodes some 70 MFS transporters and 80 ABC transporters (Blattner *et al.*, 1997), that of the gram-positive bacterium *Bacillus subtilis* 81 MSF transporters and 77 ABC transporters (Kunst *et al.*, 1997), and that of the yeast *Saccharomyces cerevisiae* 78 MFS transporters and 22 ABC transporters (Paulsen *et al.*, 1998b).

Small molecules such as water and glycerol can passively cross the plasma membrane. However, it appears that different membranes exhibit very different permeability for water and glycerol accounting for the different rates of passive diffusion observed in organisms. In fact, it has been known for many years that the permeability of specialized biological membranes for water is much higher than that of artificial lipid bilayers. This observation implied the possible involvement of channels that facilitate water flux across cell membranes (Finkelstein, 1987; Koefoed-Johnson and Ussing, 1953; Macey, 1984; Macey and Farmer, 1970; Paganelli and Solomon, 1957; Wayne and Tazawa, 1990). However, the molecular justification of this view remained elusive until the discovery of the aquaporin family of transmembrane water channels, first in mammals, subsequently in plants, and finally in microorganisms (Calamita *et al.,* 1995; Maurel *et al.,* 1993; Preston *et al.,* 1992). Similarly, the occurrence of glycerol facilitators in bacteria was proposed nearly 30 years ago (Heller *et al.,* 1980; Richey and Lin, 1972; Sanno *et al.,* 1968) and was confirmed by the cloning of *glpF,* a gene encoding the *E. coli* glycerol facilitator (Sweet *et al.,* 1990). Subsequent comparative sequence analyses revealed significant homology between glycerol facilitators and water channels (Baker and Saier, 1990). They were found to be related to the bovine lens major intrinsic protein (Gorin *et al.,* 1984) from which the family name MIP is derived.

To date, more than 200 MIP family members have been identified and their role in solute and water transport has been established both *in vitro* and *in vivo,* as described in detail in other chapters of this volume. As for the microbial MFS and ABC transporters mentioned earlier, higher organisms possess an amazing number of MIP channel isoforms expressed in different subcellular compartments and tissues, under different environmental conditions or during different developmental stages. For example, more than 30 genes encoding MIP channels have been reported in the model plant *Arabidopsis thaliana* (Kjellbom *et al.,* 1999; see also Chapter 7 by Chrispeels *et al.),* and 10 have been described in humans (Borgnia *et al.,* 1999; see also chapters by Agre, Verkman *et al.,* and Knepper *et al.).* Furthermore, nine can be recognized in the nematode *Caenorhabditis elegans* genome database (www.sanger.ac.uk/Projects/C elegans/Genomic Sequence.shtml).

Most functional studies suggest that MIP channels mediate water flux across cell membranes. In plants and animals, MIP channels appear to play a role in osmoregulation at the cellular and/or organismal level. For example, plant aquaporins are involved in stress responses and in developmental processes, and their mammalian homologs display a wide variety of roles in physiology and are consequently involved in several clinical disorders. All of these aspects are described in detail elsewhere in this volume and have been the subject of recent reviews (Borgnia *et al.,* 1999; Kjellbom *et al.,* 1999).

Like other major transport protein families, MIP channels are also widespread among microbes but the number of genes encoding MIP channels per microbial genome is not more than four. However, many new MIP channel genes continue to be identified during the sequencing of microbial genomes (see www.tigr.org/tdb/mdb/mdb.html). Table I lists the 76 MIP channels that we have located in various databases by the end of March 2000 [the table does not include the sequence of the *Thermus flavus* glpF (Darbon *et al.,* 1999), which does not appear in the databases]. These proteins are found in 52 different species belonging to Archea, Bacteria, Fungi, and Protozoa. Microorganisms constitute the biggest resource of MIP channel sequences in terms of species number. This is especially interesting for structure–function analysis because many different sequences with similar or identical function are available for comparison (Heymann and Engel, 2000).

TABLE I

Microbial MIP Family Channel Proteins

Organism	Classification and habitat	Genome sequence	Sequence source	Gene or clone or cosmid	Accession number	Phylogenic subfamily/ predicted function	Gene context (operon)	Protein size (aa)
Eukaryota								
Aspergillus nidulans	Fungus, ascomycete; saprophyte	Ongoing	GenBank/EBI	C5f02a1.r1[a]	AA783486	ND	None	118[b]
Botrytis cinerea	Fungus, ascomycete; plant pathogen	Partial	www.genoscope.cns.fr	CNS01A8X	AL112633	2/GlpF	None	239[b]
Candida albicans	Fungus, yeast; human pathogen	Ongoing	sequence-www.stanford.edu	stanford 5476	Contig4-2389	1/AQP	None	273
Dictyostelium discoideum	Protozoan, saprophyte; soil	Ongoing	www.sanger.ac.uk	wacA	U68246	1/AQP	None	277
			www.sanger.ac.uk	aqpA	AB032841	1/AQP	None	279
Neurospora crassa	Fungus, ascomycete; saprophyte	Ongoing	GenBank/EBI	NCSM1G3T3[a]	AI392589	2/AQP	None	195[b]
Saccharomyces cerevisiae	Fungus, yeast; fruits and flowers; model and industrial organism	Complete	www.proteome.com	FPS1[c]	P23900	ND/GlpF[e]	None	669
			www.proteome.com	YFL054	P43549	2/GlpF	None	646
			Laizé et al., 2000	AQY1-1[d]	AAC69713	1/AQP[e]	None	327
			Laizé et al., 2000	AQY1-2[d]	P53386	1/AQP	None	305
			Laizé et al., 2000	AQY2[d]	AAD25168	1/AQP	None	289
Schizosaccharomyces pombe	Fungus, yeast; fruits and flowers; model organism	Ongoing	www.sanger.ac.uk	SPAC977	CAB69639	2/GlpF	None	598
Trypanosoma brucei	Protozoan; human pathogen; sleeping sickness	Ongoing	www.tigr.org	RPCI93	AQ641778	2/GlpF	None	160[b]
Gram-Positive Bacteria								
Bacillus anthracis	Pathogen; anthrax	Ongoing	www.tigr.org	aapZ	gba 92	1/AQP	ND	221
			www.tigr.org		gba 1391	3/AGP	ND	273

Organism	Description	Status	Source	Gene	Accession	3/GlpF[e] (AGP)	Operon	
Bacillus subtilis	Saprophyte; model and industrial organism	Complete	www.pasteur.fr	*glpF*	P18156	3/GlpF (AGP)	Operon	274
Caulobacter crescentus	Freshwater; model organism; bacterial differentiation	Ongoing	www.tigr.org www.tigr.org	*aqp*	gcc 515 gcc 439	1/AQP 1/ND	None ND	81[b] 100[b]
Clostridium acetobutylicum	Anaerobe, industrial organism	Complete	www.genomecorp.com		AE001437	3/AGP	None	242
Clostridium perfringens	Pathogen; protein-rich foods; soil		GenBank/EBI	*glpF*	X86492	3/AGP	ND	148[b]
Corynebacterium diphtheriae	Pathogen; diphtheria	Ongoing	www.sanger.ac.uk		Contig423	3/AGP	Operon	227[b]
Deinococcus radiodurans	Natural habitat unknown; resistant to radiation	Complete	www.tigr.org	*glpF*	8796	2/GlpF	ND	271
Enterococcus faecalis	Small intestine; urinary tract; pathogen; endocarditis	Ongoing	www.tigr.org www.tigr.org www.tigr.org	*glpF* *aqpZ*	gef 6204 gef 6176 gef 6403	3/AGP ND/GlpF 1/AQP	glpF/glpO glpF/PTS None	239 236 233
Lactococcus lactis	Saprophyte, anaerobic, industrial organism	Complete	GenBank/EBI	*ydp1*	P22094	3/AGP[e]	None	289
Staphylococcus aureus	Pathogen (food poisoning) skin; meat and dairy products	Ongoing	www.tigr.org		4410	3/AGP	Operon	272
Streptococcus pneumoniae	Pathogen of the respiratory tract; pneumonia	Ongoing	www.tigr.org www.tigr.org www.tigr.org	*glpF* *aqpZ*	U12567 sp 42 sp 16	3/GlpF 3/AGP 1/AQP	Operon None None	233 289 269[b]
Streptococcus pyogenes	Pathogen of the respiratory tract; scarlet fever	Ongoing	www.genome.ou.edu www.genome.ou.edu		Contig115 Contig104	3/AGP 3/AGP	Operon ND	233 282
Streptomyces coelicolor	Soil and aquatic; producer of antibiotics	Ongoing	GenBank/EBI	*gylA*	P19255	3/AGP	Operon	80[b]
Thermotoga maritima	Extreme thermophile, marine, hydrothermal vents	Complete	GenBank/EBI	*glpF*	AAD36499	3/AGP	Operon	234

(continues)

TABLE I

(continued)

Organism	Classification and habitat	Genome sequence	Sequence source	Gene or clone or cosmid	Accession number	Phylogenic subfamily/ predicted function	Gene context (operon)	Protein size (aa)
Gram-Negative Bacteria								
Borrelia burgdorferi	Anaerobe; pathogen; lyme diseases	Complete	GenBank/EBI	*glpF*	AAC66629	2/GlpF	Operon	254
Bordetella bronchiseptica	Pathogen; respiratory disease	Ongoing	www.sanger.ac.uk		Contig 2552	1/AQP	ND	236
Brucella melitensis	Goat pathogen and parasite; milk, meat, soil	Ongoing	GenBank/EBI	*aqpZ*	AAF36396	1/AQP	ND	228
Chlorobium tepidum	Anaerobe; thermophilic green sulfur bacterium; phototroph	Ongoing	www.tigr.org	*aqp*	gct 5	1/AQP	ND	268
Escherichia coli	Facultative anaerobe; mammalian colon; model organism	Complete	GenBank/EBI GenBank/EBI	*glpF* *aqpZ*	P11244 U38664	2/GlpF[e] 1/AQP[e]	Operon None	281 231
Haemophilus influenzae	Pathogen; upper respiratory tract	Complete	GenBank/EBI GenBank/EBI	*glpF* *glpF*	P44826 U32782	2/GlpF 3/AGP	Operon None	264 213[b]
Klebsiella pneumoniae	Pathogen; respiratory tract	Ongoing	genome.wustl.edu genome.wustl.edu genome.wustl.edu genome.wustl.edu	*aqpZ* *aqp* *glpF* *glpF*	Contig1030 Contig1071 Contig848 Contig757	1/AQP 1/AQP 2/GlpF 2/GlpF	ND ND Operon Operon	155[b] 180[b] 267 293
Mycoplasma capricolum	Goat pathogen; contagious caprine pleuropneumonia		GenBank/EBI	*glpF*	Z33098	3/GlpF	Operon	89[b]
Mycoplasma gallisepticum	Fowl pathogen; anaerobe		GenBank/EBI	*glpF*	P52280	3/GlpF	ND	205[b]
Mycoplasma genitalium	Pathogen; urinary tract	Complete	GenBank/EBI	*glpF*	P47279	ND/GlpF	*GlpF/ThyK*	258
Mycoplasma pneumoniae	Pathogen; mucous membrane	Complete	GenBank/EBI	*glpF*	P75071	ND/GlpF	*GlpF/ThyK*	264

Organism	Description	Status	Database	Gene	Accession	Type	Organization	No.
Pasteurella multocida	Fowl pathogen; pasteurellosis	Ongoing	www.cbc.umn.edu	*glpF*	Contig82	2/GlpF	Operon	261
Plesiomonas shigelloides	Pathogen; food and water borne diarrhoea		GenBank/EBI	*ORF10P*	AB025970	1/AQP	Cluster	233
Pseudomonas aeruginosa	Pathogen of the gastrointestinal tract; soil	Ongoing	www.genome.washington.edu GenBank/EBI	*aqpZ* *glpF*	Contig54 Q51389	1/AQP 2/GlpF[e]	None Operon	308 279
Pseudomonas putida	Soil	Ongoing	www.tigr.org www.tigr.org	*glpF* *glpF*	All 2259 All 2406	2/GlpF 2/GlpF	ND ND	147[b] 162[b]
Pseudomonas tolaasii	Soil		GenBank/EBI	*glpF*	AB015973	2/GlpF	Operon	285
Salmonella enterica serovar typhimurium	Pathogen; gut		GenBank/EBI	*pduF*	AF026270	2/PduF	Operon	264
Salmonella typhi	Pathogen of the gut; typhoid fever; aquatic	Ongoing	www.sanger.ac.uk www.sanger.ac.uk www.sanger.ac.uk	*pduF* *glpF* *pduF*	Contig443 Contig460 Contig417	2/PduF 2/GlpF 2/PduF	ND ND ND	264 279 269
Salmonella typhimurium	Pathogen of the gut; typhus; aquatic	Ongoing	genome.wustl.edu genome.wustl.edu	*pduF* *glpF*	P37451 Contig79	2/PduF[e] 2/GlpF	Operon ND	264 288
Shewanella putrefacines	Soil and aquatic; food spoilage	Ongoing	www.tigr.org	*aqpZ*	4323	1/AQP	None	231
Shigella flexneri	Pathogen of the gut; dysentery		GenBank/EBI GenBank/EBI	*glpF* *aqpZ*	P31140 AAC12651	2/GlpF[e] 1/AQP	Operon None	281 231
Shigella sonnei	Pathogen of the gut; enteritis	Ongoing	GenBank/EBI	*ORF10S*	BAA85070	1/AQP	Cluster (plasmid)	25[b]
Snorhizobium meliloti	Root nodules; nitrogen fixation	Ongoing	cmgm.stanford.edu	Stanford 382	423050H01	ND	ND	204[b]
Synechococcus sp. PCC7942	Phototroph; aquatic		GenBank/EBI	*smpX*	D43774	ND	*smpX/pacS*	269
Synechocystis sp. PCC6803	Phototroph; aquatic	Complete	GenBank/EBI	*aqpZ*	BAA17863	1/AQP	None	247

(continues)

TABLE I

(continued)

Organism	Classification and habitat	Genome sequence	Sequence source	Gene or clone or cosmid	Accession number	Phylogenic subfamily/ predicted function	Gene context (operon)	Protein size (aa)
Thiobacillus ferrooxidans	Chemolithotroph; soil; aquatic	Ongoing	www.tigr.org	TIGR 920	3498	ND	ND	215[b]
Vibrio cholerae	Pathogen of the gut; cholera; aquatic; soil	Ongoing	www.tigr.org www.tigr.org	glpF glpF	asm814 1741	ND/GlpF 2/GlpF	Operon	261 285
Yersinia pestis	Pathogen; plague	Ongoing	www.sanger.ac.uk	glpF	Contig630	2/GlpF	Operon	282
Archaea								
Archaeoglobus fulgidus	Anaerobe; thermophilic	Complete	GenBank/EBI	glpF (aqp?)	AAB89820	1/AQP	None	246
Methanobacterium thermoautotrophicum	Anaerobe, extreme environments	Complete	GenBank/EBI	aqp	AAB84602	1/AQP	None	246

Key: ND, not determined; GlpF, glycerol facilitator/transporter; PduF, propanediol facilitator; AQP, aquaporin; AGP, aquaglyceroporin; aa, number of amino acids.
[a]Deduced from mRNA sequence.
[b]Incomplete sequence obtained from GenBank.
[c]One or both NPA motifs have a different sequence.
[d]Polymorphic form
[e]Function confirmed experimentally.

Relatively little attention, however, has yet been given to the physiological role of microbial MIP channels. In fact, functional and physiological studies are largely restricted to MIP channels from the bacterium *E. coli* and the yeast *S. cerevisiae*. This chapter attempts to summarize the available information on microbial aquaporins and glycerol facilitators with emphasis on their evolutionary relationships, molecular properties, patterns of gene expression, and physiological roles. It is anticipated that studies on microbial aquaporins will provide novel insights into the structure and function of MIP channels as well as their physiological roles. Microbial MIP channels thus provide a suitable model for understanding the role of these proteins in cellular water relations because the underlying concepts of cellular osmoregulation are conserved from bacteria to humans (Blomberg and Adler, 1992; Wiggins, 1990; Wood, 1999; Yancey *et al.*, 1982).

II. MICROBIAL AQUAPORINS AND GLYCEROL FACILITATORS

A. Classification of Microbial MIP Channels into Subfamilies

MIP channels have historically been divided into two major subgroups: (1) the aquaporins *sensu strictu*, which are specifically permeable only to water, and (2) the glycerol facilitators, which are permeable to water, glycerol, and to varying degrees, other small solutes (Agre *et al.*, 1998; Froger *et al.*, 1998; Park and Saier, 1996). In addition to substrate specificity, phylogenetic and sequence analyses (Froger *et al.*, 1998; Heymann and Engel, 2000) have revealed that certain conserved residues are distinct between putative aquaporins and glycerol facilitators. These signature residues can be used for classification (Froger *et al.*, 1998; Heymann and Engel, 2000) even when functional studies have not been conducted.

Whereas most MIP channels from plants and animals are classified as water channels, glycerol facilitators account for the majority of MIPs in microorganisms. However, functional studies have only been performed on the glycerol facilitators, GlpF, from *E. coli* (Heller *et al.*, 1980; Sweet *et al.*, 1990) and Fps1p from the yeast *S. cerevisiae* (Luyten *et al.*, 1995; Sutherland *et al.*, 1997). Hence classification of glycerol facilitators is based mainly on sequence comparison and operon organization. Although the term *glycerol facilitator* is well established, we believe that it is somewhat misleading because the *E. coli* and yeast proteins have also been shown to transport a range of other polyols and related compounds (Heller *et al.*, 1980; Karlgren and Hohmann, unpublished results, 2000, Sanders *et al.*, 1997; Sutherland *et al.*, 1997). Furthermore, even though phylogenetic analysis (Fig. 1) illustrates that microbial MIP channels can be classified as aquaporins (Fig. 1, subfamily 1) or glycerol facilitators, the latter group appears to be split further into two subfamilies (Fig. 1, subfamilies 2 and 3). Subfamily 1 constitutes the functionally characterized water channels AqpZ from *E. coli* (Calamita *et al.*,

FIGURE 1   Phylogenetic tree of microbial MIP channels. Phylogenetic analysis of the MIP channels listed in Table I. A, archeaobacterium; B, bacterium; F, fungus; P, protozoan; Y, yeast.

1995), wacA from *Dictyostelium discoideum* (Flick *et al.,* 1997) and Aqy1p from *S. cerevisiae* (Bonhivers *et al.,* 1998). The classification of all other proteins in this subfamily is based on sequence similarity only. Although the MIP channel from the Archea, *Archeaoglobus fulgidus,* has been classified as a glycerol facilitator in the databases (www.tigr.org/tdb/mdb/mdb.html), it clusters with aquaporins (Fig. 1) and shows the residues characteristic of water channels (Froger *et al.,* 1998). Hence, its initial classification may be incorrect.

Subfamily 2 constitutes the glycerol facilitators of *E. coli* (Sweet *et al.,* 1990) and *Pseudomonas aeruginosa* (Schweizer *et al.,* 1997) as well as the propanediol

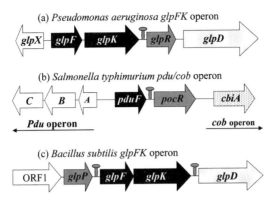

FIGURE 2 Operon structure. Operon organization of glycerol and propanediol facilitator genes in different bacteria. (a) Operon organization of the *glpFK*-containing region of the *Pseudomonas aeruginosa* chromosome. The genes encoding the glycerol facilitator and glycerol kinase are indicated as *glpF* and *glpK*, respectively; *glpX* and *glpR* encode regulatory proteins and *glpD* sn-glycerol-3-phosphate dehydrogenase (Schweizer *et al.*, 1997). (b) *Salmonella typhimurium pdu/cob* operon containing the *pduF* gene. The *pdu* operon controls the degradation of propanediol whereas the *cob* operon controls the synthesis of cobalamin, which is required for propanediol catabolism. The region between the two operons encodes two proteins, the propanediol facilitator PduF and PocR, a regulatory protein, which mediates the induction of the *pdu/cob* operon by propanediol (Chen *et al.*, 1994, 1995). The letters A, B, and C designate the first three genes in the *pdu* operon. The *pduA* gene encodes a hydrophobic protein with high similarity to the carboxysome-forming proteins of several photosynthetic bacteria, whereas *pduB* and *pduC* encode proteins of unknown function. The arrows indicate the direction of gene transcription. (c) The *Bacillus subtilis glpPFKD* region containing genes essential for growth on glycerol or glycerol-3-phosphate. The genes encoding a glycerol facilitator and a glycerol kinase are indicated as *glpF* and *glpK*, respectively. The *glpP* gene encodes a regulatory protein, whereas *glpD* encodes a glycerol-3-phosphate dehydrogenase. The four genes represent three separate transcription units (*glpP*, *glpFK*, *glpD*) and the activities of *glpFK* and *glpD* are controlled by *glpP*, the phosphoenolpyruvate:sugar phosphotransferase system (PTS) and glucose repression (Beijer *et al.*, 1993). Inverted repeats are indicated by hairpin symbols.

facilitator from *Salmonella typhimurium* (Walter *et al.*, 1997). The transport specificities of the latter two proteins have not been determined experimentally, but the genes are respectively part of the well-characterized *glp* operon in *P. aeruginosa* (Schweizer *et al.*, 1997), which is required for glycerol catabolism, and the *pdu* operon in *S. typhimurium* (Fig. 2), which is required for propanediol utilization (Walter *et al.*, 1997).

The third subfamily contains the *Lactobacillus lactis* glycerol facilitator, which has been shown to transport both water and glycerol (Froger *et al.*, unpublished data, 2000). Whether this is a general feature of the third subfamily is unknown and hence conclusions about functionality can only be speculative. To encompass the possible roles of these proteins in transporting water *and* glycerol, the term *aquaglyceroporin* (AGP) has been suggested for them.

Some microbial MIP channels do not appear to fall into any of the three subfamilies, such as the putative glycerol facilitators from *Enterococcus faecalis, Mycobacterium genitalium, Mycobacterium pneumoniae,* and Fps1p from *S. cerevisiae,* which has a number of unusual features (see further). This may reflect functional specialization as is apparent for Fps1p, which functions mainly as an export channel.

B. Distribution of MIP Channels in Microorganisms

Table I lists the known MIP channels that were found in a total of 23 complete microbial genome sequences by the end of March 2000, as well as those from ongoing sequencing projects. The data from the completed microbial genomes allows some conclusions to be drawn about the distribution of MIP channels in microorganisms. For example, there are apparently some organisms that lack MIP channels altogether, such as the Archaea *Methanococcus jannaschii* and *Pyrococcus horikoshii* and the Bacteria *Aquifex aeolicus, Helicobacter pylori, Mycobacterium tuberculosis, Treponema pallidum, Chlamydia trachomatis, Chlamydia pneumoniae, Rickettsia prowazekii,* and *Campylobacter jejuni.* The majority of these microbes are either animal pathogens or deep-sea dwellers. It is plausible that in such habitats microbes might not experience stressful osmolarity changes that would require MIP channel mediated-water/solute flux. Interestingly most microorganisms lacking a glycerol facilitator gene also do not possess a glycerol kinase gene, suggesting that these organisms might not utilize glycerol as a carbon source or as a metabolic precursor. Of course, it is possible that these organisms have other currently undefined mechanisms for water or solute transport.

In general, it appears that most bacteria whose genomes have been fully sequenced possess a glycerol facilitator homolog that is part of the same operon as the gene for glycerol kinase: an indication of a role in glycerol metabolism, as we discuss below. In addition, several genomes contain a second MIP channel, which may be an aquaporin homolog, such as in *E. coli, P. aeruginosa,* or *Shigella flexneri.* The second MIP channel may also be (1) an additional glycerol facilitator as in *Pseudomonas putida,* (2) a propanediol facilitator, as in *S. typhimurium,* or (3) a glycerol facilitator from a different subfamily, as in *Bacillus anthracis, Streptococcus pneumoniae,* and *Haemophilus influenzae.* Some bacteria appear to have more then two MIP channels, often one from each subfamily as found in *E. faecalis* and *S. pneumoniae* or from at least two different subfamilies as in *Klebsiella pneumoniae* and *Salmonella typhi.* The presence of more than one MIP channel from the same subfamily is found only in very few cases, such as in *K. pneumoniae* (two putative, quite distinct water channels and two closely related subfamily 2 members), *P. putida* (two closely related subfamily 2 glycerol facilitators), and *S. typhi* (two highly similar, putative propanediol facilitators). With the possible exception of the two *K. pneumoniae* water channels, two proteins from the same subfamily are likely to be the result of a recent gene duplication event and may fulfill the same physiological role in that microorganism. In general, however, there appears to be a tendency to maintain only one (if any) MIP channel per subfamily in a given organism. This distribution pattern supports the idea that the members of the different subfamilies may indeed exhibit different functions. This is further corroborated by the finding that in bacteria which only have a single MIP channel, the protein is in most cases found in subfamily 3, whose members may transport both water and glycerol (Froger *et al.,* unpublished data, 2000). Hence, these proteins may fulfill functions as water channels *and* glycerol facilitators. Whether this is of any physiological relevance has yet to be addressed.

From the four sequenced Archeal genomes, only two encode a MIP channel and in both instances it is a putative water channel. Hence, from this limited information it appears that glycerol is either not utilized by these organisms or that alternative uptake systems are required.

The only eukaryotic microorganism for which a complete genome sequence is available is that of *S. cerevisiae.* This genome encodes four MIP channels (André, 1995): two aquaporin homologs, 86% identical to each other, and two related glycerol facilitator homologs. The functions of only one of the aquaporins and one of the glycerol facilitators have been confirmed (Bonhivers *et al.,* 1998; Luyten *et al.,* 1995). Strikingly, most laboratory yeast strains seem to have mutations that inactivate both aquaporin genes and even industrial strains and yeasts isolated from nature appear to have mutated versions of the *AQY2* gene (Laizé *et al.,* 2000). So far, only one laboratory yeast strain, Σ1278, a derivative from an industrial isolate, has been found to have two complete aquaporin genes. *AQY1* from Σ1278 differs from that of other laboratory strains by three amino acids and, due to a frameshift mutation, the entire carboxy terminus is different. Two of the amino acid substitutions seem to be responsible for functional alteration in most laboratory strains (Bonhivers *et al.,* 1998). *AQY2* in all laboratory strains investigated so far contains an 11-bp deletion in the center of the gene leading to a premature translational stop. Various different alleles for *AQY2* have also been found in laboratory and industrial strains as well as in natural isolates (Laizé *et al.,* 2000). Although strain Σ1278 has a complete open reading frame for *AQY2,* it has not been possible to confirm in the *Xenopus* oocyte system whether *AQY2* actually encodes a functional aquaporin (Laizé *et al.,* 2000). With regard to other eukaryotic microorganisms, two aquaporins have been found in the slime mold, *Dictyostelium discoideum;* the function of one of these, WacA, has been confirmed experimentally (Flick *et al.,* 1997). Similarity searches reveal putative aquaporins in the pathogenic yeast *Candida albicans* and the filamentous ascomycetes, *Aspergillus nidulans* and *Neurospora crassa.*

Glycerol facilitators, such as that found in the parasitic protozoan *Trypanasoma brucei,* are also well represented in eukaryotic microorganisms. In particular, Fps1p from *S. cerevisiae* has been well characterized as a glycerol and polyol facilitator and will be discussed later (Luyten *et al.,* 1995; Sutherland *et al.,* 1997; Tamás *et al.,* 1999). This organism has a second open reading frame encoding a putative glycerol facilitator, *YFL054c.* Fps1p and Yfl054p are 32% identical in their six transmembrane domain cores. Like Fps1p, the protein encoded by *YFL054c* has an approximately 300 residue amino-terminal extension. Unfortunately, analysis of strains deleted for *YFL054* have not yet led to the elucidation of the protein's function (Tamás and Hohmann, unpublished observation, 2000). Recently, glycerol facilitators have been recognized in the fission yeast *Schizosaccharomyces pombe* and in the plant pathogenic ascomycete *Botrytis cinerea.* Strikingly, sequence comparison suggests that these are homologs of *S. cerevisiae* Yfl054p. Yfl054p and the homolog from *S. pombe* are 76% identical within the transmembrane core and share 30% identity even within their extensions (Tamás and Hohmann, unpublished data, 2000). However, the extensions of these two proteins and that of Fps1p appear totally unrelated. Hence Yfl054p may be the founding protein of a glycerol facilitator subfamily in fungi.

C. Origin of Microbial MIP Channels

It is generally believed that the MIP family emerged as a result of an intragenic duplication event, which probably took place 2.5 to 3 billion years ago (Heymann and Engel, 2000; Pao *et al.,* 1998; Wistow *et al.,* 1991). It has been postulated that a single gene arose in prokaryotes shortly before the emergence of eukaryotes and that subsequent gene duplication and divergence resulted in the various MIP family genes. However, the occurrence of both highly similar and dissimilar MIP channels in a single organism suggests that some MIP proteins did not arise via gene duplication, but rather have been acquired horizontally from other organisms (Park and Saier, 1996). For instance, the G+C content of *S. cerevisiae AQY1* and *FPS1* is 50% and 43%, respectively, whereas the overall G+C content of *S. cerevisiae* is 40%. This deviation lends support to the notion that *AQY1* could have been acquired horizontally and that the second highly similar yeast aquaporin gene (86% identity at protein level) was the result of a subsequent duplication event.

III. TRANSPORT PROPERTIES AND CHANNEL SELECTIVITY OF MICROBIAL MIP CHANNELS

The transport properties of MIP channels can be studied in a number of ways including the use of the heterologous *Xenopus laevis* oocyte system or the osmotic swelling of whole cells, spheroplasts, or membrane vesicles to determine water or solute transport (Hohmann *et al.,* 2000). Most available data from these types of measurements indicate that transport is very rapid, has low activation energy, and can be sensitive to mercury compounds, such as $HgCl_2$, if a cysteine residue lines the pore.

The transport properties of microbial MIP channels have been well characterized in just a few cases and in most instances functional studies have been performed in *X. laevis* oocytes. It is reasonable to assume that in such a heterologous system a different lipid environment will affect transport function or specificity (Truniger and Boos, 1993) and this could explain conflicting data obtained in some cases. The most informative transport coefficients such as osmotic water permeability (hydraulic conductivity), solute permeability, and reflection coefficient have been determined for only a limited number of channels, for example, the *E. coli* glycerol channel, GlpF, transports polyols, glyceraldehyde, glycine, and urea (Heller *et al.*, 1980) but little or no water (Maurel *et al.*, 1994; Calamita *et al.*, 1995). Consistent with a pore-type mechanism, glycerol transport via GlpF has a low activation energy ($E_a = 4.5$ kcal/mol) and is nonsaturable (Maurel *et al.*, 1994). GlpF-mediated glycerol uptake is also sensitive to the membrane lipid composition (Truniger and Boos, 1993). Similar transport properties are expected for other microbial glycerol channels because the molecular architecture of bacterial glycerol uptake systems is apparently highly conserved. With regard to microbial aquaporins, expression of the prokaryotic aquaporin gene, *E. coli aqpZ*, in *X. laevis* oocytes results in a 15-fold increase in osmotic water permeability but negligible solute transport. The observed water transport has a low activation energy ($E_a = 3.8$ kcal/mol) and is insensitive to $HgCl_2$ (Calamita *et al.*, 1995).

The water permeability of the putative *S. cerevisiae* aquaporin encoded by *AQY1*, from both wild-type and laboratory strains, has been evaluated in *X. laevis* oocytes. Only oocytes expressing *AQY1* from the strain $\Sigma1278$, which is closely related to industrial isolates, exhibit an increase in water permeability (Bonhivers *et al.*, 1998). Transport assays using yeast membranes and yeast vesicles also lead to similar observations (Coury *et al.*, 1999). In contrast, data are not yet available for *AQY2* because it could not be functionally expressed in oocytes (Laizé *et al.*, 1999, 2000).

The transport characteristics of the glycerol exporter, Fps1p, are similar to those of *E. coli* GlpF although Fps1p is known to be a regulated channel (Tamás *et al.*, 1999), which we discuss later. In contrast to most other MIP channels, the transport properties of Fps1p can be studied homologously in *S. cerevisiae*. At least three groups have also tried to functionally express Fps1p in oocytes, but this has been unsuccessful to date. Fps1p transports glycerol, erythritol, and xylitol and probably other polyols (Karlgren and Hohmann, unpublished data, 2000; Luyten *et al.*, 1995; Sutherland *et al.*, 1997). Sorbitol and mannitol seem to be transported only with very low efficiency if at all (Karlgren and Hohmann, unpublished data, 2000), and water does not seem to be transported (Coury *et al.*, 1999). Like GlpF (Sanders *et al.*, 1997), Fps1p also seems to transport antimonite (Wysocki, unpublished data, 1999), probably because the hydrated form of this ion resembles a polyol. These observations are consistent with poor substrate specificity, which is probably determined by pore size and interactions between the polyol and residues lining the channel. Although the open reading frame *YFL054* is predicted to encode a glycerol facilitator, it has not yet been possible to determine its substrate specificity.

The mechanisms dictating the commonly observed water/solute channel selectivity described above remain poorly understood. The amino acid content and length of the predicted loop region of MIP channels may play a role in determining specificity. Froger and colleagues (1998) have proposed that the molecular basis of substrate selectivity is determined by key amino acids because the substitution of two amino acids can switch the selectivity of an insect aquaporin to a glycerol channel in the *Xenopus* oocyte system (Lagree *et al.,* 1999). It is possible that these residues influence channel pore size, because this factor is also likely to be key to a complete understanding of selectivity. For example, it appears from structural data at 4.5 Å that the pore of human AQP1 is large enough to allow the passage of water, but too small for solutes such as glycerol (Mitsuoka *et al.,* 1999). However, size alone cannot explain the fact that microbial glycerol facilitators such *S. cerevisiae* Fps1p and *E. coli* GlpF transport glycerol and not water. More studies on microorganisms are thus needed to clearly elucidate the factors governing channel selectivity and determine its significance *in vivo*.

IV. FROM PRIMARY TO QUATERNARY STRUCTURE IN MICROBIAL MIPs

A. Structure-Function Analysis of Microbial MIP Channels

From an analysis of their amino acid sequences, all MIP channels are predicted to have six transmembrane domains and to share highly conserved residues. This is no less true of the microbial branch of the family. As for all MIPs, the most notable of the conserved residues are present in the presumed channel-forming loops (Heymann *et al.,* 1998), B and E, and constitute the family's signature sequences, Ser–Gly–X–His–X–Asn–Pro–Ala–Val–Thr and Asn–Pro–Ala–Arg, respectively, the so-called "NPA boxes" being underlined. However, striking differences can be observed between the sequences of microbial MIPs both within these signature motifs and at the termini. This is well illustrated in the case of the glycerol facilitator, Fps1p, from *S. cerevisiae*. Although Fps1p is clearly related to bacterial glycerol facilitators such as GlpF from *E. coli* (31% identity within the core of six transmembrane domains), it is—so far—unique in the MIP family for a number of reasons (Hohmann *et al.,* 2000). For example, neither of the family's signature NPA boxes is fully preserved, being Asn-Pro-Ser (NPS) and Asn-Leu-Ala (NLA), respectively. In fact, only four additional microbial MIP sequences contain motifs other than NPA, and it is apparent that NPA is preserved in loop B but not in loop E in most of these cases. In *E. faecalis,* the presumed glycerol facilitator (gef 6176) contains an NQA motif in loop E, in *Chlorobium tepidum,* the putative aquaporin (gct 5) has an NPV motif (Hohmann *et al.,* 2000), and the *B. cinerea* putative glycerol facilitator contains an NPS. The only deviation from NPA in loop B of a microbial MIP occurs in the *S. typhimurium* (contig 1308), which has an NLA motif. Unfortunately, these MIPs are incompletely characterized and thus their transport characteristics cannot be used to aid our understanding of the role of these atypical features. This is of particular relevance to a general understanding of MIP channels because their generic NPA motifs are believed to be integral to the formation of a continuous solute channel, the so-called "hourglass" (Jung *et al.,* 1994; see also Chapter 2 by Engel *et al.,* this volume). One possible functional consequence of these atypical motifs is that they influence MIP channel transport properties, resulting in transport specialization. Recently, it has been suggested that loop B in particular may be involved in the determination of transport direction following a comparison of the glycerol transport properties of Fps1p and *E. coli* GlpF. Physiologically, these proteins are a glycerol exporter and an uptake facilitator, respectively. Comparison of mutants where the NPA motifs were "restored" in Fps1p with those where GlpF was made more Fps1p-like by mutating NPA to NPS and/or NLA indicated that the NPS of loop B may be important in influencing Fps1p's export characteristics (Bill and Hohmann, unpublished data, 2000).

In addition to a deviation from the family's signature motifs in the channel-forming loops, Fps1p further distinguishes itself from most other microbial MIPs by having long amino- and carboxy-terminal hydrophilic extensions. This results in a protein of 669 amino acids, compared with 281 amino acids for GlpF and other typical family members. As mentioned above, the putative second *S. cerevisiae* glycerol facilitator, Yfl054p, as well as the similar *S. pombe* protein, also have long amino-terminal extensions unrelated to that of Fps1p.

Standard secondary structure predictions suggest that MIPs are rich in α helical segments. This is yet to be confirmed experimentally because to date, low-resolution structural data (at 4.5 Å) are only available for human AQP1 (Mitsuoka *et al.,* 1999). Even though the most divergent members of the MIP family are less than 20% identical (Park and Saier, 1996), it is expected that the gross structural features apparent in AQP1 will also be present in other water and glycerol channels. For example, AQP1 is functionally homotetrameric; this quaternary structure is thus anticipated for all other MIPs. A study of the crystal organization of MIP and *E. coli* AqpZ and GlpF confirms the close overall structural relationship between MIP channels (Hasler *et al.,* 1998; Ringler *et al.,* 1999) although it has been proposed recently that glycerol channels could be functionally monomeric (Lagree *et al.,* 1999).

V. PHYSIOLOGICAL ROLES

As we have already mentioned, the physiological role of microbial MIP channels has been mostly studied in the bacteria *E. coli, P. aeruginosa, T. flavus,* and *S. typhimurium,* in the yeast *S. cerevisiae,* and in the protozoan *D. discoideum.* In *E. coli, S. cerevisiae,* and *D. discoideum,* the genes for microbial MIP channels have been deleted and the phenotype of the mutant strains compared to wild-type strains. Studies on the expression of genes encoding MIP channels as well as on the location of bacterial genes in operons have provided additional information on their physiological roles. In general, the function of microbial MIP channels is in osmoregulation, metabolism—via the uptake of glycerol or related compounds as sources of carbon and energy—or disposal of metabolic end products.

A. Glycerol Facilitators in the Uptake of Substrates

Although glycerol and other uncharged small molecules are able to move across microbial membranes by simple diffusion, there is currently sufficient evidence to suggest that MIP channel proteins facilitate the uptake of these solutes. In bacteria, the organization of genes in operons, which are coexpressed and hence coregulated, usually points to their function in a common pathway. Hence, knowledge of the function and regulation of a single gene in a bacterial operon allows the function of other genes in the same operon to be predicted. Such a relationship is not observed in eukaryotic microorganisms.

Genes encoding glycerol facilitators are commonly part of the *glp* operon in both gram-positive and gram-negative bacteria (Table I, Fig. 2) The *glp* operon composes *glpF,* encoding the facilitator; *glpK,* encoding a glycerol kinase; *glpD,* encoding a glycerol-3-phosphate dehydrogenase; and two genes, *glpX* and *glpR,* which presumably encode regulators of the operon (Schweizer *et al.,* 1997). The glycerol kinase appears to be closely associated with the facilitator resulting in glycerol phosphorylation during uptake, which prevents reexport of glycerol (Voegele *et al.,* 1993). Exceptions to this operon organization have been found in a number of bacteria such as *L. lactis* (P22094), *S. pneumoniae* (SP42), *Corynebacterium acetobutylicum* (AE001437), and *H. influenzae* (U32782) where putative glycerol facilitator genes do not form part of the glycerol operon, suggesting that these MIP family proteins might also have functions other than glycerol uptake (Park and Saier, 1996).

Escherichia coli mutants lacking *glpF* grow poorly on low glycerol concentrations presumably due to insufficient glycerol permeating the cell (Voegele *et al.,* 1993). Hence it has been proposed that the facilitator is required for efficient uptake of glycerol, especially at low concentrations. The *glpF* mutant also shows altered kinetics for glycerol phosphorylation and the fact that free glycerol is undetectable in wild-type cells utilizing glycerol, supports the conclusion that transport and phosphorylation of glycerol are closely coupled (Voegele *et al.,* 1993). Substantial glycerol transport by passive diffusion through the lipid bilayer is also apparent because a *glpF* mutant does not show a glycerol-negative phenotype at high glycerol concentrations (Voegele *et al.,* 1993). Recently, a gene encoding a glycerol facilitator in *P. aeruginosa* has been cloned and a chromosomal $\Delta glpFK$ mutant isolated (Schweizer *et al.,* 1997). This mutant, which lacks both the facilitator and the glycerol kinase, does not grow on medium containing glycerol as the sole carbon source and does not transport glycerol.

The *S. typhimurium pdu* operon is required for the catabolism of 1,2-propanediol. The *pdu* operon (Fig. 2) is closely linked to the *cob* operon, which controls the synthesis of adenosyl-cobalamin (vitamin B_{12}), a cofactor required for the catabolism of propanediol. The region between the *pdu* and *cob* operons encodes two proteins, PduF, the putative propanediol transporter, and PocR, a regulatory protein that mediates the induction of the *pdu/cob* operon by propanediol (Chen *et al.,* 1994, 1995). The transport characteristics of PduF have not been determined experimentally but it is reasonable to assume that this protein is involved in the uptake of 1,2-propanediol, a compound closely related to glycerol. It is not yet known whether PduF can transport glycerol in addition to 1,2-propanediol or, indeed, whether other GlpFs can transport 1,2-propanediol. However, in addition to PduF, *S. typhimurium* has a gene encoding a glycerol facilitator that forms part of a *glpFK* operon. This suggests, in fact, that the organism possesses two facilitators, one for catabolism of propanediol and another for glycerol uptake.

Whether members of the MIP channel protein family from eukaryotic microorganisms play a role in the uptake of solutes such as glycerol is less clear. Extensive analysis of the glycerol transport characteristics of yeast wild-type and *fps1Δ* mutants has demonstrated that this protein can transport glycerol in both directions (Lages and Lucas, 1997; Luyten *et al.,* 1995; Sutherland *et al.,* 1997; Tamás *et al.,* 1999). However, mutants lacking Fps1p, Yfl054p, or the double mutant lacking both putative yeast glycerol facilitators do not show a defect in glycerol utilization (Tamás *et al.,* 1999; Tamás and Hohmann, unpublished data, 2000). Because the yeast plasma membrane seems to be relatively impermeable to glycerol (Luyten *et al.,* 1995; Tamás *et al.,* 1999), the existence of uptake proteins involved in glycerol catabolism has been suggested. In addition, it has been demonstrated that yeast cells have at least one system for the active uptake of glycerol (Lages and Lucas, 1997; Lages *et al.,* 1999; Van Zyl *et al.,* 1990). It has also been shown that yeast mutants unable to produce any glycerol themselves, and which therefore do not grow on medium containing NaCl (see below), can be rescued by as little as 5 m*M* glycerol in the growth medium, indicative of an active uptake system mediating accumulation of glycerol against a concentration gradient (Holst *et al.,* 2000). This effect has been used to identify a yeast gene, *GUP1*, which is required for rescue of the glycerol-negative mutant by low concentrations of glycerol. Gup1p is a membrane protein that either takes up glycerol or at least controls glycerol uptake, a process that appears to be closely coupled to glycerol phosphorylation. This phosphorylation is catalyzed by the glycerol kinase Gut1p. Gup1p is not a MIP channel (Holst *et al.*, 2000).

Surprisingly, deletion of the genes encoding the glycerol facilitators in *E. coli* and *S. cerevisiae* results in diminished passive diffusion of glycerol and altered cellular lipid composition (Sutherland *et al.*, 1997; Truniger and Boos, 1993). The reason for this observation is not clear, but it is possible that glycerol uptake and phosphorylation via the facilitator/kinase system provides glycerol-3-phosphate for phospholipid metabolism. Several pathways lead to the synthesis of glycerophospholipids, and the balance between different precursors seems to be critical for phospholipid biosynthesis (Daum *et al.*, 1998). In this context it is of interest that the regulation of expression of the yeast glycerol kinase gene, *GUT1*, was recently reported to be not only controlled by glucose repression and glycerol induction but also by the same regulators that control genes encoding enzymes involved in phospholipid metabolism (Grauslund *et al.*, 1999). In conclusion it appears as if there may be a connection between glycerol transport and phosphorylation on the one hand and between glycerol transport and cellular phospholipid metabolism on the other.

VI. MICROBIAL MIP CHANNELS IN OSMOREGULATION

A. Microbial Aquaporins

Most available information on the physiological roles of microbial aquaporins is derived from studies on *E. coli* AqpZ, *S. cerevisiae* Aqy1p, and *D. discoideum* WacA. AqpZ has been shown to have a direct role in the way in which *E. coli* adjusts cell turgor within the range needed for growth and survival (Calamita *et al.*, 1998). This function has been demonstrated by comparing *E. coli* cells carrying a null mutation in the *aqpZ* gene with their parental wild-type strain. Disruption of *aqpZ* is not lethal, but the viability of cells in which *aqpZ* has been knocked out is strikingly reduced when they are grown at low osmolarity. On the contrary, no significant changes in cell viability are observed when these cells are grown in high osmolarity medium. The reduced growth observed in hypo-osmotic conditions can be rescued by transforming the knockout strain with a plasmid bearing a functional *aqpZ* gene. Overall, these data are consistent with the results of regulatory studies, which show a marked increase of the *aqpZ* transcription rate when *E. coli* is grown in low osmolarity medium and a reduced expression in high osmolarity medium (Calamita *et al.*, 1998). Although questions about the physiological necessity of a water channel during prolonged hypo-osmotic stress remain to be answered, this finding clearly indicates involvement of AqpZ in prokaryotic osmoadaptation. In fact, AqpZ seems to be required both during the long-term osmoregulatory response triggered by hypo-osmotic stress and the short-term responses that occur suddenly after changes in the extracellular osmolarity. Although further investigation is required to elucidate the precise role of AqpZ, especially during the osmotic response to hypo-osmotic stress, it is likely that involvement in osmoregulation is a general feature of microbial aquaporins (Booth and Louis, 1999).

Escherichia coli wild-type and *aqpZ* mutant strains have been used in cryo-electron microscopy studies to demonstrate, *in vivo*, the ability of AqpZ to mediate rapid outward- and inward-directed water fluxes triggered by sudden up- and down-shifts of the extracellular osmolarity, respectively (Delamarche *et al.*, 1999). In addition to demonstrating the functional expression of AqpZ in *E. coli*, these studies indicate that AqpZ transports water in both directions, a property that has been also reported for many mammalian aquaporins (Meinild *et al.*, 1998). A role for AqpZ in mediating the bulk water uptake needed for cell expansion during rapid growth is suggested both by its maximal expression at the midlogarithmic phase of growth and the reduced viability characterizing the *E. coli aqpZ* mutant grown at 39°C, a temperature where the growth rate is highest. However, this function apparently contrasts with the assumption that sufficient water for cell division may be absorbed by simple diffusion across the cytoplasmic membrane during its 20- to 30-min generation time (Haines, 1994). Additional studies are therefore required to better elucidate the physiological relevance of AqpZ expression during the exponential growth phase of *E. coli*.

Interestingly, an AqpZ-like protein seems to be necessary for the expression of certain surface antigens (Kopecko *et al.*, 1980), which in turn appear to be one of the requirements for pathogenic bacteria to invade epithelial cells. In fact, it has been observed that the *Shigella sonnei* ORF10, an open reading frame with striking sequence similarity to the *aqpZ* coding region, is part of a gene cluster composed of 10 contiguous ORFs located in a plasmid (pHH201) encoding the form I antigen. It has been found that deletions of ORF10 and/or any of the other nine cluster ORFs eliminate form I antigen expression of *S. sonnei* (Houng and Venkatesan, 1998). An identical gene cluster including an *aqpZ*-like coding region (ORF10p) is also found in the pathogenic species *Plesiomonas shigelloides* (serotype 017) where in association with other genes it leads to the expression of a cell surface O-antigen (Chida *et al.*, 2000). A possible role for AqpZ in the virulence mechanisms of pathogenic bacteria is an exceedingly appealing hypothesis, which deserves investigation. We note, however, that several pathogenic organisms lack MIP channels altogether and hence the importance of aquaporins in virulence, if any, must be restricted.

Although a role in osmoregulation seems likely for the *S. cerevisiae* aquaporin Aqy1p, surprisingly, the related null mutant yeast strain tolerates osmotic changes better than the wild-type strain under laboratory conditions (Bonhivers *et al.*, 1998).

In these experiments wild-type and mutant cells were co-cultivated and repeatedly osmotically shocked. The wild-type cells did not survive this treatment and were consequently depleted in the culture, whereas the *aqy1* mutant cells survived. This evidence together with the observation that the *AQY1* gene product appears to be nonfunctional in many laboratory strains (Bonhivers *et al.*, 1998; Laizé *et al.*, 2000) led to the suggestion that functional aquaporins might have been lost in strains maintained under laboratory growth conditions (Bonhivers *et al.*, 1998). However, the interpretation of these observations is complicated by the fact that the *AQY1* gene is very poorly expressed during vegetative growth, conditions under which the above-mentioned phenotype was determined. Expression of *AQY1* is, however, strongly stimulated when diploid yeast cells enter sporulation and hence *AQY1* is clearly a developmentally regulated yeast gene, similar to *D. discoideum wacA* (see below; Chu *et al.*, 1998; Laizé and Hohmann, unpublished data, 2000). Key questions currently under study are the precise localization of Aqy1p and whether the protein has any role during sporulation, spore maturation, or spore germination.

A potential role in mediating the extrusion of water during prespore cell encapsulation has been suggested for the *D. discoideum* WacA aquaporin (Flick *et al.*, 1997). Although the *wacA* gene is expressed only in prespore cells, it has been observed that disruption of the gene does not lead to any apparent alterations in prespore cells or their ability to germinate or respond to osmotic stresses. However, the lack of an apparent phenotype could possibly be explained by the presence of unknown alternative aquaporins or the fact that the *Dictyostelium wacA* mutant was not exposed to the selective challenges of its natural habitat, the soil. According to Cotter and Raper (1968), germinating spores of *Dictyostelium* take up water very rapidly unlike prespore cells, which extrude water during encapsulation. Such rapid water fluxes imply the involvement of aquaporins. A similar argument may be used to support the involvement of Aqy1p in yeast sporulation/spore germination. Further research is required to address this intriguing question.

As outlined above, yeast *AQY2* seems to be mutated in most yeast strains (Laizé *et al.*, 2000). This is a rather unusual scenario and consequently one might speculate that there is some selective pressure against maintaining this gene. The *AQY2* ORF is complete in strain Σ1278, which consequently can express both *AQY1* and *AQY2* (Laizé *et al.*, 2000). In this strain, *AQY2* is expressed during vegetative growth and expression is stimulated after a hyperosmotic shock. This observation suggests that Aqy2p might be involved in the uptake of water during the recovery of cells from osmotic upshock. However, at present the subcellular localization of Aqy2p has not been determined and hence the protein could perhaps be located in an intracellular compartment controlling water fluxes within the cell. In fact, the two yeast aquaporins are most similar to the tonoplast aquaporins from plants and hence Aqy2p could, in fact, be a vacuolar protein. So far, attempts to assign a phenotype to cells deleted for the *AQY2* gene have been unsuccessful (Laizé and Hohmann, unpublished data, 2000).

The unresolved issues discussed in the preceding paragraphs clearly illustrate that no well-defined physiological role can yet be assigned to any microbial aquaporin. The identification and characterization of additional aquaporins in genetically tractable systems will provide further insight into the role of water transport in microbial water relations and in osmoregulation in general. Indeed, such studies certainly deserve a more widespread interest from microbiologists.

B. Yeast Osmolyte System: Control of Glycerol Metabolism

A common strategy in osmoadaptation is the accumulation of compatible solutes (Yancey *et al.*, 1982). A range of quite different compounds is employed as compatible solutes by microorganisms, such as polyols (glycerol, D-arabitol, D-mannitol, and *meso*-erythritol) in fungi (Spencer and Spencer, 1978; Yancey *et al.*, 1982) and in potassium ions, trehalose, and amino acids or their derivatives in Bacteria and Archea (da Costa *et al.*, 1998; Measures, 1975). The yeast *S. cerevisiae* employs glycerol for this purpose (Blomberg and Adler, 1992; Brown, 1978; Brown and Edgley, 1980). This is surprising, because glycerol is known to diffuse through lipid bilayers and hence one might expect that yeast cells could lose the glycerol they produce under hyperosmotic stress. However, this does not appear to be the case to any significant extent because the yeast plasma membrane is relatively impermeable to glycerol (Luyten *et al.*, 1995; Sutherland *et al.*, 1997; Tamás *et al.*, 1999). In fact, there is evidence that yeast cells can actively control the permeability of their plasma membrane to glycerol under osmotic stress, perhaps by altering the lipid composition (Sutherland *et al.*, 1997).

Glycerol is produced in two steps from the glycolytic intermediate dihydroxy acetone phosphate (Fig. 3; Blomberg and Adler, 1992) catalyzed by the enzymes glycerol-3-phosphate dehydrogenase (Gpd) and glycerol-3-phosphatase (Gpp), respectively. Both enzymes have two isoforms whose expression is differentially regulated (Ansell *et al.*, 1997; Norbeck *et al.*, 1996). The expression of *GPD1* and *GPP2* is strongly induced by hyperosmotic stress (Albertyn *et al.*, 1994; Norbeck *et al.*, 1996) and hence these two proteins appear to account for most of the glycerol production capacity under osmotic stress. The expression of *GPP1*, which is more strongly expressed under normal growth conditions than *GPP2*, is also somewhat induced under osmotic stress (Rep *et al.*, 2000). *GPD2* and *GPP1* expression are stimulated under anaerobic conditions (Ansell *et al.*, 1997) because in the absence of oxygen, glycerol production is essential for the reoxidation of NADH to NAD, which is normally performed by the respiratory chain (Hohmann, 1997; Van Dijken and Scheffers, 1986).

The mechanisms that control the induction of *GPD1* and *GPP2* under osmotic stress are being studied extensively. The high osmolarity glycerol (HOG) response pathway plays a central, though not exclusive, role in the induction of *GPD1* and *GPP2* (Rep *et al.*, 1999a, 1999b, 2000). This pathway is a prototypical MAP

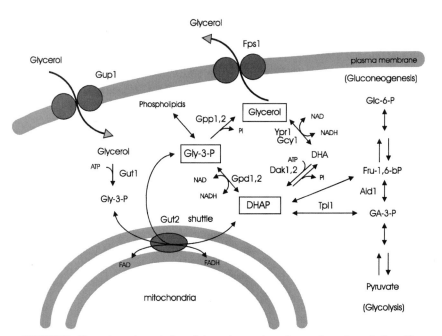

FIGURE 3 Yeast glycerol metabolism. Schematic overview of yeast glycerol metabolism. Glycerol catabolism starts with uptake, presumably through Gup1p. Glycerol is phosphorylated to glycerol-3-phosphate, Gly-3-P, by a glycerol kinase, Gut1p, and oxidized by an FAD-dependent, mitochondrial glycerol-3-phosphate dehydrogenase to dihydroxyacetonephosphate, DHAP, a glycolytic intermediate. For glycerol production DHAP is converted to Gly-3-P by an NADH-dependent, cytosolic glycerol-3-phosphate dehydrogenase, Gpd1p or Gpd2p, and subsequently dephosphorylated by glycerol-3-phosphatase, Gpp1p or Gpp2p, to glycerol. Glycerol is either accumulated within the cell or exported through the osmoregulated glycerol facilitator Fps1p, a MIP channel. DHAP and glycerol can also be interconverted via dihydroxyacetone, DHA, but the relevance of this pathway in *Saccharomyces cerevisiae* is unclear. The actions of the FAD-dependent Gut1p and the NADH-dependent Gpd1p/Gpd2p provide a shuttle for electrons into the mitochondrial electron transport chain.

(mitogen activated protein) kinase cascade (Fig. 4), as found in all eukaryotes. Recent transcriptome analysis indicates that this pathway controls the expression of more than 100 yeast genes upon osmotic shock (Rep *et al.*, 2000) and that it mediates its effects via different transcription factors such as Hot1p, Msn1p, Msn2p, Msn4p, and Sko1p. Of these, Hot1p and Msn1p are involved in controlling the glycerol biosynthesis genes (Rep *et al.*, 1999b).

Upon osmotic shock, yeast cells rapidly stimulate the production of glycerol and substantial levels of glycerol are built up in the cell within a few hours. Concentrations of up to 1 *M* of glycerol have been reported (Blomberg and Adler, 1992). In their natural environment, yeast cells are frequently exposed to high osmolarity, especially to high sugar concentrations. Equally common is exposure

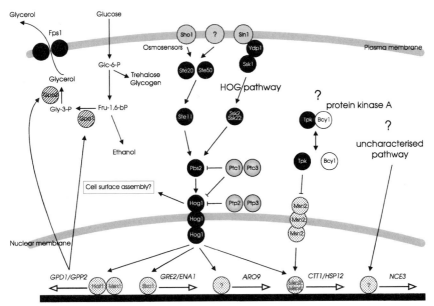

FIGURE 4 Signaling upon osmotic shock in yeast. Schematic overview of signaling upon an osmotic shock in *Saccharomyces cerevisiae*. Central to the response is the high osmolarity glycerol (HOG) pathway. Osmotic shock is sensed by at least two putative transmembrane osmosensors; one of those, Sln1p-Ypd1p-Ssk1p, forms a phosphorelay system similar to bacterial two-component systems. The signal is then transmitted through a MAP kinase cascade eventually to Hog1p, which is translocated to the nucleus. Hog1p controls and/or interacts with different transcription factors to stimulate expression of more than 100 genes. Examples are the glycerol biosynthesis genes *GPD1* and *GPP1,* the gene for the sodium pump *ENA1*, the gene *GRE2* whose product may be involved in detoxification of oxygen radicals, *ARO9,* which encodes in enzyme in amino acid metabolism, *CTT1,* which encodes a catalase, and *HSP12,* which encodes a heat shock protein of unknown function. *NCE3* encodes carbonic anhydrase and an unknown signaling pathway mediates its induction by osmotic stress. The transcription factors Msn2p and Msn4p mediate a general stress response and their subcellular localization is controlled by protein kinase A. The scheme on the left-hand side depicts glycerol metabolism. Gray, osmosensors; black with white text, protein kinases; gray fountain fill, protein phosphatases; shaded, enzymes; dotted fill, transcription factors.

to hypo-osmotic shock, for instance during rainfall. Under these conditions the cell has to rapidly dispose of accumulated glycerol in order to diminish turgor pressure. Hence, yeast has developed an efficient system to export the majority of its accumulated glycerol within a few minutes through the MIP channel, Fps1p.

C. Fps1p Solute Exporter

The *FPS1* gene was originally isolated as a multicopy suppressor of a growth defect on fermentable sugars, such as glucose, of a mutant with defective feedback

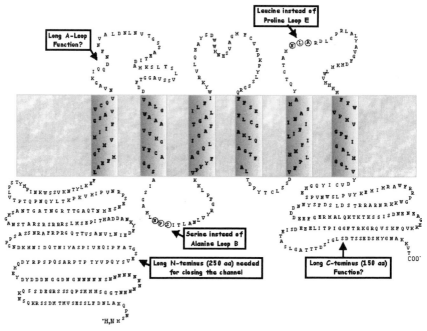

FIGURE 5 Yeast Fps1p topology. Fps1p differs from more typical members of the MIP family, such as GlpF, in a number of ways. For example, the family's signature NPA motifs are replaced by NPS and NLA in Fps1p, as indicated. The letters A to E denote Fps1p's loops of which B and E are believed to be involved in the formation of the glycerol channel by dipping into the membrane to form an "hourglass."

control of glycolysis (Van Aelst *et al.*, 1991). Subsequently it was shown that this growth defect could be partially corrected by overproduction of glycerol (Luyten *et al.*, 1995). As outlined above, Fps1p is an unusual MIP channel (Fig. 5). Its "NPA" motifs are not fully conserved, being NPS and NLA in loops B and E, respectively, and its A loop being unusually long. In addition, Fps1p is 669 amino acids long due to amino- and carboxy-terminal cytosolic extensions. Apart from two other fungal proteins, only the *Drosophila* BIB (big brain) protein has such long extensions. However, Fps1p's extensions do not show any sequence similarity to other proteins.

Fps1p has been demonstrated by direct transport assays with radiolabeled glycerol to mediate transport of glycerol into and out of the yeast cell (Luyten *et al.*, 1995; Sutherland *et al.*, 1997; Tamás *et al.*, 1999). The glycerol facilitator from *E. coli*, GlpF, when expressed in yeast, can replace Fps1p's glycerol transport function lending further support to the role of Fps1p as a glycerol transporter (Luyten *et al.,* 1995; Sutherland *et al.,* 1997; Tamás *et al.,* 1999). The phenotype associated with deletion of Fps1p clearly classifies this protein as a glycerol exporter, namely, the inability to grow under anaerobic conditions and its sensitivity to hypo-osmotic shock (Tamás *et al.,* 1999). As indicated earlier, yeast cells produce glycerol when grown in the absence of oxygen for redox balancing (Ansell *et al.,* 1997). In cells lacking Fps1p, glycerol accumulates inside the cell and inhibits growth, presumably because it leads to an osmotic imbalance (Tamás *et al.,* 1999). Under these conditions glycerol can be regarded as a metabolic end or waste product and hence Fps1p serves as a waste product exporter.

When yeast cells are grown in high osmolarity medium and then shifted to low osmolarity, they dispose of 80% of their accumulated glycerol within 5 min (Luyten *et al.,* 1995). In contrast, cells lacking Fps1p require 60 min to achieve the same low glycerol level and survive a hypo-osmotic shock in a 100-fold lower proportion than wild-type cells. Those cells that survive resume growth more slowly (Luyten *et al.,* 1995; Tamás *et al.,* 1999) and, moreover, if the *fps1Δ* mutation is combined with a mutation that weakens the cell wall, a hypo-osmotic shock is lethal (Tamás *et al.,* 1999). Because cells lacking Fps1p grow in low osmolarity medium as well as the wild type, the *fps1Δ* mutant is specifically sensitive to a hypo-osmotic shock and thus far is the only yeast mutant known to display such a phenotype (Tamás *et al.,* 1999; Ferreira and Hohmann, unpublished observations, 2000).

The role of Fps1p in osmoregulation and the control of cellular glycerol content are supported by further observations. Mutants lacking Fps1p exhibit diminished signaling through the HOG pathway, apparently because they can accumulate glycerol after a hyperosmotic shock faster than wild-type cells (Tao *et al.,* 1999; Tamás, Rep, and Hohmann unpublished observations, 1999). Strikingly, *fps1Δ* mutant cells show a defect in cell fusion during the mating process of haploid yeast cells; this defect is apparently also associated with altered osmoregulation because it can be suppressed by deletion of the *GPD1* gene and hence by reduction of glycerol production (Philips and Herskowitz, 1997). Cell fusion requires local cell wall degradation, and it appears that yeast cells have to relieve osmotic pressure at this point in order to prevent cell bursting. The observation that the presence of 1 *M* sorbitol in the medium also corrects the mating defect of *fps1Δ* mutants supports the notion that this phenotype is due to a problem with osmoregulation and cell bursting (Philips and Herskowitz, 1997). This observation also illustrates that osmotic phenomena play a role in very different cellular processes not obviously related to osmoadaptation at first sight. Finally, although there is substantial biophysical evidence for the existence of osmolyte export systems in other organisms, especially in mammalian cells (Kwon and Handler, 1995), Fps1p is to date the only eukaryotic solute exporter characterized at the molecular level.

VII. CONTROL OF THE FUNCTION OF MICROBIAL MIP CHANNELS

MIP channel function can be controlled at different levels, as also discussed in other chapters in this volume. For instance, in plant cells the expression of genes encoding aquaporins has been demonstrated to be controlled by stress (Balk and de Boer, 1999; Bohnert *et al.*, 1995; Yamada *et al.*, 1995) as well as by developmental cues (Gao *et al.*, 1999; see Chapter 7 by Chrispeels *et al.*). AQP2 and 5 are paradigms of mammalian aquaporins whose localization to their target membrane is controlled by hormonal stimuli (Deen *et al.*, 1995; Kamsteeg *et al.*, 1999; see Chapter 6 by Deen and Brown and Chapter 5 Verkman *et al.* and references therein). In addition, gating mechanisms could potentially control the function of MIP channels within the membrane as has recently been suggested for the control of mammalian AQP6 (Yasui *et al.*, 1999) and plant PM28A (Johansson *et al.*, 1998; Kjellbom *et al.*, 1999). For microbial MIP channels, control of gene expression as well as gating has been demonstrated as a means of regulation.

Control of the expression of microbial MIP genes has already been discussed along with the analysis of their physiological roles. Indeed, their expression pattern often serves as a guide toward their physiological role, as illustrated by the coregulation of bacterial glycerol facilitators with glycerol kinases in glycerol metabolism and the stimulated expression of *E. coli aqpZ* under hypo-osmotic conditions (Calamita *et al.*, 1998). In addition we have mentioned the control of yeast *AQY2* by osmotic shock and that of yeast *AQY1* and *Dictyostelium wacA* by sporulation (Chu *et al.*, 1998; Flick *et al.*, 1997), although the involvement of these proteins in osmoadaptation and development, respectively, has not yet been demonstrated. In the following section we focus on the control of protein function, which we have thus far not specifically addressed.

A. Control of Protein Activity

Regulation of microbial MIP channels at the protein level has only been well studied in *S. cerevisiae*. Although the expression of the gene encoding the glycerol facilitator, *FPS1,* does not change with growth conditions, the protein is regulated by osmotic shock: The channel apparently closes within seconds after a hyperosmotic shock and opens equally fast after a hypo-osmotic shock (Luyten *et al.*, 1995; Tamás *et al.*, 1999). This regulation ensures that glycerol can be accumulated under hyperosmotic conditions and be released after hypo-osmotic shock. However, the precise mechanism controlling Fps1p is not understood. Extensive analysis of the possible involvement of different signaling pathways in yeast, such as the HOG pathway, suggests that none of these systems is needed for gating of Fps1p (Luyten *et al.*, 1995; Tamás *et al.*, 1999). Moreover, a search for mutants that resemble the phenotype of mutants lacking Fps1p has revealed no gene other than

FPS1, suggesting but not excluding the fact that Fps1p does not need any other protein for closing (Ferreira and Hohmann, unpublished results, 2000). Gustin and co-workers have reported the presence of a mechanosensitive ion channel in the plasma membrane of *S. cerevisiae* that is activated by stretching of the membrane (Gustin *et al.,* 1988) but whether the channel activity of Fps1p could be regulated in a similar way is unknown.

A short domain within the amino-terminal extension of Fps1p apparently controls glycerol movement (Luyten *et al.,* 1995; Tamás *et al.,* 1999). Deletion of this sequence abolishes closing, thereby causing loss of glycerol from the cell during growth in high osmolarity medium and sensitivity to hyperosmotic conditions. This regulatory sequence has been narrowed down by deletion analysis to fewer than 20 amino acids and certain amino acid replacements within this sequence have been found to abolish closing. It also appears that the spacing between this domain and the first transmembrane domain of Fps1p is critical for function (Tamás and Hohmann, unpublished data, 2000). However, the sequence does not reveal any hint as to the function of this domain and it is not understood how it controls gating. Detailed mutational analyses as well as novel genetic screens have been devised to address this question and the possible involvement of other parts of Fps1p in the control of the transport function. Interestingly, the glycerol facilitator GlpF from *E. coli* can mediate glycerol transport into and out of yeast cells, as indicated above (Luyten *et al.,* 1995; Tamás *et al.,* 1999). However, its transport function is not regulated by osmotic shock in yeast and hence expression of GlpF also results in glycerol loss from the cell and in an osmosensitive phenotype, analogous to the expression of Fps1p lacking the regulatory domain (Tamás *et al.,* 1999). Hence, GlpF provides a basis to study the parts of Fps1p that are required for it to be a gated channel.

VIII. CONCLUSIONS AND FUTURE PERSPECTIVES

Microbial MIP channels are currently being identified at a rate of about one per week (Hohmann *et al.,* 2000). Although their sequences alone are a helpful source of information for generating structure–function relationships, (Heymann and Engel, 2000) the determination of their physiological roles is of fundamental importance. This field has been long neglected since MIP channels transport substrates that are able to—or thought to be able to—cross the lipid bilayer by simple diffusion. However, it has become clear that the permeability of microbial membranes for water and glycerol (and related substances) may be limiting, at least under certain conditions. Under such circumstances MIP channels may provide the means to control water and glycerol fluxes into and out of the cell. This principle is best illustrated for the glycerol channel, Fps1p, in yeast osmoregulation. The yeast plasma membrane appears to be relatively impermeable to glycerol and moreover there is evidence that yeast can actively control the permeability of glycerol through the lipid bilayer (Sutherland *et al.*, 1997; Tamás *et al.*, 1999). The presence of Fps1p, whose function is itself regulated by osmotic shock, allows yeast cells to control their intracellular glycerol content in response to external osmolarity.

The fact that glycerol and water can cross lipid bilayers has made analysis of the role of MIP channels more difficult. Phenotypes associated with the deletion of genes encoding MIP channels are not necessarily as clear as in the case of yeast Fps1p, where deletion of the gene causes a sensitivity to hypo-osmotic shock and a constitutively open channel results in sensitivity to high osmolarity. For instance, deletion of *E. coli glpF* causes only a limited inability to catabolize glycerol (Truniger *et al.*, 1992; Voegele *et al.*, 1993) and deletion of the *aqpZ* gene in this organism causes a visible sensitivity to low osmolarity only after co-cultivation in competition with the wild type (Calamita *et al.*, 1998). Hence, detection of the phenotype of a MIP channel mutant may require very careful inspection as well as knowledge of the conditions the organism experiences in its natural environment. This is hardly surprising because in nature, even subtle differences in viability under specific conditions may provide a major growth and survival advantage. Moreover, analysis of deletion mutants should be combined with studies on the expression pattern of the relevant MIP because this may provide further hints about the conditions under which a MIP channel may be functional.

The water channels Aqy1p from yeast and WacA from *D. discoideum* illustrate this latter aspect. Both proteins are developmentally regulated and hence a phenotype should be sought that is associated with the relevant developmental program: spore formation and germination in this case. It is likely that subtle phenotypes associated with mutation under conditions different from these, where the gene is not or poorly expressed, may be artefacts.

Other important principles are also illustrated by the role of yeast Fps1p. MIP channels can transport solutes and/or water in both directions and hence may serve a role in uptake and/or in efflux. In addition, MIP channels may also function in the export of metabolic end products along a concentration gradient, such as that observed for glycerol export through Fps1p under anaerobic conditions. It is at present not known whether this feature is shared by other MIP channels but is certainly worth considering.

Overall, microbial MIP channels provide an interesting field for the study of microbial physiology in processes such as osmoregulation and developmental programs, which in turn can provide models for the study of higher organisms. In addition, microbial systems themselves provide a test bed for the analysis of MIP channels from higher organisms following their heterologous expression. So far, MIP channels have mainly been studied in *Xenopus* oocytes, but because it has been demonstrated that, for instance, the function of GlpF is affected by lipid composition (Truniger and Boos, 1993), such heterologous systems may provide misleading results. Yeast expression and vesicles isolated from yeast are thus increasingly used to study aquaporin function. The yeast system has been demonstrated to be robust with respect to the routing of mutant MIPs, hence allowing their functional analysis. Undoubtedly, one power of functional expression in yeast, combined with proper test systems, is the direct investigation of transport and regulation by genetic analysis.

Acknowledgments

Work in the laboratory of the authors has been supported by the Commission of the European Union (BIO4-CT98-0024 to S. H. and FMRX-CT96-0128 to S. H. and G. C.) as well as by the National Research Foundation of South Africa (B. A. P.).

References

Agre, P., Bonhivers, M., and Borgnia, M. J. (1998). The aquaporins, blueprints for cellular plumbing systems. *J. Biol. Chem.* **273,** 14659–14662.

Albertyn, J., Hohmann, S., Thevelein, J. M., and Prior, B. A. (1994). *GPD1,* which encodes glycerol-3-phosphate dehydrogenase is essential for growth under osmotic stress in *Saccharomyces cerevisiae* and its expression is regulated by the high-osmolarity glycerol response pathway. *Mol. Cell Biol.* **14,** 4135–4144.

André, B. (1995). An overview of membrane transport proteins in *Saccharomyces cerevisiae. Yeast* **11,** 1575–1611.

Ansell, R., Granath, K., Hohmann, S., Thevelein, J., and Adler, L. (1997). The two isoenzymes for yeast NAD-dependent glycerol 3-phosphate dehydrogenase encoded by *GPD1* and *GPD2,* have distinct roles in osmoadaption and redox regulation. *EMBO J.* **16,** 2179–2187.

Baker, M. E., and Saier, M. H., Jr. (1990). A common ancestor for bovine lens fiber major intrinsic protein, soybean nodulin-26 protein, and *E. coli* glycerol facilitator. *Cell* **60,** 185–186.

Balk, P. A., and de Boer, A. D. (1999). Rapid stalk elongation in tulip (*Tulipa gesneriana* L. cv. Apeldoorn) and the combined action of cold-induced invertase and the water-channel protein gamma TIP. *Planta* **209,** 346–354.

Beijer, L., Nilsson, R. P., Holmberg, C., and Rutberg, L. (1993). The *glpP* and *glpF* genes of the glycerol regulon in *Bacillus subtilis. J. Gen. Microbiol.* **139,** 349–359.

Blattner, F. R., Plunkett, G., III, Bloch, C. A., Perna, N. T., Burland, V., Riley, M., Collado-Vides, J., Glasner, J. D., Rode, C. K., Mayhew, G. F., Gregor, J., Davis, N. W., Kirkpatrick, H. A., Goeden, M. A., Rose, D. J., Mau, B., and Shao, Y. (1997). The complete genome sequence of *Escherichia coli* K-12. *Science* **277,** 1453–1474.

Blomberg, A., and Adler, L. (1992). Physiology of osmotolerance in fungi. *Adv. Microbial Physiol.* **33,** 145–212.

Bohnert, H. J., Nelson, D. E., and Jensen, R. G. (1995). Adaptation to environmental stresses. *Plant Cell* **7,** 1099–1111.

Bonhivers, M., Carbrey, J. M., Gould, S. J., and Agre, P. (1998). Aquaporins in *Saccharomyces.* Genetic and functional distinctions between laboratory and wild-type strains. *J. Biol. Chem.* **273,** 27565–27572.

Booth, I. R., and Louis, P. (1999). Managing hypoosmotic stress: Aquaporins and mechanosensitive channels in *Escherichia coli. Curr. Opin. Microbiol.* **2,** 166–169.

Borgnia, M., Nielsen, S., Engel, A., and Agre, P. (1999). Cellular and molecular biology of aquaporin water channels. *Ann. Rev. Biochem.* **68,** 425–458.

Brown, A. D. (1978). Compatible solute and extreme water stress in eukaryotic microorganisms. *Adv. Microbial Physiol.* **17,** 181–242.

Brown, A. D., and Edgley, M. (1980). Osmoregulation in yeast. *In* "Symposium on Genetic Engineering of Osmoregulation" (D. W. Rains, R. C. Valentine and A. Hollander, Eds.) pp. 75–90. Plenum Press, New York.

Calamita, G., Bishai, W. R., Preston, G. M., Guggino, W. B., and Agre, P. (1995). Molecular cloning and characterization of AqpZ, a water channel from *Escherichia coli*. *J. Biol. Chem.* **270,** 29063–29066.

Calamita, G., Kempf, B., Bonhivers, M., Bishai, W., Bremer, E., and Agre, P. (1998). Regulation of the *Escherichia coli* water channel gene *aqpZ*. *Proc. Natl. Acad. Sci. U.S.A.* **95,** 3627–3631.

Chen, P., Andersson, D. I., and Roth, J. R. (1994). The control region of the *pdu/cob* regulon in *Salmonella typhimurium*. *J. Bacteriol.* **176,** 5474–5482.

Chen, P., Ailion, M., Bobik, T., Stormo, G., and Roth, J. (1995). Five promoters integrate control of the cob/pdu regulon in *Salmonella typhimurium*. *J. Bacteriol.* **177,** 5401–5410.

Chida, T., Okamura, N., Ohtani, K., Yoshida, Y., Arakawa, E., and Watanabe, H. (2000). The complete DNA sequence of the O antigen gene region of *Plesiomonas shigelloides* serotype O 17 which is identical to *Shigella sonnei* form I antigen. *Microbiol. Immunol.* **44,** 161–172.

Chu, S., DeRisi, J., Eisen, M., Mulholland, J., Botstein, D., Brown, P. O., and Herskowitz, I. (1998). The transcriptional program of sporulation in budding yeast. *Science* **282,** 699–705.

Cotter, D. A., and Raper, K. B. (1968). Properties of germinating spores of *Dictyostelium discoideum*. *J. Bacteriol.* **96,** 1680–1689.

Coury, L. A., Hiller, M., Mathai, J. C., Jones, E. W., Zeidel, M. L., and Brodsky, J. L. (1999). Water transport across yeast vacuolar and plasma membrane-targeted secretory vesicles occurs by passive diffusion. *J. Bacteriol.* **181,** 4437–4440.

da Costa, M. S., Santos, H., and Galinski, E. A. (1998). An overview of the role and diversity of compatible solutes in *Bacteria* and *Archaea*. *Adv. Biochem. Eng. Biotechnol.* **61,** 117–153.

Darbon, E., Ito, K., Huang, H. S., Yoshimoto, T., Poncet, S., and Deutscher, J. (1999). Glycerol transport and phosphoenolpyruvate-dependent enzyme I- and HPr-catalysed phosphorylation of glycerol kinase in *Thermus flavus*. *Microbiology* **145,** 3205–3212.

Daum, G., Lees, N. D., Bard, M., and Dickson, R. (1998). Biochemistry, cell biology and molecular biology of lipids of *Saccharomyces cerevisiae*. *Yeast* **14,** 1471–1510.

Deen, P. M., Croes, H., van Aubel, R. A., Ginsel, L. A., and van Os, C. H. (1995). Water channels encoded by mutant aquaporin-2 genes in nephrogenic diabetis insipidus are impaired in their cellular routing. *J. Clin. Invest.* **95,** 2291–2296.

Delamarche, C., Thomas, D., Rolland, J. P., Froger, A., Gouranton, J., Svelto, M., Agre, P., and Calamita, G. (1999). Visualization of AqpZ-mediated water permeability in *Escherichia coli* by cryoelectron microscopy. *J. Bacteriol.* **181,** 4193–4197.

Finkelstein, A. (1987). "Water Movement through Lipid Bilayers, Pores, and Plasma Membranes: Theory and Reality." John Wiley and Sons, New York.

Flick, K. M., Shaulsky, G., and Loomis, W. F. (1997). The wacA gene of *Dictyostelium discoideum* is a developmentally regulated member of the MIP family. *Gene* **195,** 127–130.

Froger, A., Tallur, B., Thomas, D., and Delamarche, C. (1998). Prediction of functional residues in water channels and related proteins. *Protein Sci.* **7,** 1458–1468.

Gao, Y. P., Young, L., Bonham-Smith, P., and Gusta, L. V. (1999). Characterization and expression of plasma and tonoplast membrane aquaporins in primed seed of *Brassica napus* during germination under stress conditions. *Plant Mol. Biol.* **40,** 635–644.

Gorin, M. B., Yancey, S. B., Cline, J., Revel, J.-B., and Horwitz, J. (1984). The major intrinsic protein (MIP) of the bovine lens fibre membrane: Characterisation and structure based on cDNA cloning. *Cell* **39,** 49–59.

Grauslund, M., Lopes, J. M., and Ronnow, B. (1999). Expression of *GUT1*, which encodes glycerol kinase in *Saccharomyces cerevisiae*, is controlled by the positive regulators Adr1p, Ino2p and Ino4p and the negative regulator Opi1p in a carbon source-dependent fashion. *Nucleic Acids Res.* **27,** 4391–4398.

Gustin, M. C., Zhou, X.-L., Martinac, B., and Kung, C. (1988). A mechanosensitive ion channel in the yeast plasma membrane. *Science* **242,** 762–765.

Haines, T. H. (1994). Water transport across biological membranes. *FEBS Lett.* **346,** 115–122.

Hasler, L., Walz, T., Tittmann, P., Gross, H., Kistler, J., and Engel, A. (1998). Purified lens major intrinsic protein (MIP) forms highly ordered tetragonal two-dimensional arrays by reconstitution. *J. Mol. Biol.* **279,** 855–864.

Heller, K. B., Lin, E. C., and Wilson, T. H. (1980). Substrate specificity and transport properties of the glycerol facilitator of *Escherichia coli. J. Bacteriol.* **144,** 274–278.

Heymann, J. B., and Engel, A. (2000). Structural clues in the sequences of the aquaporins. *J. Mol. Biol.* **295,** 1039–1053.

Heymann, J. B., Agre, P., and Engel, A. (1998). Progress on the structure and function of Aquaporin 1. *J. Struct. Biol.* **121,** 191–206.

Hohmann, S. (1997). Shaping up: The response of yeast to osmotic stress. *In* "Yeast Stress Responses" (S. Hohmann and W. H. Mager, Eds.), pp. 101–145. R. G. Landes Company, Austin, TX.

Hohmann, S., Bill, R., Kayingo, G., and Prior, B. A. (2000). Microbial MIP channels. *Trends Microbiol.* **8,** 33–38.

Holst, B., Lunde, C., Lages, F., Oliveira, R., Lucas, C., and Kielland-Brandt, M. C. (2000). *GUP1* and its close homologue *GUP2,* encoding multimembrane-spanning proteins involved in active glycerol uptake in *Saccharaomyces cerevisiae. Mol. Microbiol.* **37,** 108–124.

Houng, H. S., and Venkatesan, M. M. (1998). Genetic analysis of *Shigella sonnei* form I antigen: Identification of a novel IS630 as an essential element for the form I antigen expression. *Microb. Pathog.* **25,** 165–173.

Johansson, I., Karlsson, M., Shukla, V. K., Chrispeels, M. J., Larsson, C., and Kjellbom, P. (1998). Water transport activity of the plasma membrane aquaporin PM28A is regulated by phosphorylation. *Plant Cell* **10,** 451–459.

Jung, J. S., Preston, G. M., Smith, B. L., Guggino, W. B., and Agre, P. (1994). Molecular structure of the water channel through aquaporin CHIP: The hourglass model. *J. Biol. Chem.* **269,** 14648–14654.

Kamsteeg, E. J., Wormhoudt, T. A., Rijss, J. P., van Os, C. H., and Deen, P. M. (1999). An impaired routing of wild-type aquaporin-2 after tetramerization with an aquaporin-2 mutant explains dominant nephrogenic diabetes insipidus. *EMBO J.* **18,** 2394–2400.

Kjellbom, P., Larsson, C., Johansson, I., Karlsson, M., and Johanson, U. (1999). Aquaporins and water homeostasis in plants. *Trends Plant Sci.* **4,** 308–314.

Koefoed-Johnson, V., and Ussing, H. H. (1953). The contribution of diffusion and flow to the passage of D_2O through living membranes. Effect of neurohyphophyseal hormone on isolated anuran skin. *Acta Physiol. Scand.* **28,** 60–76.

Kopecko, D. J., Washington, O., and Formal, S. B. (1980). Genetic and physical evidence for plasmid control of *Shigella sonnei* form I cell surface antigen. *Infect. Immun.* **29,** 207–214.

Kunst, F., Ogasawara, N., Moszer, I., Albertini, A. M., Alloni, G., Azevedo, V., Bertero, M. G., Bessieres, P., Bolotin, A., Borchert, S., Borriss, R., Boursier, L., Brans, A., Braun, M., Brignell, S. C., Bron, S., Brouillet, S., Bruschi, C. V., Caldwell, B., Capuano, V., Carter, N. M., Choi, S. K., Codani, J. J., Connerton, I. F., and Danchin, A., *et al.* (1997). The complete genome sequence of the gram-positive bacterium *Bacillus subtilis. Nature* **390,** 249–256.

Kwon, H. M., and Handler, J. S. (1995). Cell volume regulated transporters of compatible solutes. *Curr. Opin. Cell Biol.* **7,** 465–471.

Lages, F., and Lucas, C. (1997). Contribution to the physiological characterization of glycerol active uptake in *Saccharomyces cerevisiae. Biochim. Biophys. Acta* **1322,** 8–18.

Lages, F., Silva-Graca, M., and Lucas, C. (1999). Active glycerol uptake is a mechanism underlying halotolerance in yeasts: A study of 42 species. *Microbiology* **145,** 2577–2585.

Lagree, V., Froger, A., Deschamps, S., Hubert, J. F., Delamarche, C., Bonnec, G., Thomas, D., Gouranton, J., and Pellerin, I. (1999). Switch from an aquaporin to a glycerol channel by two amino acids substitution. *J. Biol. Chem.* **274,** 6817–6819.

Laizé, V., Gobin, R., Rousselet, G., Badier, C., Hohmann, S., Ripoche, P., and Tacnet, F. (1999). Molecular and functional study of *AQY1* from *Saccharomyces cerevisiae*. Role of the C-terminal domain. *Biochem. Biophys. Res. Commun.* **257,** 139–144.

Laizé, V., Tacnet, F., Ripoche, P., and Hohmann, S. (2000). Polymorphism of yeast aquaporins. *Yeast* **16,** 897–903.

Luyten, K., Albertyn, J., Skibbe, F., Prior, B. A., Ramos, J., Thevelein, J. M., and Hohmann, S. (1995). Fps1, a yeast member of the MIP-family of channel proteins, is a facilitator for glycerol uptake and efflux and it is inactive under osmotic stress. *EMBO J.* **14,** 1360–1371.

Macey, R. I. (1984). Transport of water and urea in red blood cells. *Am. J. Physiol.* **246,** C195–C203.

Macey, R. I., and Farmer, R. E. L. (1970). Inhibition of water and solute permeability in human red blood cells. *Biochim. Biophys. Acta* **210,** 104–106.

Maurel, C., Reizer, J., Schroeder, J. I., and Chrispeels, M. J. (1993). The vacuolar membrane protein gamma-TIP creates water specific channels in *Xenopus* oocytes. *EMBO J.* **12,** 2241–2247.

Maurel, C., Reizer, J., Schroeder, J. I., Chrispeels, M. J., and Saier, M. H. Jr. (1994). Functional characterization of the *Escherichia coli* glycerol facilitator, GlpF, in *Xenopus oocytes. J. Biol. Chem.* **269,** 11869–11872.

Measures, J. C. (1975). Role of amino acids in osmoregulation of non-halophilic bacteria. *Nature* **257,** 398–400.

Meinild, A. K., Klaerke, D. A., and Zeuthen, T. (1998). Bidirectional water fluxes and specificity for small hydrophilic molecules in aquaporins 0–5. *J. Biol. Chem.* **273,** 32446–32451.

Mitsuoka, K., Murata, K., Walz, T., Hirai, T., Agre, P., Heymann, J. B., Engel, A., and Fujiyoshi, Y. (1999). The structure of aquaporin-1 at 4.5-Å resolution reveals short alpha-helices in the center of the monomer. *J. Struct. Biol.* **128,** 34–43.

Nikaido, H., and Saier, M. H., Jr. (1992). Transport proteins in bacteria: Common themes in their design. *Science* **258,** 936–942.

Norbeck, J., Påhlman, A. K., Akhtar, N., Blomberg, A., and Adler, L. (1996). Purification and characterization of two isoenzymes of DL-glycerol 3-phosphatase from *Saccharomyces cerevisiae*. Identification of the corresponding *GPP1* and *GPP2* genes and evidence for osmotic regulation of Gpp2p expression by the osmosensing MAP kinase signal transduction pathway. *J. Biol. Chem.* **271,** 13875–13881.

Paganelli, C. V., and Solomon, A. K. (1957). The rate of exchange of tritiated water across the human red cell membrane. *J. Gen. Physiol.* **41,** 259–277.

Pao, S. S., Paulsen, I. T., and Saier, M. H., Jr. (1998). Major facilitator superfamily. *Microbiol. Mol. Biol. Rev.* **62,** 1–34.

Park, J. H., and Saier, M. H. (1996). Phylogenetic characterization of the MIP family of transmembrane channel proteins. *J. Membrane Biol.* **153,** 171–180.

Paulsen, I. T., Sliwinski, M. K., and Saier, M. H., Jr. (1998a). Microbial genome analyses: Global comparisons of transport capabilities based on phylogenies, bioenergetics and substrate specificities. *J. Mol. Biol.* **277,** 573–592.

Paulsen, I. T., Sliwinski, M. K., Nelissen, B., Goffeau, A., and Saier, M. H., Jr. (1998b). Unified inventory of established and putative transporters encoded within the complete genome of *Saccharomyces cerevisiae*. *FEBS Lett.* **430,** 116–125.

Philips, J., and Herskowitz, I. (1997). Osmotic balance regulates cell fusion during mating in *Saccharomyces cerevisiae*. *J. Cell Biol.* **138,** 961–974.

Preston, G. M., Carroll, T. P., Guggino, W. B., and Agre, P. (1992). Appearance of water channels in *Xenopus* oocytes expressing red cell CHIP 28 protein. *Science* **256,** 385–387.

Rep, M., Albertyn, J., Thevelein, J. M., Prior, B. A., and Hohmann, S. (1999a). Different signalling pathways contribute to the control of *GPD1* expression by osmotic stress in *Saccharomyces cerevisiae*. *Microbiology* **145**, 715–727.

Rep, M., Reiser, V., Holzmüller, U., Thevelein, J. M., Hohmann, S., Ammerer, G., and Ruis, H. (1999b). Osmotic stress-induced gene expression in *Saccharomyces cerevisiae* requires Msn1p and the novel nuclear factor Hot1p. *Mol. Cell. Biol.* **19**, 5474–5485.

Rep, M., Krantz, M., Thevelein, J. M., and Hohmann, S. (2000). The transcriptional response of *Saccharomyces cerevisiae* to osmotic shock: Hot1p and Msn2p/Msn4p are required for induction of subsets of HOG-dependent genes. *J. Biol. Chem.* **275**, 8290–8300.

Richey, D. P., and Lin, E. C. (1972). Importance of facilitated diffusion for effective utilization of glycerol by *Escherichia coli*. *J. Bacteriol.* **112**, 784–790.

Ringler, P., Borgnia, M. J., Stahlberg, H., Maloney, P., Agre, P., and Engel, A. (1999). Structure of the water channel AqpZ from *Escherichia coli* revealed by electron crystallography. *J. Mol. Biol.* **291**, 1181–1190.

Saier, M. H., Jr. (1994). Convergence and divergence in the evolution of transport proteins. *Bioessays* **16**, 23–29.

Saier, M. H., Jr. (1998). Molecular phylogeny as a basis for the classification of transport proteins from bacteria, archaea, and eukarya. *Adv. Microb. Physiol.* **40**, 81–136.

Saier, M. H., Jr., Eng, B. H., Fard, S., Garg, J., Haggerty, D. A., Hutchinson, W. J., Jack, D. L., Lai, E. C., Liu, H. J., Nusinew, D. P., Omar, A. M., Pao, S. S., Paulsen, I. T., Quan, J. A., Sliwinski, M., Tseng, T. T., Wachi, S., and Young, G. B. (1999). Phylogenetic characterization of novel transport protein families revealed by genome analyses. *Biochim. Biophys. Acta* **1422**, 1–56.

Sanders, O. I., Rensing, C., Kuroda, M., Mitra, B., and Rosen, B. P. (1997). Antimonite is accumulated by the glycerol facilitator GlpF in *Escherichia coli*. *J. Bacteriol.* **179**, 3365–3367.

Sanno, Y., Wilson, T. H., and Lin, E. C. C. (1968). Control of permeation to glycerol in cells of *Escherichia coli*. *Biochem. Biophys. Res. Comm.* **32**, 344–349.

Schweizer, H. P., Jump, R., and Po, C. (1997). Structure and gene-polypeptide relationships of the region encoding glycerol diffusion facilitator (*glpF*) and glycerol kinase (*glpK*) of *Pseudomonas aeruginosa*. *Microbiology* **143**, 1287–1297.

Spencer, J. T. F., and Spencer, D. M. (1978). Production of polyhydroxy alcohols by osmotolerant yeasts. *In* "Primary Products of Metabolism" (H. Rose, Ed.), pp. 394–425. Academic Press, London.

Sutherland, F. C. W., Lages, F., Lucas, C., Luyten, K., Albertyn, J., Hohmann, S., Prior, B. A., and Kilian, S. G. (1997). Characteristics of Fps1-dependent and -independent glycerol transport in *Saccharomyces cerevisiae*. *J. Bacteriol.* **179**, 7790–7795.

Sweet, G., Gandor, C., Voegele, R., Wittekindt, N., Beuerle, J., Truniger, V., Lin, E. C., and Boos, W. (1990). Glycerol facilitator of *Escherichia coli*: Cloning of *glpF* and identification of the *glpF* product. *J. Bacteriol.* **172**, 424–430.

Tamás, M. J., Luyten, K., Sutherland, F. C. W., Hernandez, A., Albertyn, J., Valadi, H., Li, H., Prior, B. A., Kilian, S. G., Ramos, J., Gustafsson, L., Thevelein, J. M., and Hohmann, S. (1999). Fps1p controls the accumulation and release of the compatible solute glycerol in yeast osmoregulation. *Mol. Microbiol.* **31**, 1087–1104.

Tao, W., Deschenes, R. J., and Fassler, J. S. (1999). Intracellular glycerol levels modulate the activity of Sln1p, a *Saccharomyces cerevisiae* two-component regulator. *J. Biol. Chem.* **274**, 360–367.

Truniger, V., and Boos, W. (1993). Glycerol uptake in *Escherichia coli* is sensitive to membrane lipid composition. *Res. Microbiol.* **144**, 565–574.

Truniger, V., Boos, W., and Sweet, G. (1992). Molecular analysis of the *glpFKX* regions of *Escherichia coli* and *Shigella flexneri*. *J. Bacteriol.* **174**, 6981–6991.

Van Aelst, L., Hohmann, S., Zimmermann, F. K., Jans, A. W. H., and Thevelein, J. M. (1991). A yeast homologue of the bovine lens fibre MIP gene family complements the growth defect of a *Saccharomyces cerevisiae* mutant on fermentable sugars but not its defect in glucose-induced RAS-mediated cAMP signalling. *EMBO J.* **10,** 2095–2104.

Van Dijken, J. P., and Scheffers, W. A. (1986). Redox balances in the metabolism of sugars by yeasts. *FEMS Microbiol. Rev.* **32,** 199–225.

van Veen, H. W., and Konings, W. N. (1997). Multidrug transporters from bacteria to man: Similarities in structure and function. *Semin. Cancer Biol.* **8,** 183–191.

van Veen, H. W., and Konings, W. N. (1998). The ABC family of multidrug transporters in microorganisms. *Biochim. Biophys. Acta* **1365,** 31–36.

Van Zyl, P. J., Kilian, S. G., and Prior, B. A. (1990). The role of an active transport mechanism in glycerol accumulation during osmoregulation by *Zygosaccharomyces rouxii*. *Appl. Microbiol. Biotechnol.* **34,** 231–235.

Voegele, R. T., Sweet, G. D., and Boos, W. (1993). Glycerol kinase of *Escherichia coli* is activated by interaction with the glycerol facilitator. *J. Bacteriol.* **175,** 1087–1094.

Walter, D., Ailion, M., and Roth, J. (1997). Genetic characterization of the *pdu* operon: Use of 1,2-propanediol in *Salmonella typhimurium*. *J. Bacteriol.* **179,** 1013–1022.

Wayne, R., and Tazawa, M. (1990). Nature of the water channels in the internodal cells of *Nitellopsis*. *J. Membr. Biol.* **116,** 31–39.

Wiggins, P. M. (1990). Role of water in some biological processes. *Microbiol. Rev.* **54,** 432–449.

Wistow, G. J., Pisano, M. M., and Chepelinsky, A. B. (1991). Tandem sequence repeats in transmembrane channel proteins. *Trends Biochem. Sci.* **16,** 170–171.

Wood, J. M. (1999). Osmosensing by bacteria: Signals and membrane-based sensors. *Microbiol. Mol. Biol. Rev.* **63,** 230–262.

Yamada, S., Katsuhara, M., Kelly, W., Michalowski, C. B., and Bohnert, H. J. (1995). A family of transcripts encoding water channel proteins: Tissue specific expression in the common ice plant. *Plant Cell* **7,** 1129–1142.

Yancey, P. H., Clark, M. E., Hand, S. C., Bowlus, R. D., and Somero, G. N. (1982). Living with water stress: Evolution of osmolyte systems. *Science* **217,** 1214–1222.

Yasui, M., Hazama, A., Kwon, T. H., Nielsen, S., Guggino, W. B., and Agre, P. (1999). Rapid gating and anion permeability of an intracellular aquaporin. *Nature* **402,** 184–187.

CHAPTER 9

Future Directions of Aquaporin Research

Stefan Hohmann,[*] **Søren Nielsen,**[†] **Christophe Maurel,**[+] **and Peter Agre**[‡]

[*]Department of Cell and Molecular Biology/Microbiology, Göteborg University, S-40530 Göteborg, Sweden; [†]Department of Cell Biology, Institute of Anatomy, University of Aarhus, DK-8000 Aarhus, Denmark; [+]Biochimie et Physiologie Moléculaire des Plantes, ENSAM/INRA/CNRS/UMII Montpellier, France; [‡]Departments of Biological Chemistry and Medicine, Johns Hopkins University, School of Medicine, Baltimore, Maryland 21205

  I. Identification and Characterization of New MIP Channels
 II. Why Have So Many MIP Channels?
III. Analysis of the Physiological Roles of Aquaporins
 IV. Structure and Function
  V. Aquaporins as Targets for Treatment of Human Disease
 VI. Aquaporins as Possible Targets for Genetic Engineering and
     Crop Improvement
VII. Metabolic Engineering with Aquaporins
VIII. The Aquaporin Research Community
     References

In the preceding chapters some of the peers in research on MIP channels have provided state-of-the-art overviews on the present knowledge on structure, function, and physiological roles of aquaporins and glycerol facilitators. In this final chapter we try to address future developments and raise questions and topics for future research in this exciting field. This discussion chapter does not contain extensive reference to published articles, as those can be found in the different chapters. It does, however, take into account very recent results reported at the Molecular Biology and Physiology of Water and Solute Transport Conference, which was held in July 2000 in Gothenburg, Sweden.

Current Topics in Membranes, Volume 51

I. IDENTIFICATION AND CHARACTERIZATION OF NEW MIP CHANNELS

More than 200 MIP channels have now been identified in all organisms ranging from Archea to human. Unexpectedly, one reason for the expansion of the family is due to the realization that the canonical NPA boxes in loops B and E are not strictly conserved. Several examples have now been reported for MIP channels that fulfill all topological criteria and exhibit many conserved amino acids but have "NPA" motifs that differ in one amino acid position. For instance, NPS and NLA have been reported in several MIP channels and in some instances both NPA motifs are altered in one and the same protein. Hence, the MIP family is further expanding to proteins that are divergent from the core of the family, the aquaporins *sensu strictu* and the glycerol facilitators. This is important when genome sequences are searched for members of the family using signature sequences, and the criteria to include a novel protein in the MIP family should be carefully revised and updated with new sequences emerging. The structural and functional consequences of divergent NPA motifs are not presently understood and future functional studies should address those in more detail (Heymann and Engel, 2000).

Genome sequencing projects reveal new sequences for genes encoding MIP channels at an ever-increasing rate. Two aspects are most interesting in this respect. First, the complete sequence of a genome immediately identifies the complete set of MIP channels for that organism. For instance, although 10 MIP channels are known for humans, the complete genome sequence will almost certainly lead to the discovery of new family members. With the genome sequence of the plant *Arabidopsis* being completed in 2001, we will know all of the aquaporins present in a higher eukaryote. A total count in the range of 35 MIP genes is expected in this organism. Hence, soon we will have an overview of the total set of MIP channels for a number of model animal and plants. Bioinformatics tools will be used to classify those into subfamilies and hence into potential functional groups. This initial classification could then be a guide for functional analysis. Comparison of the complete set of MIP channels across different related species may then reveal key aquaporins that have been conserved and whose function is likely to be of crucial importance and those aquaporins that may play more specialized roles in different species. However, when comparing protein families across species border one should keep in mind that even between relatively closely related species, amplification of genes within a family occurs, as has for instance been observed for the sugar transporter family in yeasts (see further).

The second interesting aspect of genome sequencing is the identification of many orthologs, that is, a large number of genes/proteins with (presumably) the same function in different organisms. For instance, glycerol facilitators that are part of a glycerol utilization operon have been sequenced from a whole range of gram-positive and gram-negative bacteria. Comparison of those sequences aids the identification of amino acid residues that are presumably important for the specific function (Froger *et al.,* 1998; Heymann and Engel, 2000). In conclusion, with more genome sequences becoming available at an even higher pace, bioinformatics analyses of those sequences remain an attractive tool in the studies of both the putative physiological roles and the structural determination of transport function and specificity of MIP channels.

The wealth of sequence information on MIP channels illustrates that hunting for new genes by PCR approaches, which has for many years been the main source for new MIP sequences, will become less attractive, unless it is done to address a specific question in an organism whose genome has not yet been sequenced or will not be sequenced in the foreseeable future. Hence the race for new genes is basically over and the main task is the analysis of the function of the many MIP channels in different systems, that is, the "fun part" of this research.

II. WHY HAVE SO MANY MIP CHANNELS?

Why are there so many MIP channels in one organism? Rough estimates indicate that about 1 in 1000 genes in eukaryotes may encode a MIP channel and in plants it may be even more. Some clues to this philosophical question may come from comparison with the major facilitator superfamily (MFS) or the ABC transporters in microorganisms. Those families are more divergent but we could use as an example the 20 yeast hexose transporters, which belong to the MFS family and are at least as closely related to each other as are the MIP channels from a given organism. It appears that these proteins have divergent substrate specificity and affinity, and they are expressed under different growth conditions (Boles and Hollenberg, 1997). This expression pattern allows the yeast cell to adjust sugar transport specificity, affinity, and capacity to the growth conditions (sugar concentration and stress).

It is straightforward to imagine that expression of different MIP channels with different specificity under different growth conditions may serve a similar purpose. Some genes may be expressed in certain developmental stages adding an additional level of complexity. This may be especially true for plants, which have to achieve proper water and solute transport throughout their development and in response to a large variety of environmental constraints. In addition, some family members (MFS or MIP) are located in intracellular compartments and hence may control transport between the cytosol and different organelles, such as the vacuole, the ER, mitochondria, and chloroplasts. The MFS family of fungi in fact offers another interesting option: some proteins that resemble very closely hexose or amino acid transporters are in fact sugar or amino acid sensors and control via signaling pathways the expression of other transporters and even more complex regulatory systems such as a developmental switch of the growth pattern in yeast

(Boles and Hollenberg, 1997; Iraqui *et al.,* 1999). Hence, some of the MIP channels may be rather (osmo?) sensors than transporters, or a combination of both.

To complete the discussion of possible analogies between yeast hexose transporters and MIPs, although each yeast hexose transporter may have a defined function, their physiological roles are sufficiently overlapping such that deletion of a single gene does not reveal a clear phenotype. Hence, redundancy or overlapping function is certainly an aspect to consider for the many MIP channels in a single organism.

III. ANALYSIS OF THE PHYSIOLOGICAL ROLES OF AQUAPORINS

There is no doubt that aquaporins and solute facilitators have profound physiological roles but how should those be approached, which channels should be studied, and should we strive for unified, standard procedures to characterize those channels? Comprehensive answers to the following questions should lead to an understanding of the physiological role of a MIP channel (and in fact any other transport protein):

1. What is the substrate spectrum?
2. Where is the protein localized?
3. When and where is the protein expressed?
4. What is the phenotype of a knockout mutant?

There are numerous examples in the literature where the answers to at least some of those questions were different depending on the investigator (at some point this was even true for the sequence!). Such controversial results indicate that the systems and protocols for functional analyses should be chosen carefully. For instance, the standard procedure for testing MIP channel specificity is still expression in *Xenopus* oocytes, and this system has been most instrumental in studying water and glycerol transport. In addition, oocytes are very helpful tools for electrophysiology (ion transport) and intracellular pH measurements. However, several naturally occurring MIP channels, as well as mutants from certain aquaporins, are not localized to the oocyte plasma membrane and quantitative transport measurements are sometimes complicated. Hence, other systems, like growth assays with whole-yeast cells and secretory or plasma membrane vesicles from yeast cells expressing a MIP channel, are suitable alternatives being considered by more and more groups. This example illustrates that numerous approaches are around to study a key property, transport specificity, and certainly different protocols may contribute to conflicting data for a number of reasons (for instance, different lipid composition in different membranes!). It would certainly be helpful, albeit difficult to achieve, if different research groups agreed on a limited number of standard procedures in order to enhance cross-laboratory reproducibility.

In any case, recent data suggest that the substrate spectrum of MIP channels may go well beyond water, urea, and glycerol. Transport of other polyols up to C5 has been reported as well as transport of ions and even gasses (Cooper and Boron, 1998; Yasui et al., 1999). This wider and certainly not fully explored substrate spectrum of MIP channels suggests that the physiological role of many members of this family may not be limited to water transport. Future work should hence consider comprehensive studies on the spectrum of transported substrates, perhaps in fact using different, well-established test systems.

The understanding of the physiological role of a transport protein requires knowledge of its subcellular localization. The recent discovery that mammalian AQP6 is located in intracellular vesicles (Yasui et al., 1999) underscores the importance of this point also for animal cells, while location in different cellular membranes is well established in plants (Kjellbom et al., 1999). Thus, a critical role for intracellular aquaporins in plant cell osmoregulation and in nitrogen fixation by symbiotic root nodules has been proposed. In extension, it will be most important to establish in which tissue or cells within a multicellular organism a given MIP channel is expressed and under which conditions. Although taking the expression pattern for a given protein as a guide to its physiological role can also be misleading, it is of crucial importance in establishing the physiological role and also to interpret correctly the phenotype of knockout mutants: If the phenotype caused by a knockout appears in a tissue or under conditions where the gene product is not expressed, the phenotype is likely either an artifact or due to a compensatory mechanism.

Knockout models are an indispensable tool in establishing a physiological role for any protein and, of course, for MIP channels too. This is true for any biological system, mammals, animals, plants, fungi, and bacteria. As becomes clear from different chapters in this book, the interpretation of knockout phenotypes, even in "simple" organisms such as yeast or bacteria, may not be straightforward. There are several reasons for that. First of all, the large number of MIP channels in multicellular organisms suggests that several of these proteins may have redundant or partially overlapping functions. Hence, loss of one protein may not lead to a strong phenotype and the construction of mutants lacking multiple MIP channels may be necessary (raising this question: How can we find those that have redundant functions?). Secondly, many MIP channels transport substrates that penetrate the membrane even without assistance of a channel, albeit at a lower rate. However, this lower rate may under most conditions already be sufficient, thereby restricting the physiological role of MIP channels to specialized conditions. Finally, interpretation of knockout phenotypes may be complicated by incomplete establishment of the other criteria needed to assess a physiological role, that is, range of transported substrates, localization, and expression pattern.

IV. STRUCTURE AND FUNCTION

The structure of AQP1 and GlpF at atomic resolution will be available soon and this knowledge will give another boost to this field. The key questions that we hope to answer on the basis of MIP channel structure are (1) what is the mechanism of water transport, (2) is the transport mechanism for water and glycerol (and other substrates) principally different, and (3) what determines the both amazing and highly important transport specificity? Extensive studies by site-directed mutagenesis will follow to probe the interpretation of the structure. Novel systems to screen in yeast for mutants that alter transport specificity may also be helpful and their value may be enhanced by the possibility of interpreting the consequences of random amino acid changes on the basis of the structure. Such yeast systems may also allow for the stepwise evolution of MIP channels toward different transport function, a helpful tool to further understand transport mechanisms and specificity as well as for the design of MIP channels with specified properties.

The present view on the mechanism of transport focuses around the interaction of the substrate molecule with certain amino acid residues within the pore via hydrogen bonding. Hence the water molecule may tumble through the channel from one hydrogen bond to another (Zeuthen and Klaerke, 1999). Such a mechanism could explain how specificity is achieved, because interaction between the substrate and certain amino acid residues within the pore is involved. On the other hand, channels are viewed rather more or less as a hole in the membrane and their selectively could only be achieved via pore size and substrate shape. The proposed transport mechanism of MIP channels would suggest that a certain activation energy for transport is needed and it may be different for different MIP channels and different substrates. In any case, the activation energy for aquaporins is very low, close to that for passive diffusion, and this was one of the parameters used to identify aquaporins. Taken together, MIP channels may turn out as a paradigm of a class of proteins that combine low activation energy with an utmost transport specificity. Hence, apart from the major technological and scientific achievement, the structure of AQP1 and GlpF will be of fundamental biological and biophysical relevance.

Do we need more structural information to understand the function of MIP channels? That is, should more MIP channels be analyzed to atomic resolution? The answer should probably be yes. As pointed out above, different MIP channels have widely different transport specificity and even different amino acids in core positions, such as the canonical NPA motifs. It is, of course, at present unclear whether the AQP1 and GlpF structure and sophisticated tools of bioinformatics may be sufficient to interpret the function of these divergent channels. In addition, the understanding of transport of different substrates will also require knowledge of the structure of the substrate. For instance, transport of certain ions such as antimonite through glycerol facilitators may be due to the hydrated form of the ion resembling a polyol (Sanders *et al.,* 1997).

Perhaps of even higher relevance, however, is the question of the structural basis for channel regulation, that is, gating. Gated MIP channels have been reported in mammals, plants, and fungi and obviously gating is a most important means to control water and solute flux. At present, the mechanisms of gating of MIP channels are not understood and may well be different for different channels. Present evidence suggests gating of AQP6 (and AQP3) by pH (Yasui *et al.,* 1999; Zeuthen and Klaerke, 1999), of plant α-TIP and PM28A by phosphorylation (Kjellbom *et al.,* 1999), and of yeast Fps1p by osmotic shock and hence probably membrane stretching (Tamás *et al.,* 1999). A complete understanding of the detailed regulatory mechanisms will be important to fully understand the physiological role and may perhaps also aid the identification of MIP channel blockers.

V. AQUAPORINS AS TARGETS FOR TREATMENT OF HUMAN DISEASE

The studies of naturally occurring mutants of AQP2, but especially analyses of knockout models of AQP2 and other AQPs, strongly suggest that aquaporins could be potential targets for the treatment of different human disease conditions for which no treatment is currently available or available treatments cause severe side effects. Inhibitors of kidney aquaporins would function as diuretics and hence could be helpful in the treatment of hyponatremia, which is a consequence of congestive heart failure, preeclampsia, liver cirrhosis, and the syndrome of inappropriate secretion of the antidiuretic hormone. Brain edema, a fatal condition that is often a consequence of hyponatremia, stroke, accidents, and cancer, could potentially be treated by blockers of AQP4. Specific inhibition of AQP5 in sweat glands could be used as a treatment of hyperhydrosis and could be a means to control transpiration.

Certainly more work is required to verify aquaporins as suitable drug targets, and this work will include very careful analyses of the phenotypes of knockout models as well as pathophysiological studies in humans. Another most important aspect is to find suitable inhibitors for aquaporin function. The only presently known inhibitors of aquaporins are mercurials, which are of course not suitable for drug development, as well as tetraethylammonium. Several different approaches can be imagined, such as the screening of large compound libraries. This approach will require the establishment of suitable high-throughput test systems. In addition, the structure of aquaporins at atomic resolution as well as a complete understanding of the transport and gating mechanisms may aid the design of potential lead compounds. It is clear that, at present, the availability of a drug based on an aquaporin blocker is many years ahead and will require challenging research at different levels (physiology, pathophysiology, structural analysis, etc.). Given the already established and expected important roles of aquaporins in human physiology, these efforts may be justified and this is apparently also acknowledged by funding bodies in different countries.

VI. AQUAPORINS AS POSSIBLE TARGETS FOR GENETIC ENGINEERING AND CROP IMPROVEMENT

The high diversity of MIP channels in plants, with more than 30 genes already known in *Arabidopsis,* somehow reflects the intricate and crucial relationship of plants with water. The subtle control of plant aquaporin gene expression by tissue-specific and environmental cues, together with the inhibition by mercury compounds of water transport in roots, have provided the first lines of evidence about the integrated function of plant aquaporins. However, the precise physiological role of a definite MIP channel has not been established in any plant system. There is to date only one report on the analysis of a knockout model, or rather a down-regulation (antisense RNA) model, which showed a 3-fold enlargement of the root system (Kaldenhoff *et al.,* 1998). This dramatic effect conforms to well-established observations that plants adjust their development depending on the availability of water and mineral nutrients. However, this effect establishes for the first time fascinating and direct connections between membrane water transport and the plastic development of plants. This effect also suggests that future studies on knockout models will reveal many surprises although the use of single-gene knockout models may not be easy because of the reasons discussed above. Several research groups now concentrate on *Arabidopsis.* Its genome sequence is about to be completed and, we hope, large collections of knockout mutants for many or even all aquaporins will soon become available. Those will certainly be a most informative resource and their careful analysis, including the construction of plants with multiple knockouts and the analysis of the corresponding gene expression patterns, will be a most challenging task for the future.

The knowledge of plant aquaporins at the molecular and cellular levels is also somehow rudimentary and hence these proteins are at the center of key issues in plant cell physiology. The transport *in planta* of gasses such as CO_2 and NH_3 by aquaporins raises intriguing possibilities about a role for these proteins in leaves during photosynthesis or in roots during symbiotic nitrogen fixation. Aquaporins have also emerged as fine markers of plant membrane differentiation and have revealed, for instance, that plant cells can contain distinct vacuolar subtypes. They will provide valuable tools in future cell biological studies. Finally, the mechanisms that couple aquaporin functions to developmental and environmental stimuli are largely unknown. Thus, the determination of the physiological and transport roles of plant aquaporins is still ongoing, and it is difficult to assess the true potential of aquaporins as a target for genetic engineering or as markers for the breeding of novel crop varieties. However, interesting perspectives have recently emerged. The function of aquaporins in root water transport has been the object of most physiological studies and is now well established. It is also clear that aquaporins are involved in many facets of plant development from seed maturation and germination to sexual reproduction and in the response of plants to environmental factors as diverse as nutrient availability, drought, salt stress, day and night cycles, or anaerobiosis. Their role in these processes is being studied by an increasing number of public and private laboratories.

VII. METABOLIC ENGINEERING WITH AQUAPORINS

A certainly attractive possible use of aquaporins is in the engineering of solute transport and metabolism. The basic principle is illustrated by yeast Fps1p. Engineering of this protein results in constitutive efflux of glycerol from the yeast cell and hence to overproduction of glycerol. In another example, an engineered yeast strain that produces xylitol has been stimulated to produce even more xylitol (an artificial sweetener) when expressing a specific version of Fps1p. Hence it should be possible to use MIP channels with specific transport properties to stimulate passive efflux of certain compounds from microorganisms. The availability of screening and evolution systems for MIP channels with altered transport specificity could be instrumental in such approaches. The principle is not restricted to microorganisms that have been engineered to produce certain useful chemicals. One could also imagine engineering plants such that solutes are overproduced and transported in certain cells and tissues. Such effects could be helpful in the engineering of drought or cold stress tolerance. There are certainly a number of possible applications of this overriding principle and future studies will aim at demonstrating its usefulness.

VIII. THE AQUAPORIN RESEARCH COMMUNITY

The topics of the chapters in this volume illustrate that research on MIP channels is a multidisciplinary endeavor. Extensive cross-reference between chapters also indicates that researchers in different disciplines such as structure–function analysis and mammalian and plant physiology communicate or even collaborate. This makes this research field exciting and informative and, especially for the young researchers involved, of highest educational value. In July 2000, some 200 researchers gathered in Gothenburg, Sweden, and discussed the latest developments in the field of MIP channel research. This multidisciplinary though specialized conference confirmed that researchers working on very different biological systems, in medical and clinical research, in plant physiology and biotechnology, in microbiology, in biophysics, in structural biology, and so on, can find common ground to discuss their MIP channel data and pose relevant questions to each other.

This is also documented by joint research networks and, for instance, the joint publication of this volume. Even more intense collaboration, for instance, at the level of standardized test systems for the determination of substrate spectra or the exchange of biological material may advance the field at an even higher pace. A joint initiative for a common web-based information resource, updating newly identified MIP channels, analysis protocols, available tools and antibodies, and so on, could be the basis for combining research efforts in this field in an even more powerful way. Suitable domain names such as *www.aquaporin.org* have been reserved.

References

Boles, E., and Hollenberg, C. P. (1997). The molecular genetics of hexose transport in yeasts. *FEMS Microbiol. Rev.* **21,** 85–111.

Cooper, G. J., and Boron, W. F. (1998). Effect of PCMBS on CO_2 permeability of *Xenopus* oocytes expressing aquaporin 1 or its C189S mutant. *Am. J. Physiol.* **275,** C1481–C1486.

Froger, A., Tallur, B., Thomas, D., and Delamarche, C. (1998). Prediction of functional residues in water channels and related proteins. *Protein Sci.* **7,** 1458–1468.

Heymann, J. B., and Engel, A. (2000). Structural clues in the sequences of the aquaporins. *J. Mol. Biol.* **295,** 1039–1053.

Iraqui, I., Vissers, S., Bernard, F., de Craene, J. O., Boles, E., Urrestarazu, A., and Andre, B. (1999). Amino acid signaling in *Saccharomyces cerevisiae:* A permease-like sensor of external amino acids and F-Box protein Grr1p are required for transcriptional induction of the AGP1 gene, which encodes a broad-specificity amino acid permease. *Mol. Cell Biol.* **19,** 989–1001.

Kaldenhoff, R., Grote, K., Zhu, J. J., and Zimmermann, U. (1998). Significance of plasmalemma aquaporins for water-transport in *Arabidopsis thaliana. Plant J.* **14,** 121–128.

Kjellbom, P., Larsson, C., Johansson, I., Karlsson, M., and Johanson, U. (1999). Aquaporins and water homeostasis in plants. *Trends Plant Sci.* **4,** 308–314.

Sanders, O. I., Rensing, C., Kuroda, M., Mitra, B., and Rosen, B. P. (1997). Antimonite is accumulated by the glycerol facilitator GlpF in *Escherichia coli. J. Bacteriol.* **179,** 3365–3367.

Tamás, M. J., Luyten, K., Sutherland, F. C. W., Hernandez, A., Albertyn, J., Valadi, H., Li, H., Prior, B. A., Kilian, S. G., Ramos, J., Gustafsson, L., Thevelein, J. M., and Hohmann, S. (1999). Fps1p controls the accumulation and release of the compatible solute glycerol in yeast osmoregulation. *Mol. Microbiol.* **31,** 1087–1104.

Yasui, M., Hazama, A., Kwon, T. H., Nielsen, S., Guggino, W. B., and Agre, P. (1999). Rapid gating and anion permeability of an intracellular aquaporin. *Nature* **402,** 184–187.

Zeuthen, T., and Klaerke, D. A. (1999). Transport of water and glycerol in aquaporin 3 is gated by H(+). *J. Biol. Chem.* **274,** 21631–21636.

Index

381

ISBN 0-12-153351-4

GLP cluster

GLPB3
GLPFmge
GLPFmpn
GLPFmga
GLPB2 GLPFlac
GLPFstr
GLPFbac
GLPY1 GLPY1 GLPY2 GLPY2
AQPA AQPsco AQPS
GLPB1 GLPFsal AQParc
GLPFpse
GLPFhae
GLPFeco
AQP2
AQP5
AQP7 AQP0
AQP9 AQP6
GLPA AQP3 AQP1 AQP0
AQP4
AQPhin
GLPcel2 AQPcic
GLPcel4
GLPcel3 AQPbib
GLPcel1

TIPG
TIPic
TIPD TIP
TIPA
AQPscy
AQPZ AQPZ
PIP1
PIP2 PIP
NIPnic AQPcel
NIPory
NIPgly
NIP NIPara AQP8 AQP8
AQPdic
AQY AQPD
AQPY **AQP cluster**

CHAPTER 2, FIGURE 1 Phylogenetic analysis of the aquaporin superfamily suggests a classification into two clusters, AQP and GLP, 16 subfamilies, and 46 types. The types are considered to be representative of the whole family of more than 160 sequences obtained from Genbank, SWISS_PROT, EMBL, and the genome databases (Heymann and Engel, 1999).

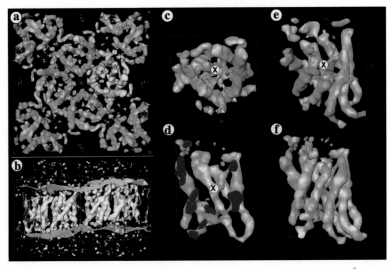

CHAPTER 2, FIGURE 16 The unit cell of 2D AQP1 crystals has a side length of 96 Å and houses eight asymmetric units that form two tetramers integrated into the bilayer in opposite orientations. (a) The view along the 4-fold symmetry axis (♦) shows the cytoplasmic surface of the central tetramer, with one monomer colored in gold. Tetramers adjacent to the central tetramer are seen from the extracellular side, which exposes the connecting loop between monomers (*). Molecular boundaries are reflected by narrow gaps marked by arrows in the vertical slice displayed in (b). This slice, with a width of 152 Å and a thickness of 60 Å, contains four monomers and has been cut as outlined in (a). The overlaid surface reconstruction determined by metal shadowing and atomic force microscopy (Walz *et al.*, 1996) illustrates that AQP1 protrudes significantly from the membrane on the extracellular side. In addition, the surface extends down to the lipid bilayer between tetramers as indicated by a two-headed arrow. (c) The AQP1 monomer as seen in (a) from the cytosolic side and (d) cut open to expose the central density **X**. The slice shown in (d) is outlined in (c), and it is 16 Å thick. (e) The monomer after a clockwise rotation around the *x* axis by 45° and (f) after another 45° clockwise rotation around the *x* axis. (♦) marks the 4-fold axis in (a) and (c), while the horizontal line in (e) indicates the *x* axis.

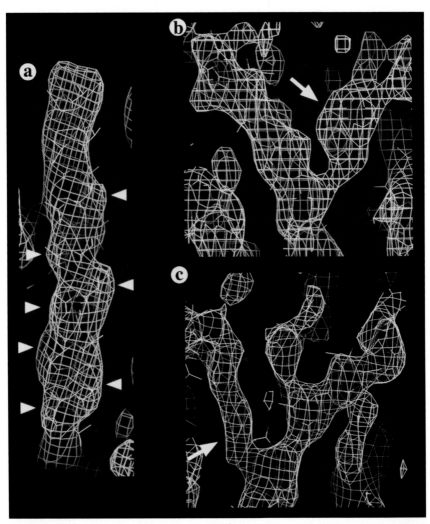

CHAPTER 2, FIGURE 18 Views of the 4.5-Å resolution potential map contoured at 1.0 σ. (a) The potential map around a transmembrane rodlike structure shows protrusions corresponding to side chains of the transmembrane helix (*arrowheads*). Using the protrusions as markers, the poly-Ala helix (represented by the stick model) was manually built and subsequently refined. (b, c) The potential maps defining the two pore helix loop structures forming density X are shown in separate panels. The membrane surface is at the top of panels (b) and (c). Arrows indicate the densities of the pore loops while the stick models represent the pore helices.

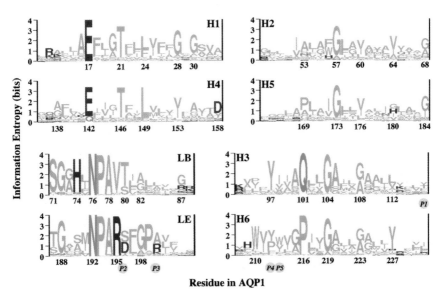

Residue in AQP1

CHAPTER 2, FIGURE 19 Sequence logos reveal the conservation of residues at particular positions in the sequence. The core segments of the sequence alignments were converted to sequence logos (Schneider and Stephens, 1990) and shown with the residue numbers for AQP1. The five positions (P1–P5) that were found by Froger *et al.* (1998) to be different between the two clusters are shown in italics in circles. The scale gives the certainty of finding a particular amino acid type at each position, and is related to the entropy as $R_{seq} = \log_2\{20 \cdot H \cdot c(m)\}$, where $c(m)$ is a correction factor for small sample size m. Colors: Gray, hydrophobic; light-blue, polar; green, amide; red, acidic; blue, basic.

```
               5              15              25              35              45
                                                        H1 ────────────────────────
AQP0   - - - - - - - - -   - - - - M M E L R S   A S F W R A I F A E   F F A T L F Y V F F   G L G S S L R W A P
AQP1   - - - - - - - - -   - - - M A S E F K K   K L F W R A V Y A E   F L A T T L F Y F I   S I G S A L G F K Y
AQP2   - - - - - - - - -   - - - - - - - - - -   - - M F R K L A A E   C F G I F W L Y F G   G C G S A Y L A A G
TIP1   M A T S A R R A Y G   F G - R A D E A T H   P D S I R A T L A E   F L S I F Y F Y F A   A E G S I L S L D K
GLPF   - - - - - - - - -   - - - - - - M S Q T   S T L K G Q C I A E   F L G I G L L I F F   G V G C V A A L K Y

              55              65              75              85              95
                                H2 ────────────────────────
AQP0   G - - - - - - - -   - - - - - P L H Y L   Q V A M A F G L A L   A T L Y Q S V G H I   S G A H V N P A V T
AQP1   P Y G N N Q T - - -   - - - - A Y Q D N Y   K V S L A F G L S I   A T L A Q S V G H I   S G A H L N P A V T
AQP2   F P E L G - - - -   - - - - - - I G F A   G V A L A F G L T Y   L T M A F A V G H I   S G G H F N P A V T
TIP1   L Y W E H - - A A H   A G T N T P G G L I   L Y A L A H A F A L   F A A Y S A A I N Y   S G G H V N P A V T
GLPF   A G - - - - - - -   - - - - A S F G Q W   E I S Y I W E L G Y   A M A I Y L T A G Y   S G A H L N P A V T

             105             115             125             135             145
                    H3 ────────────────
AQP0   F A F L Y G S Q M S   L L R A F C Y M A A   Q L L G A V A G A A   Y L Y S V T - - - -   - - - - - - P P A
AQP1   L G L L L S C Q I S   I F R A L M V I I A   Q C Y G A I V A T A   I L S G I T - - - -   - - - - - - - S S L
AQP2   I G L W A G G R F P   A K E Y Y G V Y I A   Q Y Y G G I V A A A   L L Y L I A S G K T   G F D A - - A A S G
TIP1   F G A L Y G G R V T   A I R A I Y Y W I A   Q L L G A I L A C L   L L R L T T - - - -   - - - - - - - - - -
GLPF   I A L W L F A C F D   K R K Y I P F I Y S   Q V A G A F C A A A   L Y Y G L Y Y N L F   F D F E Q T H H I Y

             155             165             175             185             195
                                        H4 ────────────────
AQP0   Y R G - - - - - -   N L A L N T L H P A   V S Y G Q A T T V E   I F L I L Q F Y L C   I F A T Y D - E R R
AQP1   T G N - - - - - -   S L G R N D L A D G   Y N S G Q G L G I E   I I G T L Q L Y L C   Y L A T T D - R R R
AQP2   - F A S - - - - - -   N G V G E H S P G G   Y S M L S A L Y Y E   L Y L S A G F L L Y   I H G A T D - - - K
TIP1   - N G M - - - - - -   R P Y G F R L A S G   Y G A V N G L Y L E   I I L T F G L V Y Y   V Y S T L I D P K R
GLPF   R G S Y E S Y D L A   G T F S T Y P N P H   I N F Y Q A F A V E   M V I I A I L M G L   I L A L T D D G N G

      H5 ────────────────
             205             215             225             235             245
AQP0   N G Q L G S V A L A   Y G F S L A L G H L   F G M V Y T G A G M   N P A R S F A P A I   L T G N - - - - -
AQP1   R D L G G S A P L A   I G L S Y A L G H L   L A I D Y T G C G I   N P A R S F G S A V   I T H N - - - - -
AQP2   F A P A G F A P I A   I G L A L T L I H L   I S I P Y T N T S Y   N P A R S T A V A I   F Q G - - - - - -
TIP1   G S L G I I A P L A   I G L I V G A N I L   Y G G P F S G A S M   N P A R A F G P A L   Y G W R - - - - -
GLPF   V P R G P L A P L L   I G L L I A Y I G A   S M G P L T G F A M   N P A R D F G P K Y   F A W L A G W G N Y

             255     H6 ──────────────────────────────────
                          265             275             285             295
AQP0   - - - - - - F T - N   H W V Y W V G P I I   G G G L G S L L Y D   F L L F P R L K S I   S - - - - - E R L S
AQP1   - - - - - - F S - N   H W I F W V G P F I   G G A L A Y L I V D   F I L A P R S S D L   T - - - - - D R Y K
AQP2   - - - - G W A L E Q   L W F F W V V P I Y   G G I I G G L I V R   T L L E K R D - - -   - - - - - - - - - -
TIP1   - - - - - - - W H D   H W I V W V G P F I   G S A L A A L I V E   Y M V I P - - T E P   P T H H A H G Y H Q
GLPF   A F T G G R D I P Y   F L V P L F G P I Y   G A I Y G A F A V R   K L I G R H L P C D   I C V Y E E K E T T

             305             315             325
AQP0   V L K G A K P D V S   N G Q P E Y T G E P   Y E L N T Q A L
AQP1   V W T S G Q Y E E Y   D L D A D D I N S R   Y E M K P K - -
AQP2   - - - - - - - - -   - - - - - - - - - -   - - - - - - - -
TIP1   P L A P E D Y - - -   - - - - - - - - - -   - - - - - - - -
GLPF   T P S E Q K A S L -   - - - - - - - - - -   - - - - - - - -
```

CHAPTER 2, FIGURE 47 Comparison of five aquaporin sequences aligned according to Heymann and Engel (2000). Helices are indicated by solid lines. The highlighted residues are similar to those highlighted in Fig. 19. Remarkable variations are observed in the surface loop regions, explaining the differences of surface topographies (Fig. 49). Functional loop B is equal in length in all proteins and highly conserved in the first half, but loop E is by 10–12 residues longer in GlpF than in the others.

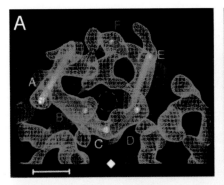
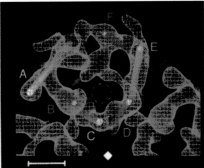
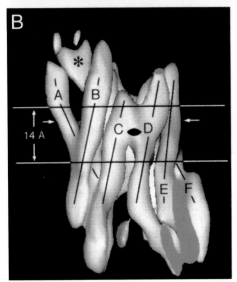

CHAPTER 5, FIGURE 14 Electron crystallography of AQP1 crystals in reconstituted proteoliposomes.(A) Stereo pair of 3D density maps of frozen-hydrated AQP1 viewed approximately perpendicular to the bilayer. The rods trace the approximate paths of the centers of six tilted helices packed with a right-handed twist. These helices surround a vestibular region which narrows to a diameter of ~8 Å (*indicated by dashed circle*). The arrow indicates density assigned to the vertically apposed NPA sequences. (B) Surface-shaded representation of the 6-helix barrel viewed parallel to the bilayer. In-plane molecular pseudo-two-fold symmetry (*black ellipse*) is strongest within the demarcated transmembrane region. (Adapted from Cheng *et al.*, 1997.)

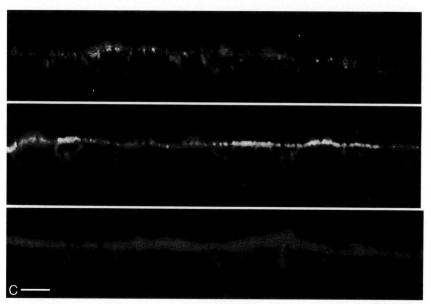

CHAPTER 6, FIGURE 2 Localization of AQP2 in transfected LLC-PK$_1$ and MDCK cells. LLC-PK$_1$ cells stably expressing an AQP2-c-*myc* fusion protein were treated (A) without or (B) with vasopressin, fixed and stained for c-*myc*. Without stimulation, AQP2 is mainly located on intracellular, perinuclear vesicles, whereas after vasopressin stimulation, AQP2 is mainly located on the plasma membrane. (See text page 241 for parts A and B). (C) Transfected MDCK cells stably expressing AQP2 (WT10 cells; upper, middle panel) or native MDCK cells (lower panel) grown on coverslips were treated without (upper panel) or with forskolin (middle, lower panel), fixed, and apically labeled with biotin LC-hydrazide. After permeabilization, cells were incubated with affinity-purified rabbit anti-AQP2 antibodies, rat anti-mouse E-cadherin, and TRITC-coupled extravidin, and after washing, with affinity-purified goat anti-rabbit IgG coupled to FITC and Cy-5-coupled goat anti-rat IgG. Following washing, dehydration and embedding, *x-z* axes images were obtained with a Bio-Rad MRC-1000 laser scanning confocal imaging system. E-cadherin (blue), apical glycoproteins (red), and AQP2 (green) are indicated. Note the redistribution of AQP2 from intracellular vesicles to the apical membrane upon forskolin treatment. (Originally published in Deen *et al.*, 1997a.)

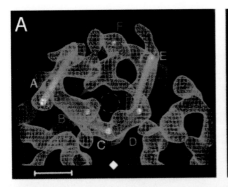
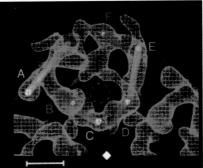

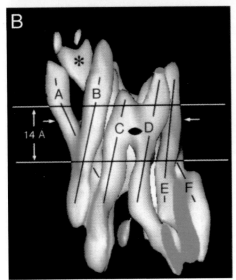

CHAPTER 5, FIGURE 14 Electron crystallography of AQP1 crystals in reconstituted proteoliposomes.(A) Stereo pair of 3D density maps of frozen-hydrated AQP1 viewed approximately perpendicular to the bilayer. The rods trace the approximate paths of the centers of six tilted helices packed with a right-handed twist. These helices surround a vestibular region which narrows to a diameter of ~8 Å (*indicated by dashed circle*). The arrow indicates density assigned to the vertically apposed NPA sequences. (B) Surface-shaded representation of the 6-helix barrel viewed parallel to the bilayer. In-plane molecular pseudo-two-fold symmetry (*black ellipse*) is strongest within the demarcated transmembrane region. (Adapted from Cheng *et al.*, 1997.)

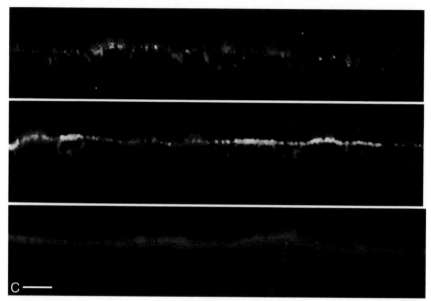

CHAPTER 6, FIGURE 2 Localization of AQP2 in transfected LLC-PK₁ and MDCK cells. LLC-PK₁ cells stably expressing an AQP2-c-*myc* fusion protein were treated (A) without or (B) with vasopressin, fixed and stained for c-*myc*. Without stimulation, AQP2 is mainly located on intracellular, perinuclear vesicles, whereas after vasopressin stimulation, AQP2 is mainly located on the plasma membrane. (See text page 241 for parts A and B). (C) Transfected MDCK cells stably expressing AQP2 (WT10 cells; upper, middle panel) or native MDCK cells (lower panel) grown on coverslips were treated without (upper panel) or with forskolin (middle, lower panel), fixed, and apically labeled with biotin LC-hydrazide. After permeabilization, cells were incubated with affinity-purified rabbit anti-AQP2 antibodies, rat anti-mouse E-cadherin, and TRITC-coupled extravidin, and after washing, with affinity-purified goat anti-rabbit IgG coupled to FITC and Cy-5-coupled goat anti-rat IgG. Following washing, dehydration and embedding, *x-z* axes images were obtained with a Bio-Rad MRC-1000 laser scanning confocal imaging system. E-cadherin (blue), apical glycoproteins (red), and AQP2 (green) are indicated. Note the redistribution of AQP2 from intracellular vesicles to the apical membrane upon forskolin treatment. (Originally published in Deen *et al.*, 1997a.)

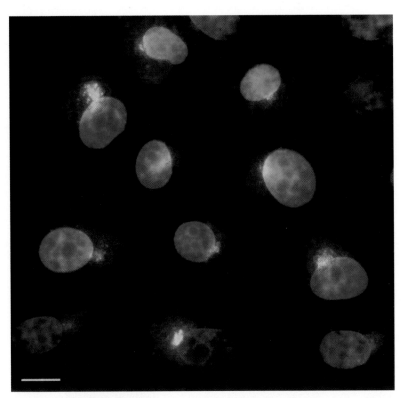

CHAPTER 6, FIGURE 3   Localization of AQP2 in LLC-PK$_1$ cells on treatment with bafilomycin. Treatment of AQP2-expressing LLC-PK$_1$ cells with the H$^+$-ATPase-inhibitor bafilomycin for 2 h causes an accumulation of AQP2 (orange/red) in a dense perinuclear patch, coresponding at least partially to the trans-Golgi network. Nuclei are counterstained with DAPI (blue). Very few AQP2-positive vesicles are detectable in the rest of the cytoplasm after bafilomycin treatment, indicating that virtually all of the intracellular AQP2, both newly synthesized and recycling, are trapped in this perinuclear patch. A similar perinuclear accumulation is seen after exposure of cells to low temperature (20°C) for 2 h. Bar = 5 μm (Courtesy of D. Brown.)

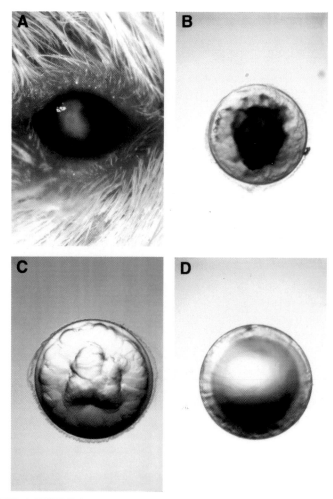

CHAPTER 6, FIGURE 6   Phenotype of the lens of a CAT mouse. (A) The eye of a mouse homozygous for the CAT mutation in the MIP protein. (B–D) Dissected lenses from homozygote (B) or heterozygote (C) CAT mice or a wild-type mouse (D). (Courtesy of A. Shiels, Department of Opthalmology, Washington University School of Medicine, St. Louis, Missouri.)